T0249765

Reaction–Diffusion Computers

Reaction–Diffusion Computers

Andy Adamatzky
Faculty of Computing, Engineering and Mathematical Sciences
University of the West of England
Bristol, UK

Ben De Lacy Costello
Faculty of Applied Sciences
University of the West of England
Bristol, UK

Tetsuya Asai
Graduate School of Information Science and Technology
Hokkaido University
Sapporo, Japan

ELSEVIER

Amsterdam – Boston – Heidelberg – London – New York – Oxford
Paris – San Diego – San Francisco – Singapore – Sydney – Tokyo

ELSEVIER B.V.	ELSEVIER Inc.	ELSEVIER Ltd	ELSEVIER Ltd
Radarweg 29	525 B Street, Suite 1900	The Boulevard, Langford Lane 84	Theobalds Road
P.O. Box 211,	San Diego, CA 92101-4495	Kidlington, Oxford OX5 1GB	London WC1X 8RR
1000 AE Amsterdam	USA	UK	UK
The Netherlands			

First edition 2005

Library of Congress Cataloging in Publication Data
A catalog record is available from the Library of Congress.

British Library Cataloguing in Publication Data
A catalogue record is available from the British Library.

ISBN-13:   0-978-0-444-52042-5
ISBN-10:   0-444-52042-2

♾ The paper used in this publication meets the requirements of ANSI/NISO Z39.48-1992 (Permanence of Paper).
Printed and bound in the United Kingdom
Transferred to Digital Print 2010

Preface

What is a reaction-diffusion processor? A reaction-diffusion processor in the context of this book is a real chemical medium, usually composed of a thin layer of solution or gel containing chemical reagents, that in its space–time dynamics transforms data to results in a sensible and programmable way. Data, to be processed, can be represented by the concentration of certain reagents and spatial structures, e.g. diffusive or excitation waves, spread from these initial data points. The spreading patterns interact to produce either stationary structures, e.g. a precipitate concentration profile, or dissipative structures, e.g. oscillating patterns. The final state, or even just a particular spatial state of the whole medium, represents a result of the reaction-diffusion computation.

The spreading of waves is analogous to information transfer. And, the interaction of diffusive or phase waves realises the computation. An important attribute of this mode of computation is that there is an absence of a rigid hardware-like structure. Essentially, the 'liquid' processor has an 'amorphous' structure which may be considered as a layer of micro-volume reaction-diffusion chemical processors capable of massive parallelism.

Characteristic advantages of reaction-diffusion processors include parallel input of data (usually, via the spatial distribution of the reactant concentrations), massively parallel information processing (typically, via spreading and interaction of either phase or diffusive waves) and parallel output of results of the computation (commonly, the results are represented by patterns of reactants or a coloured precipitate that enables the use of optical reading devices).

These features together with the relative ease of laboratory experiments (most reactions occur at room temperature and do not require any specialist equipment), constructional simplicity of formal design (all reaction-diffusion systems are well simulated in two-dimensional cellular automata) and the pleasure of parallelism *per se* make reaction-diffusion chemical processors an invaluable tool for developing advanced unconventional parallel computing architectures.

Recently, a great deal of attention has been paid to the study of the computational properties of spatially extended chemical systems. To date, it has been proved experimentally that reaction-diffusion chemical processors are capable of computing shortest paths, image processing, computational geometry, pattern recognition and logical computation. In the last ten years enough results have been obtained to demonstrate that reaction-diffusion chemical processors are not simply curiosities invented by theoreticians but promising — and somewhat revolutionary — computing architectures offering an alternative to the as yet unchallenged domination of the current silicon designs. This is because spatially extended reaction-diffusion processors are equivalent to massively parallel computers.

A two-dimensional reaction-diffusion processor, implemented in a thin-layer liquid phase in a Petri dish, consists of millions of micro-volumes, nearly 10^{19}. The concentrations of reactants in each micro-volume are changed in parallel depending on reagent concentrations in neighbouring micro-volumes. Therefore, a thin layer of a chemical medium could be seen as an (ir)regular array of elementary few-bit processors. The great number of elementary processing units makes chemical computers tolerant to impurities of reaction-diffusion media while local connectivity allows for localisation of spatial inhomogeneities in the reacting medium. The 'amorphous' structure of the chemical medium guarantees that a massively parallel chemical processor will self-reconfigure and restore its original architecture after some parts of the physical processing medium are removed.

Non-linear chemistry and non-classical computation share more than just a prefix: these two fields of science flourished on fertile soil — the analysis of emerging phenomena of pattern formation, dissipative structures, somewhat unpredictable space–time dynamics and, basically, unconventional views of existing scientific norms, laws and rules. **Chapter 1** informally introduces the basic concepts of self-organisation and controllability in spatially distributed chemical systems, and considers examples of excitable, oscillating, precipitating chemical systems. In the chapter we also discuss previously established results in the implementation of basic computational operations in chemical systems: mass-flow chemical gates, logics of excitation impulses in a geometrically constrained medium, chemical diodes, image processing and dynamical memory. We analyse how well reaction-diffusion computers are positioned amongst other types of novel and emerging computing devices and paradigms.

In **Chap. 2** we design chemical processors that solve a classical problem of computational geometry — Voronoi diagrams, tessellation of a plane from a given finite planar set. Basically, we represent the planar data set by a configuration of droplets of one reactant and another reactant is contained within the substrate-gel. The 'data' reactant diffuses from the sites of the initial application and forms a coloured precipitate when reacting with the reactant in the substrate. When wave fronts of 'data' reactant, originating from different sources, meet up they exhaust reactant from the substrate and no precipitate is formed. Thus, uncoloured sites of the medium represent edges of the Voronoi diagram. In chemical systems front initiation, propagation and interaction are the primary mechanisms for Voronoi-diagram formation. On the basis of cellular-automata models the general concept of the formation of Voronoi diagrams is explained and related mechanisms are simulated.

From Voronoi diagrams we logically pass to skeletons of planar shapes. We produced an experimental implementation of a massively parallel reaction-diffusion processor which performs one of the most essential parts of shape recognition — construction of a skeleton. A skeleton is a unique, stable and invariant representation of a shape; therefore, computation of the skeleton is an essential tool of computer vision. Skeleton computation is a typical 'natural' spatial problem that can be solved with the use of biological, chemical or physical phenomena. One possible approach — a reaction-diffusion-based computation — is explored. A contour of data of planar shape is represented by a concentration profile of one reagent; a planar substrate is mixed with another reagent. The reagent, representing the original contour, diffuses to form a coloured phase in a reaction with the substrate-reagent. However, at sites where two diffusion wave fronts meet no coloured phase is formed and the substrate retains its uncoloured state. These loci of the computation space represent a skeleton of the given contour.

Is there any chance to process two data sets independently in a reaction-diffusion processor? To find an answer we study the possibility of designing a multitasking chemical processor that independently and simultaneously computes Voronoi diagrams of two different planar data sets. We define a two-tasking chemical processor as two distinct reactant–substrate couples within a reaction-diffusion processor that solve separate tasks but share the same physical space. A micro-volume of the physical space is an elementary processor of a massively parallel chemical processor; therefore, two reaction-diffusion systems occupying the same space are considered to be a single chemical processor. The outcomes of the experiments albeit in a simple chemical system are significant because the system constitutes the first class of a synthetic chemical parallel processor capable of at least two computations at the same time.

Can reaction-diffusion chemical media carry out logical operations? In **Chap. 3** we discuss several implementations of reaction-diffusion logical circuits.

Firstly, we employ the particulars of diffusive wave-front interactions in a two-reactant reaction-diffusion medium to construct a laboratory prototype of an XOR gate. In the design, the values of the logic variables are represented by the presence or absence of a precipitate, 'wires' are constructed of a substrate-loaded gel and the computation is based on diffusive wave dynamics. We also discuss implementation of an AND gate and study a three-valued composition, derived from the gate dynamic, and discuss possible logics that could be derived from this composition. From the architecture-based computing in a geometrically constrained medium we move further to collision-based computing, computation with mobile self-localisations in a homogeneous medium.

So, secondly, we study a photosensitive subexcitable Belousov–Zhabotinsky medium which exhibits propagating wave fragments that preserve their shapes during substantial periods of time. In numerical studies we show that the medium is a computational universal architecture-less system, if the presence and absence of wave fragments are interpreted as truth values of a Boolean variable. When two or more wave fragments collide they may annihilate, fuse, split or deviate from their original paths; thus, values of the logical variables are changed and certain logical gates are realised as a result of the collision. We demonstrate exact implementation of basic operations with signals and logical gates in Belousov–Zhabotinsky dynamic circuits. The findings provide a theoretical background for subsequent experimental implementation of collision-based, architecture-less, dynamical computing devices in homogeneous active chemical media.

Thirdly, the chapter finishes with computational studies of a non-trivial reaction-diffusion cellular automaton, which exhibits rich glider dynamics in its evolution. We specify particulars of glider interaction and exemplify logical operations and gates realised in glider collision.

Chapter 4 brings an insight into chemical controllers for robots, a kind of on-board 'liquid brain' capable of path planning, decision making and controlling a robotic hand. Here we employ all the advantages of reaction-diffusion chemical computing — massive parallelism, amorphous structure, high tolerance to damage, parallel inputs and outputs — that could be employed as a 'liquid brain' for a new class of robots, i.e. to control robot behaviour.

There are few chemical computers that can — potentially — control a robot's behaviour off-line, if some interface between a stationary reactor and a mobile robot is developed. They compute a shortest collision-free path in a bounded space with obstacles — the obstacles can be physically cut out of the excitation substrate, represented by local changes in the chemical species' concentrations, or by singularities in the medium's illumination. Unrealistic omniscience is the main disadvantage of these approaches — in all cases the whole robotic arena must be projected onto a reaction-diffusion chemical medium; there is no interaction between the robot, physical reality and chemical computer. In the chapter we develop an idea of chemical controllers for robots a little further and discuss experimental implementations of a chemical reaction-diffusion processor for robot-motion planning, finding a shortest collision-free path for a robot moving in an arena with obstacles. In some cases we do indeed map an image of the robot arena, in which the robot is to navigate, onto a thin-layer chemical medium using a method that allows obstacles to be represented as local changes in the reactant concentrations. Disturbances created by the 'objects' generate diffusive and phase wave fronts. The spreading waves approximate a repulsive field generated by the obstacles. This repulsive field is then inputted to a discrete model of an excitable reaction-diffusion medium, which computes a tree of shortest paths leading to a selected destination point. Two types of chemical processors are discussed: a disposable palladium processor, which executes arena mapping from a configuration of obstacles, given before an experiment, and a reusable Belousov–Zhabotinsky processor, which allows on-line path planning and adaptation for a dynamically changing configuration of obstacles.

Is it possible to navigate a robot with an on-board chemical controller? To answer the question we put a chemical reactor filled with Belousov–Zhabotinsky chemical mixture on-board a mobile wheeled robot. When a point-wise part of the medium is stimulated with silver wire, waves of excitation spread from the initial source of stimulation, 'looking' at the spatial patterns of excitation in the chemical reactor; the robot extracts from the topology of the excitation wave fronts a position of stimulation and rotates and moves in the direction of stimulation. Thus, by subsequent stimulation of relevant parts of the on-board chemical reactor we can guide a robot in the robotic arena.

To study particulars of direct interaction between a robotic device and liquid-phase chemical medium, we designed a closed system where a Belousov–Zhabotinsky thin-layer chemical reactor was linked to the robotic hand via an array of photosensors and the fingers of the hand were able to stimulate the excitation dynamics in the medium via the local addition of an activator species. A principal working loop of the chemo-robotic system is that oxidation wave fronts travelling in the medium are detected by photosensors and cause (via a micro-controller) the fingers of the hand to bend. When a finger bends, it is set up to apply a small quantity of colloidal silver to the reaction and thus causes an additional excitation wave. The travelling and interacting waves stimulate further movements of the fingers and patterns of behaviour are observed.

Implementations of reaction-diffusion computers discussed in the book suffer from a deficiency of 'specialisation' — from a common-sense point of view they are 'hard wired' to execute a strictly limited set of tasks and to solve just a few problems. By making a reaction-diffusion processor programmable we would open new horizons for future theoretical findings and wide practical applications of reaction-diffusion computing. Therefore, in **Chap. 5**, we search for answers to the question 'What properties of a chemical medium can be used to (re-)program reaction-diffusion computers?'. To start with, we briefly analyse experimental results, available so far, on controlling spatio-temporal dynamics of reaction-diffusion media with temperature, electric field, substrate's structure and illumination, and then derive several theoretical models based on these experimental findings. In a cellular-automaton model of a two-reagent chemical medium we show how computational properties — the ability to subdivide space by a Voronoi diagram — emerge in a parametric space of reaction rates. Then, we show how to re-program three-valued logical gates, and to switch operations implemented in geometrically constrained media between various types of three-valued logics, by varying the local excitability of the medium. There, we realise several types of logical gates, including Łukasiewicz conjunction and disjunction, Sobociński conjunction in cellular automata and FitzHugh–Nagumo models of T-shaped excitable media.

Chemical reaction-diffusion processors — at least in their current experimental implementations at macro-scales — are typically very slow; thus, it would take half-an-hour for a chemical computer to calculate a skeleton of a planar shape or for an on-board chemical controller to 'make a decision' which stimulation source the robot should follow. This does not pose any significant limitations on future applications of reaction-diffusion computers, because they can do well on the micro- or nano-scale, subject to the fabrication technology. Principles of reaction-diffusion computation *per se* are quite extraordinary in their simplicity, high level of parallelism and intrinsic closeness to existing massively parallel hardware architectures. This means that designing silicon-based analogues of a reaction-diffusion chemical processor is not a 'step back' from advanced wet-ware to archival hardware but simply taking an opportunity to get immediate benefits from excellent ideas of non-classical computation. **Chapter 6** presents an overview of the semiconductor implementation of reaction-diffusion computers in large-scale integrated (LSI) circuits. There, we show how to model reaction-diffusion processes in LSI circuits and discuss several designs of reaction-diffusion digital chips, based on cellular-automaton models of reaction-diffusion and excitable systems. Feasibility of a reaction-diffusion digital chip is demonstrated in the construction of a Voronoi diagram and decomposition of images. The chapter concludes with analogue reaction-diffusion chips, where closer to physical reality non-linear characteristics of chemical systems are employed. We propose designs of reaction-diffusion chips based on Oregonator, Turing and Lotka–Volterra systems and Wilson–Cowan oscillator systems. We exemplify functionality of analogue reaction-diffusion chips in edge detection, feature extraction and fingerprint re-construction tasks.

By making reaction-diffusion computers smaller we can increase the number of elementary processors, and thus the degree of parallelism per time step, and also make the processors work faster. So, the next two chapters invite readers to the field of micro- and nano-scale reaction-diffusion computing. **Chapter 7** deals with p-n-p-n devices, couplings of transistors which can be switched into a high-current, low-voltage state, capable of 'accidental' triggering and thus can act analogously to 'excitable' elements. When arranged in arrays p-n-p-n devices 'transmit' their excitation via leakage of electrons, exciting neighbouring devices. We analyse operational characteristics of reaction-diffusion p-n-p-n-based processors and assess their potential for implementation of reaction-diffusion algorithms.

Can reaction-diffusion computers get smaller and smaller? – this is a subject of **Chap. 8**, where we offer designs of single-electron reaction-diffusion processors: arrays of single-electron diffusively coupled oscillators. In the chapter we tackle the mathematical background and practical designs of the non-linear oscillators and their coupling links, and then provide examples of space–time excitation dynamics in the oscillator array, including expanding circular patterns, rotating spirals and multiplying patterns. The chapter concludes with blueprints of potential manufacturing methods for the single-electron reaction-diffusion devices.

As may be expected, the overwhelming majority of the publications pertaining to chemical computers discuss predominantly the advantages of these novel prototypes of spatially extended chemical media. However, now we consider that it is very important to look at the subject critically and to highlight the limitations of these reaction-diffusion processors, which can include inaccurate or even incorrect computations. It is our belief that in order to develop better prototypes of chemical-based computers we must understand the limitations of the already existing prototypes. In addition, from a theory of computation point of view, it would be useful to be aware of at least one problem that can be considered unsolvable by chemical computers. Therefore, in **Chap. 9**, we evaluate the limits of a class of reaction-diffusion processors in undertaking two test problems: construction of a Voronoi diagram and inversion — re-construction of a planar data set — of a Voronoi diagram. We study how a Voronoi diagram is constructed and how it is inverted in a planar chemical processor. We demonstrate that a Voronoi diagram is computed only partially in the chemical processor. We also prove that given a specific Voronoi diagram it is impossible to re-construct the planar set — from which the diagram was computed — in the reaction-diffusion chemical processor. In this chapter we open the first ever line of inquiry into the computational 'disability' of reaction-diffusion chemical computers.

*   *   *

The book brings together results of a decade-long study into designing experimental and simulated prototypes of reaction-diffusion computing devices for image processing, path planning, robot navigation, computational geometry, logics and artificial intelligence. The book is unique because it gives a comprehensive presentation of the theoretical and experimental foundations and cutting-edge computation techniques, chemical laboratory experimental setups and hardware-implementation technology employed in the development of novel nature-inspired computing devices. This timely volume presents a detailed overview and original analysis of information processing in spatially extended amorphous non-linear media, and then applies these theoretical findings in describing the fabrication of working prototypes of wet-ware. The text is self-contained and requires a minimal knowledge of computer science, chemistry, physics and electronic engineering to understand all aspects discussed within the book.

Acknowledgments

Some findings, results and implementations discussed in the book were obtained in collaboration with Hiroshi Yokoi, University of Tokyo (who designed a robotic hand and its interface with the Belousov–Zhabotinsky chemical medium and undertook a series of experiments on the interaction between the robotic hand and a chemical processor), Ikuko Motoike, Future University Hakodate (who performed a numerical simulation of three-valued logical gates in an excitable medium), Chris Melhuish, University of the West of England (who participated in all robotics experiments and enabled things to run smoothly), Ian Horsfield and Chris Bytheway, University of the West of England (who adapted a mobile wheeled robot for experiments with chemical controllers for robot navigation), Norman Ratcliffe, University of the West of England (who took part in initial discussions of the 'wet-brains for robots' subject), Andy Wuensche (who discovered the beehive cellular-automaton rule which formed a basis for logical computation in reaction-diffusion cellular automata), Peter Hantz (who was a visiting researcher to the University of the West of England and who carried out experiments on the formation of Voronoi diagrams in unstable systems), Yoshihito Amemiya, Hokkaido University (for most valuable discussions and suggestions on semiconductor devices and circuits), Yusuke Kanazawa, Sharp Corporation (who designed and measured analogue reaction-diffusion chips) and Takahide Oya, Hokkaido University (who performed a numerical simulation of single-electron reaction-diffusion devices). Thanks are also due to Leon Chua, University of California at Berkeley (for his constant support and encouragement) and Nicolas Rambidi, Moscow State University (for his promotion of the field of chemical computation). Many thanks to Michael Jones for copyediting the manuscript.

Contents

Chapter 1

Non-linear chemistry meets non-classical computation

In this chapter the idea of what is meant from a chemical perspective by the term reaction-diffusion processor (or chemical processor) will be discussed. This will be backed up with some examples of real chemical systems. The second major section of this chapter will provide a brief overview of what has already been accomplished in the field of reaction-diffusion-based chemical processing. This will be backed up by examples from our own work and other groups working in the field of non-linear chemical reactions. The overview will predominantly focus on work containing actual experimental results (chemical computations) and not merely on the results obtained from theoretical studies utilising reaction-diffusion-based models. However, if theoretical work identifies new areas which could be exploited to produce chemical processors it will be discussed. The third section aims to identify experimental research that has not yet been linked to information-processing applications but that we feel is potentially useful. Often the study of non-linear reactions such as say the Belousov–Zhabotinsky (BZ) reaction is carried out in order to model biological systems, for example the formation of spiral waves in neural tissue [76] or cardiac muscle [94] or merely to study interesting chemical systems *per se*. However, there are probably many experimental systems already in existence which could be re-assessed and/or developed in order to realise useful chemical-processing prototypes. The same chemical systems may also provide information in order to construct new types of reaction-diffusion-based silicon devices, another focus of the book. In the fourth section some alternative types of non-standard or unconventional computing architectures will be discussed with particular emphasis on biologically inspired systems such as DNA computing and computing with neurons. Questions about how these methods currently compare and differ to the discussed chemical-based reaction-diffusion processors will be tackled. Also discussed will be the central theme of the book: how do these chemical processors compare to our conventional silicon processors? The final section will discuss the future for chemical-based reaction-diffusion processors and chemically inspired computation. Where does the field go from here? What if despite their natural parallelism and other perceived advantages their limitations which will be discussed openly during the book prove insurmountable?

1.1 What is a chemical processor?

1.1.1 Apology

But, before we get started, hopefully these sections will be self-explanatory; however, because of the diverse nature of the subject area we apologise if some concepts are touched on very briefly with insufficient explanation. Hopefully, in this case there are at least appropriate references to guide the interested reader deeper into the subject area.

1.1.2 About conventional and unconventional processing

When touching on the subject of conventional and unconventional processing we could not resist quoting Tomasso Toffoli again (as we did before in [7]): "... a computing scheme that today is viewed as unconventional may well be so because its time hasn't come yet — or is already gone" [280]. Thus, just over half-a-century ago analogue computers were considered obsolete (at that time digital ones were unconventional) but nowadays they top the charts of modern unconventional computing devices.

So what does a conventional processor do from a layman's point of view? When surfing the rich tapestry of web pages or typing on a keyboard it may be easy to forget that a conventional processor receives all these inputs converted to binary numbers (0s and 1s in the form of electrical pulses) and from then on its function involves the movement and transformation of these pulses in simple electrical circuits. Its advantages lie in the fact that it has millions of such circuits operating at high speed and can thus 'compute' outputs very quickly.

In a computer-dominated society and one with a great degree of reliance on such conventional computing devices it might be easier to assign some greater significance to what the processor does rather than look at any limitations. We are sure that if the basic technology behind this conventional computation had endless development potential then any limitations might be easier to ignore. However, there is a consensus that current methods will indeed reach a threshold and this has led to an explosion in research into unconventional methods of computation.

So what are these limitations? Some are detailed in the table below, adopted and slightly modified from Jonathan Mills's duality of digital and analogue computers, outlined in [181]:

Conventional	Unconventional
Hardware, predominantly silicon	Wet-ware, predominantly non-silicon
Sequential processor	Parallel processor
Restricted by CPU and memory	Computation occurs everywhere
Algorithms	Analogies
Precision increases temporally	Precision increases spatially
Modular and temporal programming	Holistic and spatial programming
Lexical structure	Visual structure
Explicit error correcting	Implicit error correcting

The limitations really seem to stem from the fact that conventional processors compute in a serial manner whereas biological and natural information processing seems to be predominantly via parallel mechanisms [297]. Conventional processors are hard wired while unconventional ones are soft-, chemical- and molecular-based devices. Conventional computers are fragile, in a sense that damaging one component will usually halt the work of the whole machine, and unconventional ones are 'self-healing', re-constructible, due to the behaviour of the physical matter they are built of. There are areas where conventional computers far exceed the capabilities of humans — but conversely there are areas where even simple biological entities seem to outperform computers as they interact seamlessly with their environments. For this reason there is great interest in the mechanisms underlying biological methods of information processing. As these in part can be traced to complex chemical reactions, reaction-diffusion processors fit well in this general field.

We can complete our brief discussion by outlining the subjects commonly addressed in unconventional computing research:

- cellular automata (mainly due to the 'low entrance fee', i.e. ease of designing models, topological conformity to natural spatially extended systems and huge potential to exhibit all types of complex behaviour with simple local rules);

- biological and molecular computing (conformation-based computing, DNA computing, information processing in micro-tubules, molecular memory, biochemical computing, artificial chemistry);

- chemistry-based computing (amorphous computing, implementation of logical functions, image processing and pattern recognition in reaction-diffusion chemical systems and networks of chemical reactors);

- hybrid and non-silicon computation (plastic computers, organic semiconducting devices, neuronal tissue–silicon hybrid processors);

- logics of unconventional computing (logical systems derived from space–time behaviour of natural systems, non-classical logics, logical reasoning in physical, chemical and biological systems);

- physics-based computation (analogue computation, quantum computing, collision-based computing with solitons);

- stigmergic and population-based computing (optimisation in cellular cultures, computing in societies of social insects, ecological computing);

- smart actuators (molecular motors and machines with computational abilities, intelligent arrays of actuators, molecular actuators, coupling unconventional computing devices with arrays of molecular or smart-polymer actuators).

1.1.3 Order within disorder

If we consider that all chemical reactions are really somewhat efficient molecular-based computations, in this context we control the inputs to the system (reactant

Figure 1.1: The reaction of cobalt and manganese salts with ammonia ($CoCl_2$ 0.05 M, $MnCl_2$ 0.25 M, agar gel 1%). From left to right ammonia solution concentration (4.5 M, 9 M, 13.5 M, 18 M).

concentrations, reaction conditions, etc.) and the outputs or products result via a set of complex parallel interactions.

To emphasise this point Fig. 1.1 shows an example of a chemical reaction. Prior to the start of the reaction two simple salts (cobalt chloride and manganese chloride) were mixed with water and a gelling agent. The resulting solutions were then poured into test tubes and allowed to set. Then, a different concentration of another salt (ammonia solution) was added to the top of the gel column to start the reaction. The permanent output of this chemical reaction is shown in Fig. 1.1 (see also Fig. 9.11 in colour insert). What can be observed are bands of precipitate separated by precipitate-free regions. These bands of precipitate are commonly referred to as Liesegang bands. The Liesegang phenomenon is defined as oscillatory precipitation in the wake of a moving diffusion front. This is an interesting phenomenon thought to be responsible for the patterns in rocks such as agate and readers are referred to a book by Henisch [126] which deals with the subject in more detail. Even a century after the phenomenon was first discovered an exact mechanism of how the bands form does not exist. This is even the case for a simpler system containing just cobalt chloride reacted with ammonia, so the situation is further complicated by the addition of an extra salt in our example. The result of our experiment is two

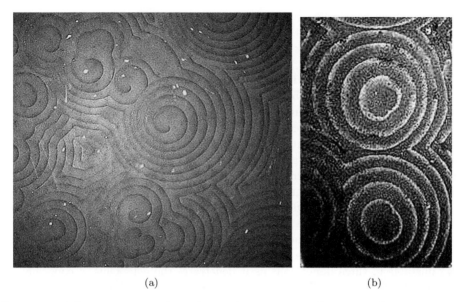

(a)	(b)

Figure 1.2: Microscopic spirals and complex structures formed in the reaction between copper chloride (0.75 M) in agarose gel (1% by weight) and potassium hydroxide (8 M): image sizes (a) 1×1 mm^2, (b) 1×0.5 mm^2).

sets of interpenetrating coloured bands. The effect of changing the concentration of ammonia ('programmability') is also obvious with the number and spacing of bands subtly changed.

In very simple terms the ammonia diffuses into the gel and reacts with both salts simultaneously (in parallel) to form intermediates and products which then interact in parallel to give the transient (bands can form and re-dissolve, especially cobalt) and finally the permanent output of the system. The macroscopic output we observe is the result of these millions of parallel molecular interactions in a constrained environment. In some respects we do not need to understand every detail of these processes — however, for the design of useful chemical processors, as much mechanistic information as possible is required.

1.1.4 Self-organisation in simple chemical systems

The previous example highlighted just how complex seemingly simple chemical processes can be. The next example just serves to emphasise this fact. Figure 1.2 shows the complex structures that can be formed when certain concentrations of a simple salt (copper chloride) immobilised in thin gel sheets are reacted with potassium hydroxide (8 M). It is amazing that such seemingly simple chemical reactions give rise to such complex patterns [120, 121, 81]. The spirals, double spirals, target waves and other complex structures are more commonly associated with more complex chemical reactions such as the BZ reaction [248] or self-organisation in natural systems such as the aggregation of slime moulds [201].

The exact mechanisms of this reaction are not clearly established; however, the patterns appear to arise from the primary product having two stable phases. The first phase is amorphous (bulk colloidal precipitate) and the second phase more crystalline. When potassium hydroxide is added to the copper chloride-loaded gel a diffusion front is initiated and behind this front the chemical reaction takes place. This advancing front becomes unstable when the product concentration exceeds a critical threshold. Small imperfections can then cause the unstable front to switch suddenly from the colloidal to the crystalline phase. A thin layer of the crystalline phase grows out from this point of instability as the front sweeps through the substrate-loaded gel. As the crystalline phase acts as a temporary barrier to diffusion, the gel layer is split into reacted and unreacted zones. However, in this reaction it is possible for the diffusing reactant to overcome this crystalline barrier, re-starting the reaction in an unreacted zone of the gel. At this stage the concentration of the product is reduced so the colloidal form is again precipitated. The whole cycle is repeated leading to the observed structures.

It should be noted that the complex structures shown are a result of the front being unstable in three dimensions as it penetrates the gel. Thus, a cone-shaped region bordered by crystalline precipitate is formed (Fig. 1.2a) and the repeated process forms a series of nested cones which when viewed from below give the target-like patterns (Fig. 1.2b).

The reason for spiral formation is unclear but this follows for a number of chemical, biological and physical systems. What should be reinforced by this example are the complex parallel chemical or molecular processes that occur even in the simplest chemical reactions under certain conditions. In both the reactions there is also an element of self-organisation or self-assembly in evidence. The ultimate aim of producing a chemical processor is to exploit this natural richness in the system's behaviour. However, currently the highly complex and non-linear nature of the reactions discussed discounts their use in constructing useful chemical processors.

So, here we have our first current limitation — our understanding of chemical systems (and molecular mechanisms) *per se* is currently not at a sufficiently high level to exploit all the natural features of a given reaction in order to construct useful chemical processors.

Having said that, by understanding enough about the mechanisms or by exploiting a natural characteristic of chemical reactions it is possible to construct a family of somewhat efficient chemical processors and some of these will be discussed below and in the following chapters.

1.1.5   So, what is a reaction-diffusion computer?

Reaction-diffusion chemical systems are well known for their unique ability to efficiently solve combinatorial problems with natural parallelism. In reaction-diffusion processors, both the data and the results of the computation are encoded as concentration profiles of the reagents. The computation *per se* is performed via the spreading and interaction of wave fronts. The reaction-diffusion computers are parallel because the chemical medium's micro-volumes update their states simultaneously, and molecules diffuse and react in parallel. Architectural and operational bases of reaction-diffusion computers are based on three principles:

- Computation is dynamical: physical action measures the amount of information;

- Computation is local: physical information travels only a finite distance;

- Computation is spatial: the nature is governed by waves and spreading patterns.

This can be summarised as follows:

Architecture/operation	Implementation
Component base, computing substrate	Thin layer of reagents
Data representation	Initial concentration profile
Information transfer, communication	Diffusive and phase waves
Computation	Interaction of waves
Result representation	Final concentration profile

Potentially, a reaction-diffusion computer is a super-computer in a goo. Liquid-phase chemical computing media are characterised by

- massive parallelism: millions of elementary — 2–4 bit — processors in a small chemical reactor;

- local connectivity: every micro-volume of the medium changes its state depending on the states of its closest neighbours;

- parallel I/O: optical input — control of initial excitation dynamics by illumination masks, output is parallel because concentration profile representing results of computation is visualised by indicators;

- fault tolerance and automatic re-configuration: because if we remove some quantity of the liquid phase, the topology is restored almost immediately.

1.2 Overview of chemical processors

1.2.1 Precipitation reactions

So far all the examples we have covered have been based on precipitating reactions. In the vast majority of these reactions salts dissolved in a hydrogel are reacted with another salt to form a third insoluble product that is often highly coloured. The diffusion of the reacting salt is limited by the hydrogel and, thus, depending on the concentrations of the reagents and other experimental parameters, the precipitation may evolve in a highly non-linear way giving rise to distributed patterns.

These reactions have advantages such as a permanent output of the results but disadvantages such as non-re-usability and extremely slow speed.

Processors of this type have been predominantly developed and utilised during our work and will be discussed in greater depth in the chapters ahead. They are suitable for solving simple specialised computational tasks such as construction of a Voronoi diagram [282, 81], construction of logic gates, pre-processors for shape recognition [19] and pre-processors for shortest-path problems [14].

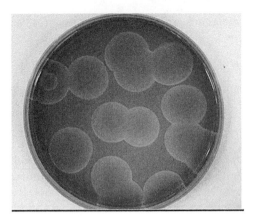

Figure 1.3: Excitation waves formed in a thin layer of the ferroin-catalysed BZ reaction. A wave at each location was initiated using an array of silver wires.

1.2.2 Active chemical media

A more dynamic class of chemical reactions exists and these have been used to produce a range of chemical processors. Currently these are predominantly based on the BZ reaction. The major types of active chemical media are listed below.

Excitable chemical media

The major type of active media utilised in the construction of chemical processors to date are excitable media. An excitable chemical medium has a single steady state that is stable to small perturbations, but responds with an excitation event — for the BZ system, this excitation event is the oxidation of the redox catalyst — if the perturbation is above a critical threshold level [284]. After excitation the system enters a refractory period during which time it is insensitive to further excitation before returning to its steady state.

When the various locations of these spatially extended systems are coupled via diffusion, the excitation event will spread as a wave of excitation leading to spatio-temporal patterns. If the perturbation is continuous at a certain point (as in the top left corner of Fig. 1.3), then a series of concentric waves will be generated (a wave train). When these waves of excitation collide (this can be observed in Fig. 1.3, where circular waves have collided and fused) they annihilate and the system returns to the steady state at that point. So, from a chemical-processor point of view, the system is thus re-usable.

Excitable systems in nature include neuronal and heart tissue [148].

Subexcitable media

The so-called subexcitable regime lies between the excitable state and the unexcitable state. In excitable chemical media, waves initiated by perturbations of a sufficient size propagate throughout the media. In an unexcitable medium no perturbation is large enough to trigger a wave. In a subexcitable medium with higher excitability, wave segments with free ends are formed and either expand or contract, depending on their size and the medium's excitability. In a subexcitable medium with lower excitability, waves with free ends contract and eventually disappear. This point where wave segments fail to propagate marks the unexcitable/subexcitable threshold [250]. From the perspective of constructing chemical processors, this feature is desirable as it allows fine control over wave evolution and dynamics.

Oscillating chemical reactions

In most chemical reactions the intermediate compounds increase on initiation but approach zero towards the end of the reaction. In oscillating chemical reactions [202], several of the intermediates oscillate periodically. This is commonly caused where a set of parallel chemical processes exist where at least one is autocatalytic (promotes the rate of its own production) and one is inhibitory. In the BZ reaction and other oscillating reactions these bulk oscillations, usually occurring in well-stirred systems, can be followed via an accompanying change in colour. From the point of view of constructing chemical processors, oscillating systems are interesting as oscillations can be controlled with external forcing and the coupling of such reactions can result in complex behaviour.

Bistable and multistable reactions

In bistable systems there are two steady states and possibly some unstable states. Large enough perturbations can cause transitions between these stable states and small perturbations may cause transitions between the unstable and steady states. Bistable and multistable systems can only exist in open systems subject to a flow of reactants, as closed systems have a unique equilibrium state [119].

Often, chemical systems displaying multistable behaviour are chemically coupled, i.e. they consist of two bistable or oscillating reactions linked via a common intermediate [29]. Reactions may also display more complex dynamical phenomena if they are physically coupled together [159].

Turing patterns

Turing-type active media usually lead to spatially periodic patterns that are stationary in time. This occurs due to an activator–inhibitor system whereby the inhibitor must diffuse more rapidly than the activator. Such patterns have been found in the chlorite–iodide–malonic acid [68] reaction and more recently in the BZ-AOT system [288]. Lee *et al.* [162] also showed experimentally in the ferrocyanide–iodate–sulphite reaction the existence of self-replicating spots.

Turing patterns are widely accepted by theoretical biologists as a model for pattern formation in living organisms [195].

1.2.3 The BZ reaction

As the Belousov–Zhabotinsky (BZ) reaction is by far the most commonly studied active chemical medium for prototyping chemical processors, it is worth describing

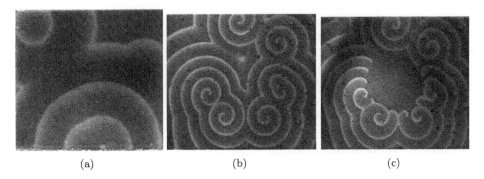

(a)   (b)   (c)

Figure 1.4: Effects of light on waves in the light-sensitive BZ reaction: (a) natural target wave formation in ambient light, (b) spiral wave formation where light from a blue LED was used to break the original waves in (c). Wave-free area caused by illumination with a blue LED at one spot ($\sim$ 10 sec). The formation of a number of free ends can be observed — these will naturally generate additional spirals.

some of the key features of the reaction in more detail. Depending on experimental conditions the BZ reaction exhibits many of the states described above. The BZ reaction involves the oxidation of an organic acid such as malonic acid with a solution of acidified bromate in the presence of a one-electron transfer redox catalyst, such as ferroin $[Fe(phen)_3^{2+}]$ or a light-sensitive ruthenium bipyridyl complex $[Ru(bipy)_3^{2+}]$. The change in the redox behaviour of the catalyst is usually accompanied by a change of colour; e.g. in Fig. 1.3 the ferroin (orange) is oxidised to ferriin (blue).

In the light-sensitive BZ reaction the ruthenium catalyst is excited by 460-nm light (blue) and reacts with bromomalonic acid to produce bromine, an inhibitor of autocatalysis. Thus, by varying the light levels the excitability of the system can be controlled.

From the perspective of constructing chemical processors the photosensitive nature of the ruthenium catalyst is attractive. It affords the possibility of perturbing and controlling either the temporally oscillating stirred BZ reaction or the spatially patterned unstirred (thin-layer) BZ reaction with a high degree of precision. Figure 1.4 (see also Fig. 9.12 in colour insert) shows an example of the effects of light on the BZ reaction. In the experiment a thin silica gel sheet (0.2 mm $\times$ 1.4 mm $\times$ 1.4 mm) was immersed in a solution of the light-sensitive catalyst (0.0002 M Ru(bpy)$_3$ Cl$_2$). The gel was then soaked in water and placed in a Petri dish containing the reagents of the BZ solution (NaBrO$_3$ [0.28 M], NaBr [0.098 M], malonic acid [0.25 M] and H$_2$SO$_4$ [0.4 M]). The effects of illumination are demonstrated by using a blue light emitting diode (LED) to locally perturb the reaction.

1.2.4   External control

In utilising these active media for constructing chemical processors we and other workers in the field exploit the natural features of the media such as travelling waves, periodic oscillations and other dynamic features in order to transfer information.

In addition to just exploiting the natural features of the reactions, experiments have shown that these active media can be dynamically guided to some output state (a controlled pattern formation or type of dynamic behaviour) via the application of external fields.

External fields which have been used to control the evolution of active chemical reactions (particularly but not exclusively in the BZ reaction) include temperature [295, 174], gravitational field [198], electric field [146, 206, 268, 223, 255], magnetic field [59] and light field [240, 315, 164, 106].

Various effects have been induced including wave-front (de)stabilisation, changing the wave velocity, altering the wave direction, generation of new waves and patterns, controlling spiral drift and frequency of rotation and suppressing or inducing oscillations.

Using these and other techniques, experimentalists are now able to exhibit a degree of control over the behaviour of non-linear systems. This is allied to more general work which often comes under the heading 'controlling chaos'. Although this field in general is relevant to the aim of producing reaction-diffusion processors from highly non-linear chemical reactions, an in-depth discussion is not possible. However, interested readers are referred to the following papers which give an overview of this subject relating to many areas: [56, 219, 220, 217, 262, 245, 207, 108, 85].

Specific examples of the use of external forcing, particularly the use of light, to create and dynamically 'program' reaction-diffusion processors are discussed in the following sections. An overview of programming is also given in Chap. 5.

1.2.5 Logic gates constructed from chemical media

There have been a number of experimental approaches to constructing logical gates in chemical media. This work on the whole exploits some natural feature of the chemical reactions such as wave dynamics. The experiments have usually involved the construction of specific channels or other geometrically constrained setups. These rather artificial experimental approaches which constitute a basic copy of the logic circuits found in conventional processors do not really exploit the complexity or natural parallelism of the chemical systems, but they do serve as a simple example of how chemical systems can be used to implement computation.

Using coupled reactors

Over 30 years ago the construction of logic gates in a bistable chemical system was described by Rossler [238]. Hjelmfelt *et al.* [130, 131, 127, 129, 128] produced a theoretical construct suggesting the use of 'chemical' reactor systems coupled by mass flow for implementing logic gates, neural networks and finite-state machines.

In other work Lebender and Schneider [161] described methods of constructing logical gates using a series of flow rate coupled continuous flow stirred tank reactors (CSTRs) containing a bistable chemical reaction. The minimal bromate reaction involves the oxidation of cerium(III) (Ce^{3+}) ions by bromate in the presence of bromide and sulphuric acid. In the reaction the Ce^{4+} concentration state is considered as '0', or FALSE ('1', or TRUE) if a given steady state is within 10% of the minimal (maximal) value. The reactors were flow rate coupled according to rules given by a feed-forward neural network run using a PC. The experiment is started by feeding in two TRUE states to the input reactors and then switching the flow rates to generate TRUE–FALSE, FALSE–TRUE and FALSE–FALSE. In this three-coupled reactor system the AND (output TRUE if the inputs are both TRUE), OR (output TRUE if one of the inputs is TRUE), NAND (output TRUE if one of the inputs is FALSE) and NOR gates (output TRUE if both of the inputs are FALSE) could be realised. However, to construct XOR and XNOR gates two additional reactors (a hidden layer) were required. These composite gates are solved by interlinking AND and OR gates and their negations.

The authors noted that the suggested experiments suffered from low-speed processing as linear reactions show only gradual changes between initial and final states, but they postulated that if fast-switching chemical reactions were used then the limiting step would be the physical flow rate of the reactants. In their work coupling was implemented by computer, but they suggested that true chemical computing of some Boolean functions may be achieved by using the outflows of reactors as the inflows to other reactors, i.e. serial mass coupling.

Chemical waves in capillaries

In experiments Tóth and Showalter showed that logic gates could be constructed using the interaction of chemical waves propagating through capillary tubes [283]. In original work [284] thin layers of excitable BZ mixtures were connected by precision-bore capillary tubes of differing internal diameters. A wave initiated on one side of an impenetrable barrier (between the reactors) enters and travels through the capillary tube forming a hemisphere of excited solution at the exit. When the tube diameter is greater than some critical value the excitation is transferred to the second compartment, otherwise the hemisphere collapses and no wave is transferred. Thus in effect they produced a chemical analogue of the on–off switch with a binary output, as a single input wave can give rise to an output wave (binary 1) or no output wave (binary 0). They also observed complex resonance patterns for successive input waves.

In further work [283] they utilised the above observations to construct arrays of two or more tubes able to implement logic gates based on input/ouput signals in the form of chemical waves.

In this case AND — inputs of 0 (no wave in input tube 1) and 1 (wave in input tube 2) yield 0 (no wave in output chamber) and inputs 1, 1 yield 1 (wave in output chamber) — and OR — inputs 1, 1; 0, 1 or 1, 0 yield 1 (wave in output chamber) — gates were realised experimentally using two coaxial capillary tubes bridging barriers separating a central output chamber from two input chambers. The reaction-mixture composition (changing excitability) could be used to convert the assembly from an OR to an AND gate.

The authors suggest methods to construct other gates via combinations of the aforementioned gates. The key element of all the gates (except the OR gate) was the union of two subcritical regions of excitation to initiate a wave in the output chamber. They noted that the construction of logic gates was based on the geo-

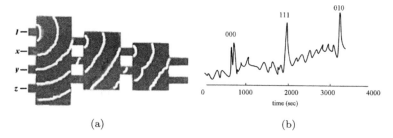

(a)   (b)

Figure 1.5: Example of a three-layer network representing an OR gate: (a) wave evolution is shown as superimposed snapshots at time intervals of 110 s for input (0, 0, 0), (b) grey level intensity signals at output channels in response to inputs (0, 0, 0), (1, 1, 1) and (0, 1, 0). Single and double peaks correspond to TRUE and FALSE, respectively. Modified from [267] with kind permission of O. Steinbock.

metrical configuration, temporal synchronisation of the waves and the ratio of the tube radius to the critical nucleation radius of the excitable medium.

Printed logic

In a distinct type of experimental implementation, Steinbock *et al.* [267] constructed logic gates based on wave propagation in a geometrically constrained excitable BZ system. In their experiments the catalyst for the reaction (bathoferroin) was ink-jet printed onto membranes to give specific pre-determined patterns with geometries designed to provide logic gates.

Figure 1.5 shows an OR gate constructed using this experimental approach. In the figure the dark areas are catalyst-loaded regions of the membrane. The gate has four initiation sites on the left and two outputs on the right. The input (x, y or z) corresponds to wave initiation in the channel at a specified time (TRUE, or 1, FALSE, or 0). The input 1 is a 'clock' input to establish the case for zero input $(x, y, z) = (0, 0, 0)$. The output is defined as TRUE or FALSE by synchronous or asynchronous waves exiting from the output channels. The overall output was obtained by summing the grey-level intensities of an area in each output channel.

In addition to more simple logical gates, by varying the geometries and connected layers a set of complex logical gates could be realised. The authors also showed that the operation of gates could be changed dynamically by applying an electrical potential for a short period of time to channels linking rectangular domains.

A chemical diode

Agladze *et al.* [26] were able to construct a 'chemical diode' using an excitable BZ medium. Their experimental system consisted of two square glass plates loaded with the BZ catalyst ferroin and placed in a Petri dish with a small gap (60 νm) between them. One plate was orientated to the contact area via the plane border (P-side) and the other orientated via the corner (C-side). The glass plates were

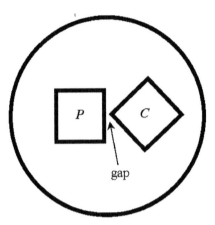

Figure 1.6: Schematic of the experimental setup used to implement the chemical diode. A wave initiated on P will propagate to C but a wave initiated on C will not propagate to P [26].

immersed in the BZ solution and waves were stimulated alternately on each plate in turn. The waves propagate across the gap from P to C but not in the opposite direction from C to P. In this case the unidirectional propagation of the wave results from the asymmetric geometry of the medium (Fig. 1.6).

Waves must have a critical size (or frequency) to allow them to propagate across gaps, and when propagating on the C plate in this experimental setup the wavefront geometry is systematically reduced below this level. It should be noted that the critical gap width controlling diode functionality is dependent on the excitability of the system, and therefore the propagation could in theory be controlled via the application of an external field such as light.

Light-controlled logic

In further work Agladze et al. [27] used the photosensitive BZ reaction to study wave propagation and annihilation at the boundaries of excitable and inhibitory regions. Depending on the degree of excitability of the two areas (which was controlled via illumination — dark = excitable, light = inhibitory) waves can either penetrate into the inhibitory region or they collapse in the excitable zone. Thus, the layer adjacent to the boundary between the inhibited and non-inhibited regions of the active medium can either decrease or increase the effective size of the excitable element compared to its original geometry. Thus, whilst not implementing any experimental gates the authors speculate that this externally controlled inhibition would allow the shift of the boundary layer in either direction, which in turn should enable them to re-align the switching of logic gates in excitable media without changing the actual geometry of the gates.

In a related theoretical paper Motoike and Yoshikawa [190] showed that a travelling wave can be transmitted from one excitable field to another excitable field via an intervening passive diffusion field in a characteristic manner depending on the

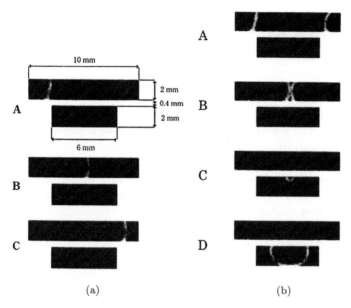

Figure 1.7: Experimental implementation of a T-shaped coincidence detector: (a) time evolution of input pulse propagation which travels from left to right without generating a signal in the output (A–C correspond to times 0, 45 and 90 sec, respectively), (b) time evolution of two input pulses that coincide in the signal channel resulting in an output (A–D correspond to times 0, 40, 60 and 80 sec, respectively).

spatial geometry of the excitable fields. In this way they could design various logic gates and a time-sequential memory device. This study was extended to include computation with multiple excitable pulses, i.e. an output pulse is only generated when the time difference between certain input pulses falls within a certain range [189].

In related experimental studies Gorecki *et al.* [112] produced a T-shaped coincidence detector; see Fig. 1.7. This consists of an active area in the form of a small rectangle with another active horizontal bar above it. These two shapes are separated by an inactive/passive (non-excitable) gap of 0.4 mm. Active areas are defined by low light intensity and passive areas by high light intensity. If a wave is initiated at one end of the horizontal bar then it will travel past the small rectangle and the excitation will not spread (Fig. 1.7a). However, if waves are started at each end and they converge above the small rectangle then excitation will spread and a wave is formed in the output channel (Fig. 1.7b). Thus, a so-called coincidence detector is created which also acts as a logical AND gate. The authors also suggested theoretical setups for counting elements involving three chemical diodes, a T-shaped coincidence detector and a chemical memory cell used in conjunction.

Ichino *et al.* [138] were able to construct the same chemical diode as in [26] but simply using the projection of the diode image onto a light-sensitive BZ reaction. The inactive areas (inhibitory) around the diode including the inactive gap corresponded to a high light intensity. They also showed experimentally that by altering the light intensity they could perform different operations using the same overall geometry of the projected circuit.

Chemical transformers

There is much interest in the propagation of chemical waves across passive barriers of various widths. The passive gap can take the form of a physical area without catalyst or can be actively controlled via the application of high light intensity. A periodic input signal of waves may be modulated by the barrier into a complex output signal depending on the gap width and frequency of the input [261]. It is easy to see how this may have applications in information processing, especially if many gaps of differing widths/spacing are employed [278].

The operation of the 'T'-shaped coincidence detector depends on the differential transmission of waves across non-excitable gaps. Gorecka and Gorecki [111] discussed its use as a band filter of chemical signal frequency.

Interestingly, in an analogous natural example the transmission of calcium waves across inactive gaps (cell-free) is observed during extracellular communication in glial cells [124].

1.2.6   Image processing using active chemical media

Kuhnert *et al.* [154] in original experiments demonstrated that it was possible to use a thin layer incorporating the light-sensitive BZ catalyst $(Ru(bpy)_3^{2+})$ and an oscillating BZ solution for basic image-processing operations. In the experiments halftone images were projected onto the thin layer using differential light intensities. The resultant photo-induced phase shift in the chemical oscillations allowed them to implement contrast enhancement, contour detection and other image-processing operations. The authors also pointed out that their reaction was a chemical realisation of an associative memory and suggested that it may be possible to implement learning networks via chemical means.

In such an oscillating reaction a periodic change of colour (orange to green) is observed corresponding to the bulk oscillations of the system. By using appropriate filters the contrast can be improved so that any image is effectively black and white. If an image is projected onto a thin layer of this photosensitive oscillating reaction then each point is perturbed dependent on the intensity and time of illumination, resulting in a proportional delay of the next oscillation. This effective photo-inhibition is caused chemically by a local increase in the inhibitor (bromide ions) [25].

In a continuation of this work Rambidi and Yakovenchuk [229, 230, 232] utilised a similar system to study the range of possible image-processing operations. In this work Rambidi and Yakovenchuk compared the responses of the pseudo-two-dimensional oscillating chemical systems used to those of shunted feedback neural networks [118].

The chemical medium is considered as a network of coupled cells whereby the cell size is linked to the diffusion length [96] of the reaction and the coupling is via diffusion. As cells are connected via diffusion then each cell should be connected

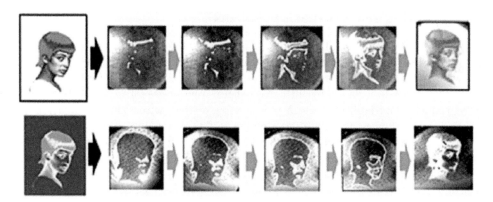

Figure 1.8: Processing of positive (top) and negative (bottom) half-tone pictures using a light-sensitive oscillatory BZ medium. Reproduced with kind permission of N. G. Rambidi.

to every other cell; however, in reality there is a time delay proportional to the distance between cells: therefore the interaction decreases proportional to separation distance. Thus, any control of a cell in the network via the application of light will in turn affect the neighbouring cells via diffusive coupling. In general, this analogy between excitable media and neural networks held true for networks with short-range coupling; however, the more complicated image-processing operations achievable with networks having long-distance coupling were not observed in real experiments. Thus, it must be assumed that the effective coupling of excitable-media cells does not seem to extend far beyond the adjacent nearest neighbours.

In further experiments with a thin reagent layer (0.25 mm, silica gel) placed in a reaction vessel with an oscillating BZ reaction, the following basic image-processing operations were observed for a given black and white input image: (i) contour enhancement, (ii) alternation of positive and negative images and (iii) disappearance or enhancement of small features.

Where images having several levels of brightness were used as projected (using a computer-controlled video projector) inputs, the areas having different brightnesses were enhanced at different stages of the reaction's evolution suggesting that more subtle image-processing operations could be possible with chemical systems (see Fig. 1.8).

Methods for fine control of the basic image-processing operations included control of temperature, reaction composition (acidity), exposure and intensity of light and exposing the negative or positive form of the initial image (see differences in Fig. 1.8 top and bottom). It was noted that the chemical medium is capable of short-term memory as the input image is stored by the medium and subjected to image transformation. The time for which the image remains clear in the processor ranges from 5 min to an hour dependent on the medium and the contrast of the input image.

1.2.7   Solving shortest-path problems

Steinbock *et al.* [269] were able to show that propagating waves in an excitable BZ medium could be used to find optimal paths through complex labyrinths. They utilised a membrane soaked in an excitable BZ solution with rectangular areas removed in order to form a labyrinth. No natural waves were formed but they could be reproducibly initiated at certain points using a silver wire. The propagation of the waves was then analysed using time-lapse digital imaging techniques. When a wave is initiated it will split at each junction and will therefore reach all points connected to the starting point. Some wave fragments propagate into blind channels and disappear on collision with the boundary. Image sequences give the path length from any location in the maze to a given target point. A colour-coded map was produced which corresponded to the time elapsed from the wave initiation to the local maxima of excitation. Thus, the minimum path length from any point in the maze to the target point was given by the product of the elapsed time assigned to that point and the constant wave velocity (2.41 ± 0.18 mm/min). Collisions of waves that were temporarily separated by obstacles marked the boundary lines between different paths with the same absolute distance. The authors noted that the transit time and distance for every location in the labyrinth are determined in parallel by a single propagating wave.

Agladze *et al.* [28] were able to demonstrate how the optimal path between two points in a two-dimensional vector field could be calculated with the aid of a chemical wave. They utilised a thin liquid layer of the BZ reaction and monitored wave dynamics with a CCD camera, a VCR and frame-grabbing software. Obstacles to increase the complexity of the path finding were added to the chemical medium via the application of a strong light spot or via the addition of a drop of chemical inhibitor KCl. If a path between two points was to be found, a wave was initiated at A using a silver wire and the wave monitored as it passed through the obstacles to point B. A wave was then initiated at point B and a compound image produced of the forward- and backward-propagating waves. The contour of the joined intersection points of the waves at each stage as they travelled from A to B and from B to A through the obstacles represents the optimal path between A and B.

Rambidi and co-workers suggested another methodology for finding the shortest paths in labyrinths using reaction-diffusion media. Instead of using trigger waves (waves formed usually in excitable systems) they proposed a scheme whereby light-controlled phase waves are used. Phase waves are much faster than trigger waves as they propagate independently of diffusion along a phase gradient in an oscillatory medium. In this setup a light-sensitive BZ reaction in oscillating mode is used and a composite image of a labyrinth and a pre-determined non-uniform background is projected onto it. The non-uniform background causes a phase wave to propagate from the input towards the output of the labyrinth. To determine the shortest path a number of further steps were required including a high degree of conventional processing. The first was to record time-lapse images corresponding to the steps of the phase wave spreading through the labyrinth. The second was to test the connections between each labyrinth fragment at every branching point of the spreading wave (conventional image processing). The final step was to subtract unconnected fragments from the initial labyrinth image again using conventional digital image processing. The authors noted that the methodology was applicable to labyrinths with multiple inputs and outputs but not to labyrinths with cyclic structures where the path determined would consist of two coupled routes.

1.2.8 Memory on an excitable field

Motoike *et al.* [191] were able to show experimentally how a memory could be implemented in an excitable chemical reaction (see Fig. 1.9). In their experiments they used a Nafion membrane soaked in the BZ catalyst ferroin immersed in an excitable BZ reaction. The membrane assembly consisted of a circular track with a small gap of 0.168 mm (the only break in the circle) and an input channel to the track (Fig. 1.9a). The input channel was separated from the ring by the same gap of 0.168 mm and placed at right angles to the ring just above the gap (Fig. 1.10). When a wave is initiated in the input channel it travels along until it reaches the gap. At this point, due to the critical gap width, it can transfer across the gap onto the ring. However, once on the ring the second gap dictates that it may only travel in the anti-clockwise direction. This is because the effective diffusion length of the wave perpendicular to the boundary is greater than that parallel to the boundary. Thus, by using a suitable asymmetric spatial arrangement of excitable fields a diode can be created between adjacent excitable fields even where these fields exhibit the same excitability.

When the pulse travels full circle it approaches the gap and is able to overcome it due to its geometry being parallel to the boundary. Thus, all pulses inputted to the ring continue to propagate on the ring within the life of the reaction. Thus, the inputted pulses are retained as 'dynamic memory'. If the input channel is shifted downwards (Fig. 1.9b) so that it is just below the gap in the circle, then the input pulses will travel in a clockwise direction thus erasing the counter-clockwise waves. The authors equate this to a real-time memory which can be erased and re-written (Fig. 1.9c). A more detailed discussion of possible experimental geometries for implementing memory on an excitable field is discussed including output channels from the ring and two input channels.

1.2.9 Pattern recognition and chemical neural networks

Hjelmfelt *et al.* [129] simulated a pattern-recognition device constructed from large networks of mass-coupled chemical reactors containing a bistable iodate–arsenous acid reaction. They encoded arbitrary patterns of low and high iodide concentrations in the network of 36 coupled reactors. When the network is initialised with a pattern similar to the encoded one then errors in the initial pattern are corrected bringing about the regeneration of the stored pattern. However, if the pattern is not similar then the network evolves to a homogenous state signalling non-recognition.

In experimental work Laplante *et al.* [159] used a network of eight bistable mass-coupled chemical reactors (via 16 tubes) to implement pattern-recognition operations. They demonstrated experimentally that stored patterns of high and low iodide concentrations could be recalled (stable output state) if similar patterns were used as input data to the programmed network. This highlights how a pro-

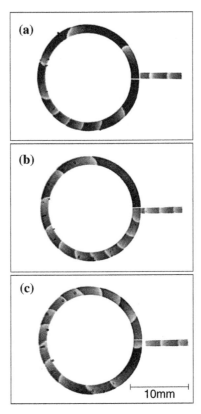

Figure 1.9: Real-time memory on an excitable field (BZ reaction): (a) successive input signals are introduced into the ring and only propagate in an anti-clockwise direction, (b) input channel is shifted to the lower position and waves now rotate in a clockwise direction, (c) the input channel is removed and the waves continue to propagate in the clockwise direction [191]. Reproduced with kind permission of I. Motoike and S. Nakata.

grammable parallel processor could be constructed from coupled chemical reactors. This chemical system has many properties similar to parallel neural networks.

As yet no large-scale experimental network implementations have been undertaken, mainly due to the complexity of analysing and controlling so many reactors. That said, there have been many experimental studies carried out involving coupled oscillating and bistable systems [272, 75, 44, 43, 74, 132]. The reactions are coupled together either physically by diffusion or an electrical connection or chemically, by having two oscillators that share a common chemical species. The effects observed include multistability, synchronisation, in-phase and out of phase entrainment, amplitude or 'oscillator death', the cessation of oscillation in two coupled oscillating systems, or the converse, 'rhythmogenesis', in which coupling two systems at steady state causes them to start oscillating [86].

An interesting development in this area is the use of micron-sized BZ catalyst-

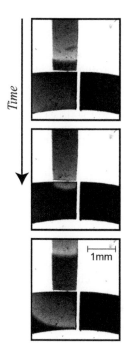

Figure 1.10: Close-up of the 'diode' junction (used in Fig. 1.9) showing wave propagation in only one direction around the ring [189]. Reproduced with kind permission of I. Motoike and S. Nakata.

loaded ion-exchange beads [309] which exhibit uniform global oscillations (below a critical bead size of 0.62 mm) when immersed in the BZ solution at certain experimental parameters. In experiments, beads in diffusive contact were shown to exhibit in-phase synchronisation. In recent experiments Fukuda *et al.* [105] demonstrated global synchronisation in a 10 by 10 array of coupled oscillators (beads) using the application of light noise at an optimal intensity.

These chemical systems act as simple models for networks of coupled oscillators such as neurons, circadian pacemakers and other biological systems [147]. For example, biological systems are assumed to process information via a phase coincidence of oscillations representing the firing of neurons [115].

In previous sections we have seen the use of excitable and oscillating chemical systems to solve a number of computational tasks. However, to some degree the lack of compartmentalisation in these systems limits the domain of solvable tasks, thus making it difficult to realise general-purpose computing. This proposed methodology of utilising networks of coupled oscillating chemical reactions may provide a solution. The fact that these coupled oscillators can be controlled via the application of external fields (data input) such as light provides a possible method for undertaking a number of complex computations provided an effective methodology for realising large-scale networks can be found.

1.3 Other chemical systems

This section will include discussion of some fairly recently discovered chemical reactions or techniques that show promise for use in constructing reaction-diffusion processors. Although no published work to date discusses directly their use in information processing, some of the experiments can be interpreted in these terms.

1.3.1 The BZ reaction in a micro-emulsion (the BZ-AOT system)

Recently, Vanag and Epstein [288, 289, 290, 291, 292, 293] identified a modified BZ system consisting of a water-in-oil reverse micro-emulsion with nanometre-sized droplets of water stabilised by the surfactant sodium bis(2-ethylhexyl)sulfosuccinate (AOT) dispersed in octane.

This system results in a remarkable array of pattern formations many of which were not obtainable in the BZ system previously, including Turing patterns. Also observed were standing waves and clusters which oscillate in time but remain stationary in space [288, 293]. Another phenomenon observed was accelerating waves [288] that accelerate as they approach each other and then travel at right angles after undergoing collision (normal BZ waves travel with constant velocity and annihilate on collision).

Packet waves were also identified [288]; these waves can reflect off obstacles and collide with other packet waves of the same amplitude leading to the formation of standing waves. Alternatively, when packet waves collide they can pass through each other like solitons, after experiencing transient interference in the region of overlap. Thus, these packet waves give an example of 'chemical interference' and exhibit behaviour similar to electromagnetic waves.

Anti-spirals which propagate towards their centres [289] and dash waves [291] and segmented spirals [292], which consist of alternating regions of excitation separated by non-excited media, are also observed.

The versatility of this system arises from its controllability, whereby the size and spacing of the water droplets can be controlled by varying the oil, water and surfactant concentrations. The diffusion properties of the main intermediates in the conventional BZ reaction (activators and inhibitors) are also altered relative to each other due to their differential solubilities in water and oil. Communication may also be as a result of mass exchange during droplet collision fission/fusion.

This great diversity of dynamic and tuneable patterns in the BZ-AOT system and the chance to control these additionally via external forcing make it a promising system to investigate further in terms of implementing computations. The natural compartmentalisation of the system to create a massively parallel network of nano-sized chemical reactors (volume $\sim$ 10–22 L) interacting via diffusion would also seem to lend itself to information-processing applications.

1.3.2 Waves supported by noise

Wang *et al.* [300] showed that waves supported by applied noise could propagate in a subexcitable medium of the light-sensitive BZ reaction that was otherwise unable to support sustained wave propagation. This is similar to a phenomenon known as stochastic resonance, where the detection of weak signals of non-linear dynamical systems may improve with increasing noise up to a maximum noise level where the signal is overwhelmed. This phenomenon attracts particular interest due to the possibility of noise-supported transmission in excitable biological media [303].

In their experiments they had a thin layer of silica gel with immobilised light-sensitive BZ catalyst and this was bathed with fresh catalyst-free reagents (open reactor). The gel was exposed to a uniform light intensity transmitted from a video projector through a 460-nm band-pass filter. The light intensity was adjusted to a reference level so as to keep the illuminated area below the excitability threshold (subexcitable) [142]. During the experiment, the projected illumination field consisted of an array of square cells with the intensity in each adjusted at equal time intervals to random values above or below the reference level (Gaussian distribution about a mean centred on the reference value). A dark area at one side of the projected intensity was used to initiate waves. They found that wave propagation was enhanced with increasing noise amplitude [143], and sustained propagation was achieved at an optimal level. Above this optimal level the waves split and fragmented waves were observed.

In related work waves were initiated and modulated via the application of noise to a subexcitable BZ reaction [300]. In the absence of noise no waves propagate but when noise is applied waves are randomly initiated in cells where there is an accumulation of subthreshold perturbations, i.e. the cell is locally excited. The wave then starts to expand but some adjacent cells are biased to lower excitability by their noise history, thus inducing the wave to break. Other cells in the neighbourhood are biased to higher excitability, inducing further wave propagation and initiation. Thus, once wave nucleation occurs the result is a proliferation of wave segments on wide space and time scales (avalanche behaviour). The fragments may also undergo collisions either annihilating or forming new daughter fragments.

Interestingly, calcium waves induced in networks of cultured glial cells [141] display similar features to the ones identified in this chemical system, which the authors postulated may provide a possible mechanism for long-range signalling and memory in neuronal tissues.

1.3.3   Waves controlled by light modulation

In this work, using a similar experimental setup, sustained propagation of small wave fragments was achieved in a subexcitable BZ system by the periodic modulation of a homogeneous light field [250]; see Fig. 1.11. This was due to the generation of excitabilities above and below the subexcitable reference state via the application of certain periods and amplitudes of light intensity. A primary feedback loop increases or decreases the excitability of the medium as the wave becomes larger or smaller, thus stabilising the size. The data is calculated by the real-time processing of video images of the wave, which in turn updates the projected illumination intensity.

In an extension of this work, they were able to get particle-like waves that propagated in user-defined patterns using feedback-regulated excitability gradients [240]. Thus, the waves were stabilised using the light-modulation method described above

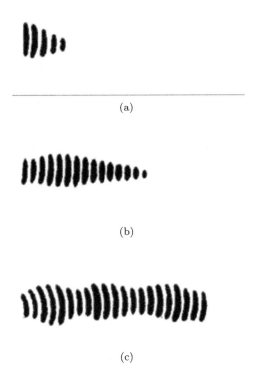

(a)

(b)

(c)

Figure 1.11: Waves supported by light modulation in a subexcitable BZ reaction: (a) wave evolution with no modulation, (b) partial support of propagation with a period of modulation (light intensity: $I = I_o \pm A/2$, $I = 3.87$ mW/cm^2 and $A = 0.22$ mW/cm^2) of 66 sec, (c) sustained propagation with a period of modulation of 133 sec. Modified from [250].

but their trajectories were altered by a secondary feedback loop. This was achieved by applying a simple linear light gradient perpendicular to the direction of propagation at the waves' centre of mass. The gradients are dynamically varied according to the waves' current and target trajectories enabling a wide variety of wave patterns to be realised. The secondary wave guiding excitability gradient represents a truly spatio-temporal feedback.

Using the described method two or more waves can be independently controlled via a localised feedback loop and can be made to interact. Waves could also be confined to specific areas by having gradient-free fields surrounded by areas where a gradient is applied perpendicular to the propagation, making the wave turn back on a circular trajectory until it re-enters the gradient-free field. There is much scope for the design and implementation of spatio-temporal feedback of this type with the linear form implemented in the above experiments only representing the simplest case. The applications to information processing of the production and fine control of particle-like waves will be discussed in Chap. 3 in terms of collision-based computing, although this may only be one area in which the experiments could be exploited.

1.4 Current state of reaction-diffusion processors

The preceding sections have attempted to define a reaction-diffusion processor and sought to give an overview of the development of experimental prototypes. This section will attempt to set a current context for chemical reaction-diffusion processors in terms of competing unconventional and conventional technologies and explore possible future directions.

1.4.1 Biologically inspired computing

There exists the possibility of drastically improving conventional digital devices by utilising molecular primitives. However, to date although some relatively successful experiments have been implemented [183], it remains a distant prospect. However, the chemical systems discussed in the preceding sections adhere far more closely to a non-von Neumann paradigm of biologically inspired computation [177, 71]. Thus, we will attempt to compare and contrast existing unconventional biological computing paradigms with the chemical reaction-diffusion computing.

Differences?

Before we do discuss the specific performance of some biological computing architectures, what really is the difference (if any) between chemical computing (reaction-diffusion computing) and biologically inspired computing?

As mentioned above, the chemical-based processors we have described operate in a manner more akin to processors of biological origin rather than digital computing devices. The similarities between active chemical media and biological/natural systems are also emphasised by the great parallels in self-organised pattern formation and signal transmission mechanisms. For these reasons, as has been highlighted many active chemical systems have served as robust models of biological systems.

The major difference is that biological computing paradigms are predominantly based on the action of complex molecules, enzymes, proteins, cells and DNA. Their chemical counterparts discussed in the preceding sections are in contrast based on very simple molecular processes. These large biological molecules are highly evolved and somewhat task specific, although with a degree of adaptability. However, if their information-processing capabilities are assessed they are predominantly based on a hierarchical series of chemical reactions.

DNA computation

If you search for DNA computing on an Internet search engine, you will find thousands upon thousands of relevant hits. The same is not true for chemical computing or reaction-diffusion computing or some other unconventional computing paradigms, so why this huge interest? The interest was sparked by the paper of Adelman [24], who showed how DNA coupled with biochemical techniques could be used to solve a Hamiltonian path problem (to find the shortest route between a number of joined points only visiting each point once). This problem is NP-complete; thus, it is considered unsolvable by conventional computers in polynomial time.

The problem was solved using the reaction of oligonucleotides (one for each point) leading to the formation of DNA molecules encoding all random paths through the graph. These were then amplified, separated and purified to obtain DNA fragments coding the right input/output, correct number of points and ones containing all possible points. However, this process took a number of days and there were concerns about some techniques used that could be prone to errors. Although some improvements and adaptations of the original techniques have been achieved [63], there still remain some issues which are unclear with regard to the practical use of DNA computation for solving combinatorial problems of higher dimensions. These include the number of laboratory steps, durations of the laboratory steps and energy considerations [229]. As DNA computation is effectively based around chemical reactions between large molecules (oligonucleotides), there are also mass and chemical kinetic considerations. For solving more complex problems the number of nucleotides would have to increase massively. Mixing larger numbers of oligonucleotides would also reduce reaction rates, thus increasing computation times [48].

However, work continues to try and overcome these problems [49] and whilst DNA is not currently viable for information processing it cannot be ruled out in some form in the future. What is good about such applied research in both a biological and a chemical context is that it highlights fundamental gaps in our understanding and stimulates further research. If a conclusion had to be drawn about the relative merits of reaction-diffusion and DNA approaches it would be that currently the reaction-diffusion approach provides comparatively simple implementations of real-time processing prototypes albeit for a different class of problems.

Returning to the question about popularity of DNA computing, we can suggest an answer — DNA computers are popular because they bear a huge potential to formalisation. If you are a 'pure' theoretician, you can build a lot of theories on top of a DNA computing paradigm. However, most problems solved by DNA computers are only solved potentially; also, experimental implementations are tricky, unstable and hardly verifiable.

Neuronal computation

For the purposes of this brief overview we will be concerned with computational approaches that utilise networks of real cultured neurons. This approach naturally arises from the drive to mimic the brain as a computer; thus, why not just use the raw materials direct. The limitations are probably obvious — having no firm understanding of how the brain does what it does, particularly how it extracts useful information from the basic elements, makes experimental approaches difficult.

As we are comparing reaction-diffusion and neuronal systems it is worth mentioning the commonality that exists. For example, the rules governing the spiral wave dynamics in the BZ reaction and calcium waves in the brain were shown to be the same mathematically [160]. Additionally, the diffusion of simple neurotransmitters such as nitric oxide (NO) from one neuron to many adjacent neurons appears to play a significant role in information-transfer processes [221]. This so-called volume signalling operates on a different spatial and temporal scale to the electrical signalling.

In terms of information processing, there have been some relatively successful implementations albeit at an early stage of development. Shahaf and Mahom [256] were able to demonstrate learning in networks of cortical neurons. This was achieved by stimulating cultured neurons with a pattern of electrical inputs and monitoring for recurring patterns in action potentials. De Marse *et al.* [82] used the same approach to control a simulated mobile robot; see also [42].

As mentioned, a major limitation presently is how to utilise the neuronal networks effectively — how to input data, how to make optimal connections, etc. Thus, for it to develop there are major engineering considerations as well as biological ones.

If this approach were to be compared to the reaction-diffusion approach we might consider that presently the neuronal approach may disappoint due to the natural comparison with the brain, whereas we might be inclined to be more pleased if we could process information with a dish of simple chemicals!

Membrane computing

Membrane computing is a branch of natural computing whose aim is to abstract computational models and paradigms from the structure and functioning of the living cell (neurons as a particular case); see the overview in [213, 214]. After all, membrane computing is a science of formal languages being very far from experimental science. We are not aware of any laboratory implementations of membrane computers, and may envisage that the paradigm will remain in the kingdom of theory for the time being.

Evolutionary computing

Evolutionary computing or evolutionary algorithms are a family of computer-based problem-solving systems using computational models relating to known mechanisms of evolution as key elements in their design and implementation. The major types are genetic algorithms, classifier systems, evolutionary programming and genetic programming. The basic concept is to simulate the evolution of an individual structure by processes of selection, mutation and reproduction, where the processes depend on the perceived performance of the individual structures [125]. Thus, it can be considered complementary with rather than competitive to experimental reaction-diffusion systems.

1.4.2 Physical systems

Gas-discharge systems

Many complex and dynamic patterns can be observed in planar dc- and ac-driven gas-discharge systems. Patterns such as solitary waves [31, 57], spirals and targets [271], periodic and Turing patterns [30] plus other more complex patterns are commonly observed. It is interesting that these patterns and their interactions can be modelled in a qualitative manner via a three- component reaction-diffusion model [58].

Experiments have shown that these systems can solve computational tasks such as construction of a Voronoi diagram [313] and shortest-path problems [236]. The information-processing power of such systems could theoretically outperform reaction-diffusion-based media and would be close to competing with contemporary digital systems for certain tasks. However, a difficulty in reliably inputting information currently hampers their development.

Quantum computation

Quantum computation is a methodology proposed to enhance conventional methods of processing, where logical circuits are based on quantum interference effects rather than the laws of classical physics [84]. For certain applications such as factoring large numbers (useful for encryption, etc.) these computers have the potential to outperform conventional processors [260]. However, although quantum computers have been constructed they are at a pioneering state of development in practical terms.

1.4.3   Reaction-diffusion processors vs. conventional processors

In most obvious aspects reaction-diffusion processors suffer when compared to conventional silicon-based processors. We have touched on some of their perceived advantages, massive parallelism, self-organisation, etc., but the problem is how to bring these to bear on real computational situations. To some extent it is not surprising that unconventional methods such as the reaction-diffusion approach do not compare favourably, as silicon technology has been developed to the nth degree whereas reaction-diffusion processors are in their infancy. Who can deny that conventional processors have served us well? However, if silicon technology is reaching a plateau then the time is right to explore new computing methods. Really, the time has always been right because of the limitations of the current approach — thus the huge interest in natural computing methods, neural networks, etc.

Agladze *et al.* [28] provided an estimation of the current state of silicon vs. chemical computation. They postulated that a single processor in the BZ reaction was defined by the diffusion length of the order of 0.1 mm. Thus, a 9-cm Petri dish contains 600 000 processors, working in parallel. The processors are very slow, corresponding to a wave-propagation speed (0.05 mm/sec) that yields one switch per 2 sec. Thus, despite the large number of processors, the maximum number of operations is 300 000 per second, still hundreds of times slower than a modern PC. Thus, the conclusion is that for this type of processor to be viable reactions with extremely fast excitation dynamics must be found or designed. This leads nicely onto a discussion of the future of reaction-diffusion-based processors in the next subsection.

1.4.4 The future

It may seem strange to finish the introduction with a discussion about the future of reaction-diffusion-based processors. However, hopefully it will set the scene for the remaining chapters of the book. It should be pointed out that the rest of the book is mainly set aside for examples of reaction-diffusion processors tackling real-life problems. These will be actual experimental processors, models based on reaction-diffusion processes and finally a new breed of silicon-chip device inspired by reaction-diffusion processes.

In some cases the experiments may seem very simplistic but that reflects to some degree the current knowledge of this class of reactions. In order to improve the current processors and indeed design new processors, we need a better understanding of the mechanisms and of molecular processes *per se*. At the present time we are only able to exploit some natural dynamic and exert minimal control over a reaction's evolution. It will also be apparent that many examples discussed utilise a fair degree of conventional processing power in order to yield the final computation. Although this alone is not a reason against this type of processor, it does raise questions about possible paybacks from these techniques. We will attempt to be objective and look at both the positive aspects and the limitations.

However, this might be to paint a negative picture, which is not the intention. The devices we discuss are not meant to compete with conventional processors but hopefully supplement them in areas where they are currently limited. Also, the study of these reaction-diffusion systems serves as a good model for information processing in biological entities and natural systems. We will also discuss the possibility of these reaction-diffusion systems serving as models for the synthesis of new silicon technologies. If we are to extend this point then maybe there exists the possibility of producing hybrid silicon–reaction-diffusion systems and even hybrid reaction-diffusion–biological systems [157].

The natural dynamics of reaction-diffusion systems may also be exploited to make novel sensors [301] and actuators. There are already examples emerging where reaction-diffusion systems have been utilised to make functional biomimetic materials. For example, a BZ oscillator has been used to direct a periodic self-assembly process of gold nanocrystals [89]. Self-oscillating nanometre-sized gel beads were synthesised by incorporating a BZ catalyst into a polymer of N-isopropylacrylamide [239]. Prior to this a polymeric micro-array of cilia had been constructed capable of spontaneous ciliary motion when placed in the BZ solution [275]. These types of materials could serve as actuators in nano-technology applications.

In general terms, a better understanding of non-linear chemical processes and reaction-diffusion processes would benefit nearly every facet of science and engineering not just the field of computation.

And, let's face it but without the processes of evolution and design, brains and silicon processors might still occupy a chemical soup, so what better place to begin a search for a relative!

Chapter 2

Geometrical computation: Voronoi diagram and skeleton

Geometrical constructions are based on the topology and neighbourhood and, at least in Euclidian space, a 'natural' configuration of objects and intuitive ways of their analysis. So, at first sight, the majority of locality-based geometrical problems could in principle be solved in reaction-diffusion processors. In this chapter we restrict ourselves to just two problems — space tessellation, or Voronoi-diagram construction, and planar shape skeletonisation — because these two problems have explicit natural parallelism, and they are also widely present in many subfields of image processing and pattern recognition such as virtual reality and robotics.

2.1 Voronoi diagram

Let $\mathbf{P}$ be a non-empty finite set of planar points. A planar Voronoi diagram of the set $\mathbf{P}$ is a partition of the plane into such regions that, for any element of $\mathbf{P}$, a region corresponding to a unique point p contains all those points of the plane which are closer to p than to any other node of $\mathbf{P}$. A unique region

$$vor(p) = \{z \in \mathbf{R}^2 : d(p, z) < d(p, m) \forall m \in \mathbf{R}^2, \, m \neq z\}$$

assigned to point p is called a Voronoi cell of the point p. The boundary of the Voronoi cell of a point p is built up of segments of bisectors separating pairs of geographically closest points of the given planar set $\mathbf{P}$. A union of all boundaries of the Voronoi cells determines the *planar Voronoi diagram*:

$$VD(\mathbf{P}) = \cup_{p \in \mathbf{P}} \partial vor(p).$$

Examples of planar Voronoi diagrams are shown in Fig. 2.1. The first example (Fig. 2.1a) just shows that a bisector separating two points forms an edge of their Voronoi cells. Three open Voronoi cells are shown in the second example (Fig. 2.1b). A closed Voronoi cell, contacting four open Voronoi cells, is shown in the Voronoi diagram of five planar points in the third example (Fig. 2.1c). A 'full-scale' diagram of many points is constructed in Fig. 2.1d. We refer the reader

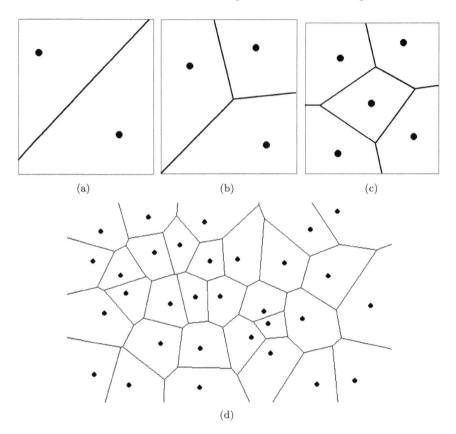

(a)   (b)   (c)

(d)

Figure 2.1: An example of a bisector of two points (a), Voronoi cells of three (b) and five points (c) and a Voronoi diagram of many points (d).

to the basic texts [224, 150, 205], where almost all types of Voronoi diagrams and most conventional algorithms of their construction are discussed.

Voronoi diagrams tend to be used in situations where space needs to be partitioned into 'spheres of influence'. Therefore, as a modelling tool they find applications in fields such as astronomy (identifying star and galaxy clusters [212], modelling gravitational influence [88]), biology (models of cells [133], tumour-cell growth [52]) and related areas such as zoology (modelling territories [45], e.g. slime-mould cultures [193]), ecology (competition models and spatial distributions [66]), chemistry (modelling crystal growth [123], modelling of composite and heterogeneous materials [185]), molecular modelling (e.g. protein volume evaluation [222]), animal markings [298, 178], geography (cartography and geographical modelling of surfaces) [65], finding the best facility location [41] and market analysis [205].

This is not to ignore their uses in pure mathematics, statistical data analysis (cluster analysis) and the many applications in computer sciences. These include practical problem-solving tools such as pattern recognition (via computation of a skeleton [19]), path planning in the presence of obstacles [90] and computer graphics or computer-generated images [83]. The Voronoi tessellation is also extensively used in the field of computational geometry, where it is applied to solve variations on the nearest-neighbour problem [113, 224].

In addition to having applications in all areas of scientific modelling, Voronoi diagrams can be observed in nature from animal markings [210] to geological structures and still further in universal structures such as gravitational caustics. Therefore, they form an important class of natural pattern formation [270].

2.2   Time-to-space mapping

Voronoi cells of a planar set represent the natural or geographical neighbourhood of the set's elements; therefore, the computation of a Voronoi diagram based on the spreading of some 'substance' from the data points is usually the first approach of those trying to design massively parallel algorithms.

The 'spreading substance' can be represented by a potential or an oscillatory field, or diffusing or phase waves, or simply by some travelling inhomogeneities or disturbances in the physical computing medium. This idea was firstly explored by Blum, Calabi and Hartnett in their 'grass-fire' transformation algorithm [54, 55, 67]. A fire is started at planar points of given data set, the fire spreads from the data points on a substrate (that constitutes the computing medium) and the sites where the fire fronts meet represent the edges of the Voronoi diagram of the planar data set.

Quite similarly, to construct the diagram with a potential field one puts the field generators at the data points and then detects sites where two or more fronts of the field waves meet. This technique is employed in the computation of a Voronoi diagram via repulsive potential and oscillatory fields in homogeneous neural networks [136, 145, 163]. These approaches are 'natural' and intuitive and their simplicity of implementation made them an invaluable tool for massively parallel image processing, particularly where VLSI implementation is concerned. However, the algorithms bear a weakness — no stationary structure is formed as a result of the computation and the meeting points of the colliding wave fronts must be detected by some 'artificial' or extrinsic methods. This disadvantage is eliminated in the reaction-diffusion technique.

2.2.1   The chemical approach

How do you go about constructing a bisector of two planar points in a chemical processor? If we drop reagents at the two points the diffusive waves spread outward from all edges of the drop with the same speed (assuming that the medium is uniform). Thus, the wave fronts travel the same distance from the sites of origination before they meet. The points where the waves meet constitute the bisector points of the two given points. An assumption was made in pioneering papers [2, 4] concerning the reaction-diffusion-based construction of a Voronoi diagram that when two wave fronts meet a precipitate is produced and that the concentration profile of the precipitate represents the edges of the Voronoi diagram. Unfortunately, this was not realised in laboratory experiments. In practice a reagent reacts with a substrate and a precipitate is formed immediately and uniformly in the resulting reaction [282, 18]. However, the precipitate is not formed at the sites where two wave fronts meet.

So, in experimental implementations of reaction-diffusion chemical processors all internal sites of the Voronoi cells are coloured by precipitate but the edges of the Voronoi cells remain uncoloured.

2.3   Cellular-automaton Voronoi diagram

We have used cellular automata as computationally fast prototypes of reaction-diffusion computers in order to model the systems and test our ideas in order to shape the further development of experimental implementations. Why cellular automata? Assuming that the computational space is homogeneous and locally connected, and every node is coupled to its closest neighbours by the same diffusive links, we can easily draw a parallel between distance and time, and thus put our wave-based approach into action.

Throughout the book we assume that every non-edge cell x of a cellular automaton updates its state depending on the states of its eight closest neighbours ○:

$$u(x) = \begin{matrix} \circ & \circ & \circ \\ \circ & x & \circ \\ \circ & \circ & \circ \end{matrix} \, .$$

The neighbourhood u determines that all processes in the cellular-automata model are constrained to the discrete metric L_∞. So, when talking about automata models we operate with a discrete Voronoi diagram, which is what a planar (continuous) Voronoi diagram is transformed to.

A discrete Voronoi diagram is usually defined on lattices or arrays of cells where one uses a two-dimensional lattice $\mathbf{Z}^2$; the distance $d(\cdot, \cdot)$ is calculated not in Euclidean but in one of the discrete metrics, e.g. L_1 or L_∞. A 'typical' discrete bisector of nodes x and y of $\mathbf{Z}^2$ is determined as

$$B(x, y) = \{z \in \mathbf{Z}^2 : d(x, z) = d(y, z)\}.$$

Not surprisingly, applying this definition of the bisector one can generate bisectors that fill a quarter of the lattices or produce no bisectors at all (see [4, 7, 168] for discussions). All of these phenomena are perfectly correct in the metrics L_1 and L_∞; however, they are not acceptable, i.e. 'do not look right', from a common-sense point of view.

When considering the term 'discrete', it would be desirable to have a common-sense discretisation of a planar Voronoi diagram. Quite reasonably, nodes contacting nodes of the bisector could form a discrete image of a continuous line segment separating the points x and y. Thus, we get a definition of a 'proper' discrete bisector:

$$B(x, y) = \{z \in \mathbf{Z}^2 : |d(x, z) - d(y, z)| \le 1\}.$$

The re-defined bisector will comprise edges of Voronoi diagrams constructed in discrete, cellular-automaton models of reaction-diffusion and excitable media.

To give the reader a sense of this type of reaction-diffusion-inspired cellular-automaton-based computing, we introduce below several versions of wave-based computations of Voronoi diagrams. These range from a naive model, where the number of reagents grows proportionally to the number of data points, to a minimalist implementation with just one reagent. We will not discuss excitable-media approaches, but details of these can be found, together with all the necessary background, in [7].

2.3.1 $O(n)$-reagent model

In this naive version of a Voronoi-diagram computation using a reaction-diffusion-based approach one needs two reagents and a precipitate to mark a bisector separating two points, i.e. n reagents are required to approximate a Voronoi diagram of n points. We place n unique reagents on n points of the given set $\mathbf{P}$; waves of these reagents spread and interact with each other where they meet.

When at least two different reagents meet at the same or adjacent sites of the space, they react and form a precipitate. Sites that contain the precipitate represent the edges of the Voronoi cells, and therefore construct the Voronoi diagram of the original points of reagent application.

In 'chemical reaction' equations the idea looks as follows (α and β are different reagents and $\#$ is a precipitate):

$$\alpha + \beta \to \#.$$

In a cellular-automaton interpretation this looks as follows:

$$x^{t+1} = \begin{cases} \rho, & \text{if } x^t = \bullet \text{ and } \Psi(x)^t \subset \{\rho, \bullet\} \\ \#, & \text{if } x^t \neq \# \text{ and } |\Psi(x)^t/\#| > 1 \\ x^t, & \text{otherwise} \end{cases} \tag{2.1}$$

where $\bullet$ is a resting state (a cell in this state does not contain any reagents), $\rho \in \mathbf{R}$ is a reagent from the set $\mathbf{R}$ of n reagents and $\Psi(x)^t = \{y^t : y \in u(x)\}$ characterises which reagents are present in the local neighbourhood $u(x)$ of the cell x at time step t. The first transition of rule (2.1) symbolises diffusion: a resting cell takes the state ρ if only this reagent is present in the cell's neighbourhood. If there are two different reagents in the cell's neighbourhood then the cell takes the precipitate state $\#$ and stays in this state forever. An example of a cellular-automaton simulation of the $O(n)$-reagent chemical processor is shown in Fig. 2.2.

Because of the stoichiometry of the cellular-automata 'chemical' equations, the precipitate $\#$ prevents the reagents from diffusing any further and the precipitate is confined to the bisector zones (diffusion is stopped 'automatically' because the formation of the precipitate reduces the number of 'vacant' resting cells). The $O(n)$-reagent model is demonstrative; however, it is computationally inefficient. Clearly we can reduce the number of reagents to four (using map-colouring theorems) but this requires planning of which sites to put the various reagents in prior to implementing the computation. This can make the pre-processing more complicated

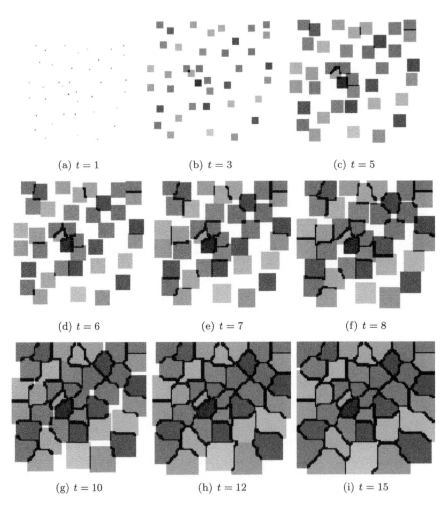

(a) $t = 1$ (b) $t = 3$ (c) $t = 5$

(d) $t = 6$ (e) $t = 7$ (f) $t = 8$

(g) $t = 10$ (h) $t = 12$ (i) $t = 15$

Figure 2.2: Computation of a Voronoi diagram in a cellular-automaton model of a chemical processor with $O(n)$ reagents. Precipitate is shown in black.

than the processing itself and totally unrealistic when dealing with real chemical processors.

Actually, the number of reagents can be reduced (literally to just two automaton states) in the cellular-automaton models when the topology of the spreading waves of excitation is taken into account [7] and in real experimental chemical processors [282, 13].

2.3.2   Minimalist model: one reagent

Now we go from one extreme to another: let us consider a model with just one reagent. The reagent α diffuses from sites corresponding to the elements of a given planar set **P**. When two diffusing wave fronts meet a super-threshold concentration of reagents prevents waves from spreading further. A cellular-automaton model represents this as follows.

Every cell has two possible states: $\bullet$ (resting) and α (reagent). If the cell is in state α it remains in this state forever. If the cell is in state $\bullet$ and between one and four of its neighbours are in state α, then the cell takes the state α; otherwise, the cell remains in the state $\bullet$ (this can be interpreted as a 'super-threshold inhibition' of the precipitation reaction). A cell state transition rule is as follows:

$$x^{t+1} = \begin{cases} \alpha, & \text{if } x^t = \bullet \text{ and } 1 \leq \sigma(x)^t \leq 4 \\ x^t, & \text{otherwise} \end{cases} \tag{2.2}$$

where $\sigma(x)^t = |y \in u(x) : y^t = \alpha|$. Some examples of Voronoi-diagram computation in cellular-automata models of a one-reagent chemical processor can be found in [13]. Some of the Voronoi diagrams computed using this approach contain segments that are incomplete or disconnected. A more realistic computational modelling approach in terms of the resulting Voronoi diagrams (with parallels to real systems) is discussed in the next section.

2.3.3   One diffusing reagent and one precipitate

A reagent β is placed onto a substrate in drops pertaining to the initial points of **P** and forms a precipitate in the reaction

$$m\beta \to \alpha,$$

where $1 \leq m \leq 4$. A cellular-automaton model assumes that every cell takes three states: $\bullet$ (a resting cell, no reagents), α (coloured precipitate) and β (reagent). The cell state transition rule is as follows:

$$x^{t+1} = \begin{cases} \beta, & \text{if } x^t = \bullet \text{ and } 1 \leq \sigma(x)^t \leq 4 \\ \alpha, & \text{if } x^t = \beta \text{ and } 1 \leq \sigma(x)^t \leq 4 \\ x^t, & \text{otherwise} \end{cases} \tag{2.3}$$

where $\sigma(x)^t = |y \in u(x) : y^t = \beta|$.

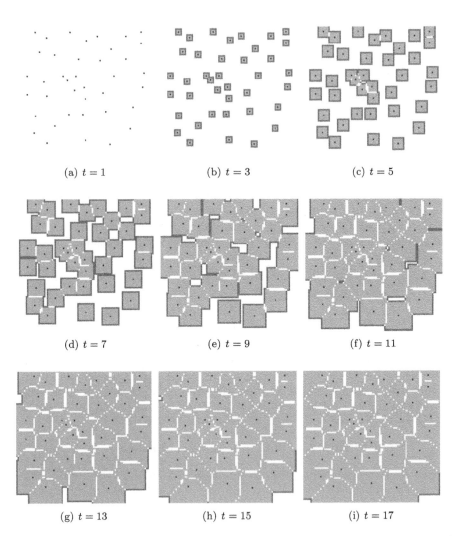

(a) $t = 1$ (b) $t = 3$ (c) $t = 5$

(d) $t = 7$ (e) $t = 9$ (f) $t = 11$

(g) $t = 13$ (h) $t = 15$ (i) $t = 17$

Figure 2.3: An example of a Voronoi-diagram computation in an automaton model of a reaction-diffusion medium with one reagent and one substrate. Reactive parts of the wave fronts are shown in red. Precipitate is shown in grey and the edges of the Voronoi diagram are white.

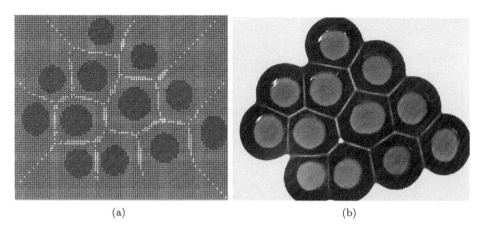

(a)   (b)

Figure 2.4: An example of Voronoi diagrams computed, for the same data discs, in a three-state cellular-automaton model of a reaction-diffusion processor (a) and in an experimental reaction-diffusion processor (b). In (a) the white pixels represent cells in the empty state; grey pixels represent cells in the precipitate state. Dark-grey pixels are artificially marked to show the initial positions of the data discs. The palladium processor shown in (b) shows how straight 'precipitate-free' bisectors are constructed between opposing reaction fronts when they collide (note that the computation is still in progress, so some bisectors are incomplete).

The development of a cellular-automaton (reaction-diffusion medium with one reagent and one substrate) dynamics during the computation of a Voronoi diagram is shown in Fig. 2.3.

An example of the cellular-automaton approximation of a Voronoi diagram of several discs is shown in Fig. 2.4.

2.3.4 Consequences of discrete simulation

The one-reagent model (2.2) defines a semitotalistic cellular automaton (assuming that • is arithmetic 0 and α is 1). Each cell of a semitotalistic cellular automaton updates its state depending on its own current state and the arithmetical sum of its neighbours' states:

$$x^{t+1} = f\left(x^t, \sum_{y \in u(x)} y^t\right).$$

In [2] we considered a cellular automaton, called a generalised semitotalistic cellular automaton, where each cell updates its state depending on its current state and a distribution of the sums of the different states in its neighbourhood (we assume q cell states): $x^{t+1} = f(x^t, \mathbf{w}_x^t)$, where $\mathbf{w}_x^t = (w_{x\alpha}^t)_{1 \leq \alpha \leq q}$, $w_{x\alpha}^t = \sum_{y \in u(x)} \chi(y^t, \alpha)$ and $\chi(y^t, \alpha) = 1$ if $y^t = \alpha$ and $\chi(y^t, \alpha) = 0$ otherwise.

A Voronoi diagram is not constructible in a cellular automaton if there are a pair of neighbouring cells from the given set $\mathbf{P}$ which are separated by incomplete bisec-

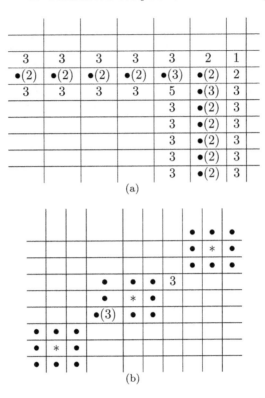

Figure 2.5: (a) A fragment of a discrete wave front generated from a single-cell source. Non-resting cells are shown by •. We have also shown the numbers of non-resting states in the eight-cell neighbourhood of resting and non-resting cells in '()' brackets. (b) An example of the interaction of wave fronts, originating from three separate sources (∗). In the interaction of two wave fronts at the bottom left-hand corner of the lattice we can see a non-resting cell (that must take the state of the precipitate at the next time step) that has three non-resting neighbours. The top right-hand part of the picture also shows a situation when a resting cell (that must take the state of the precipitate at the next time step) has three non-resting neighbours.

tors; meaning that neither 'proper' discrete or discretised planar Voronoi diagrams are calculated.

A Voronoi diagram is not constructible in a generalised semitotalistic cellular automaton with two or three cell states.

It is enough to demonstrate (see Fig. 2.5) that the local sums corresponding to approaching wave fronts can be found in the local sums corresponding to a singular expanding wave front. Therefore, there is no deterministic transition that uniquely determines the formation of precipitate states, i.e. bisectors.

By analogy, one could demonstrate the following: there is no finite-state generalised semitotalistic cellular automaton that can construct a Voronoi diagram. These propositions may indicate that a substrate-competition mechanism of bisector construction in experimental chemical processors may not work.

2.4 Chemical processors for Voronoi-diagram computation

Based on the phenomenology of the cellular-automaton construction of a Voronoi diagram, we fabricated several types of chemical processors, all of which approximate the construction of a Voronoi diagram equally well (within certain concentration ranges). Some nuances of the implementation are discussed in the following sections.

Firstly, a number of recipes are given which will allow the easy re-construction of these fascinating yet beautifully simple phenomena. For the construction of simple Voronoi diagrams of the type shown in Fig. 2.4b the lower concentration ranges are favoured. At higher concentration ranges the formation of stable Voronoi diagrams can be disrupted, but more about this later.

2.4.1 Palladium iodide processor

A gel of agar (1.5% by weight)[1] containing palladium chloride (palladium(II) chloride 99%) in the range 0.2–0.9% by weight (0.011–0.051 M) was prepared by mixing the solids in warm deionised water. The mixture was heated with a naked flame until it boiled with constant stirring to ensure full dissolution of palladium chloride and production of a uniform gel (on cooling). Aliquots ($\sim$ 6 ml) of the liquid were then transferred to Petri dishes and left to set in order to give a final gel thickness of approximately 1 mm. The use of Petri dishes was favoured as the reaction process is relatively slow and drying of the gel (syneresis) can occur if kept open to the atmosphere. The unreacted gel processors were then kept for 30 min, although they remained stable for in excess of 24 hours provided drying was controlled. For the displayed images the lower range of 0.2% was favoured.

A saturated solution (at 20°C) of potassium iodide was used as the outer electrolyte for the reactions. Drops of the outer electrolyte were applied to the surface of the gel to initiate the reaction process. A circular precipitating reaction front was initiated around each drop and this moves outwards from the source into the Petri dish (diffuses) until it interacts with an opposing reaction front of a neighbouring drop.

The result (computation) is a Voronoi diagram pertaining to the original pattern of the outer electrolyte drops; see Fig. 2.4b.

2.4.2 Ferric ferrocyanide (Prussian blue) processor

The second chemical system displaying Voronoi-diagram formation employs the ferric ion and ferrocyanide couple, where the primary product formed is ferric ferrocyanide (Prussian blue, $Fe_4(Fe(CN)_6)_3$).

Potassium ferrocyanide ($K_4Fe(CN)_6 \cdot 3H_2O$ 2.5 mg/ml (5.91 mM) (99+%), BDH Chemicals Ltd., Poole, Dorset, UK) was mixed with a 1.5% gel of agar. The preparation method used was the same as for the palladium processor detailed above.

---

[1]All chemicals listed were purchased from Sigma-Aldrich Chemical Company Ltd., Poole, Dorset, UK unless otherwise stated.

A ferric ion solution was used as the outer electrolyte (300 mg/ml, 0.74 M, iron(III) nitrate nonahydrate, but any highly pure soluble ferric ion source will suffice, 99.99+%).

2.4.3 Copper ferrocyanide processor

This reaction was carried out on the same potassium ferrocyanide gel films using copper(II) chloride as the outer electrolyte (300 mg/ml, 2.23 M, or another source of cupric ions). The resulting precipitate was rust coloured (copper ferrocyanide, $Cu_2Fe(CN_6)$).

For studies of extended pattern formation, substrate concentrations were varied between 2.5 mg/ml and 100 mg/ml (5.91 mM–0.23 M) whilst keeping the same reactant concentration, gel thickness and concentration.

Copper(II) chloride can be replaced with other salts such as nickel(II) chloride, manganese(II) chloride and cobalt(II) chloride.

2.4.4 Copper hydroxide processor

In this system copper(II) chloride was utilised as the substrate and potassium hydroxide used as the outer electrolyte (8 M). An analogous system was found to exhibit a diverse range of pattern formations at both low and high substrate concentrations including interacting chemical waves, Liesegang patterns, target waves, spiral waves and cardioid structures [120].

We investigated the system at low substrate concentrations of between 2.5 mg/ml and 20 mg/ml (0.0186–0.149 M) copper(II) chloride in a 1.5% gel of agar. The system was also investigated at high substrate concentrations of 40 mg/ml, 60 mg/ml, 80 mg/ml, 100 mg/ml and 150 mg/ml to study the extended pattern formations.

For construction of a Voronoi diagram copper(II) chloride can be replaced by copper(II) bromide, copper(II) nitrate or other salts such as nickel(II) chloride, manganese chloride, cobalt(II) chloride or silver nitrate.

2.4.5 Copper ferricyanide processor

This details a 'Liesegang-like method' for the construction of unstable/controllable processors as discussed before.

Agarose gel of 2% by weight was prepared by adding agarose powder to high-purity water at room temperature. Solubilisation was achieved by stirring for 1–2 min at 70–80°C. The inner electrolyte concentration was set by adding appropriate amounts of concentrated inner electrolyte solution at 50–60°C to an agarose solution of 100 ml (final concentrations 0.03–0.15 M $K_3[Fe(CN)_6]$). Then, the mixtures were made up to 200 ml. After shaking, the mixtures were poured into Petri dishes (gel layer 6 mm) and left to set. The reactions were started by pouring a 5-mm layer of the outer electrolyte $CuCl_2$ [2.93 M] onto the gel.

2.5 Voronoi diagrams in chemical processors

2.5.1 Mechanism of primary Voronoi diagram formation

In all the systems mentioned a stable Voronoi diagram pertaining to the original drops of the reactant solution was constructed as the primary output of the chemical processor. This is true only up to a certain system-dependent concentration (discussed in more detail in the following sections).

Figure 2.6 shows stable Voronoi diagrams constructed in the copper hydroxide and Prussian blue processors. In these reaction-diffusion systems droplets of the outer electrolyte (Fig. 2.6a, sodium hydroxide and Fig. 2.6b, ferric chloride) are put at distinct points on the surface of the gel containing the inner electrolyte (Fig. 2.6a, copper chloride and Fig. 2.6b, potassium ferrocyanide). A precipitation reaction is initiated around each droplet.

The term precipitate is used here to indicate a change in colour of the gel. However, there is no evidence to suggest that actual physical precipitation occurs, rather that a proportion of the substrate is converted to the primary product (or direct precursor to the product) within an area of the gel.

Because the process is diffusion limited, ahead of the reaction fronts there is a thin (but expanding) region where the inner electrolyte is depleted. Just before the reaction fronts initiated at distinct points meet, the depleted regions overlap, and the precipitation ceases or is significantly reduced. Since the speed of the fronts started from different points varies with time in an identical manner, the resulting pattern, consisting of precipitate regions and empty bands, represents the Voronoi diagram of the droplets. Thus, a mechanism based on substrate competition is proposed.

In this respect the chemical systems appear to mirror natural processes such as crystal growth (see Fig. 2.6b) and the growth of fungal colonies where substrate competition appears to be a plausible mechanism for tessellation of the plane.

The widths of the bisectors are controlled by the speed of diffusion of the reactant through the gel and are directly proportional to the distance between the original reactant sources. Therefore, sources in close proximity to each other annihilate to give thin sharp bisectors, whereas sources an appreciable distance apart give thick diffuse bisectors. This is apparent in Fig. 2.6: the Petri dish is split into distinct areas by wide bisectors formed between three distinct clusters of points. A special case is where sources are very close together and weak or even no bisectors are formed.

In the chemical processors of this type low precipitate concentration occurs at two points: firstly, below the original reactant solution (see Fig. 2.6a) and, secondly, at the point of annihilation between the advancing fronts.

We postulate that this is due to the action of both chemical- and physical-based mechanisms. The gel beneath the original reactant drops corresponds to the points of maximum concentration and residence time of the reactant. Thus, at this point in the reaction, there is a relatively low concentration of substrate and a precipitate-depleted region results in the gel layer beneath the drop. So, possibly the mechanism of bisector formation is also based on the reactant exceeding a critical concentration at a specific site of the gel? However, if this were the case then large bisectors would be formed between colliding fronts emanating from drops a minimal distance apart, as the collision point would serve as a secondary maximum of reactant concentration. However, this is the opposite of what is observed experimentally. Thus, to explain the experimental observations of weak bisectors at minimal drop separations and thick diffuse bisectors at maximal drop separations we propose the 'substrate-leaching' mechanism. This mechanism is based predominantly on substrate competition between the fronts. However, it appears to act over relatively long distances (given enough time). Thus, in the initial phase of the reaction the diffusion is at a maximum, as is the chemical affinity for the substrate. However, as the reaction fronts proceed both the speed of diffusion and chemical affinity for the substrate are reduced. We postulate that from the initiation of the reaction the substrate moves actively towards the reactant drop. Thus, a concentration gradient is established. This gradient acts over an increasing area of the gel (in advance of the diffusion front) as the speed of diffusion decreases. Therefore, weak bisectors are formed due to the initial inertia of the substrate in the gel (its resistance to transport is high). At this point in the reaction the diffusion speed of the reactant is close to the maximum and the concentration gradient is minimal (it acts over a negligible area of the gel), so the diffusion fronts collide and annihilate at an area of gel relatively high in substrate. Sharp but thin bisectors are formed above some critical drop separation as the mechanism acts for a short time over a relatively small area of gel but with maximum 'intensity'. At increasing drop separations more diffuse bisectors are formed between opposing reactant sources as the 'attractive forces' act over increasingly long distances.

It is apparent that the bisectors in the real chemical systems are complex continuous concentration profiles of the precipitate (or lack thereof). The centre of the bisector represents a minimal precipitate point and then the areas either side of this represent increasing levels of precipitate. There are also differences in the precipitate concentration along the length of the bisector. The minimal precipitate area is at the initial collision point of the circular fronts and/or the intersection points of three or more fronts, whereas the adjoining parts of the bisector have higher concentrations of precipitate. These observations are all consistent with the mechanisms proposed.

Further evidence for this 'substrate-leaching' mechanism comes from experimental observations of different systems. For example, in the potassium ferricyanide system where the substrate-loaded gel is dark yellow in colour a clear gel region ('halo') can be observed in advance of the precipitation front [79]. This 'halo' increases in size as the precipitation reaction proceeds. Also, in other work [81] involving approaching diffusion fronts where the fronts do not chemically compete for the same substrate, bisectors are still formed between these fronts.

2.5.2   Voronoi diagrams of planar shapes

Voronoi diagrams of other geometric shapes can be constructed by substituting the drops of outer electrolyte for pieces of absorbent materials soaked in the electrolyte solution and applying these to the gel surface (see Figs. 2.6 and 2.7).

Note that the bisectors formed when the sources are not drops but geometric shapes are curved and not straight. These particular Voronoi diagrams are related to the planar type and are referred to as generalised Voronoi diagrams.

Actually, as theoretically the sources should be points, then all Voronoi diagrams constructed in chemical processors are of this generalised type as the drops equate to circular not point sources.

It should also be noted that Voronoi diagrams of this generalised type especially where the sources are geometric shapes are more difficult to compute using conventional algorithms. The chemical systems are very easily adapted to different tasks and, whilst they do not compute things perfectly in all cases (see discussion in Chap. 9), they provide a close approximation to the computationally ideal results.

2.5.3 Constructive chemical processors

In the copper ferrocyanide reaction-diffusion processor a primary Voronoi diagram was constructed only at concentrations between 2.5 mg/ml and 5 mg/ml. In the case of the ferric ferrocyanide reaction-diffusion processor a Voronoi diagram was constructed at concentrations up to 20 mg/ml.

Although the copper ferrocyanide and ferric ferrocyanide (Prussian blue) reactions are different chemically and this exerts an influence on the behaviour at higher substrate concentrations, at lower concentrations (2.5 mg/ml potassium ferrocyanide) and for the purpose of computing a Voronoi diagram the reactions are effectively the same. This is emphasised by the creation of a mixed system whereby ferric and cupric ions are reacted on the same substrate (potassium ferrocyanide gel) to compute a single Voronoi diagram (see Fig. 2.8a, and Fig. 9.13 in colour insert).

The fact that two distinct chemical species can be utilised to construct a single Voronoi diagram reinforces the earlier assertion that the simple underlying mechanism involved in the computation of these primary Voronoi diagrams is predominantly based on substrate competition.

In reality, although the computation appears to be indistinguishable from the results computed individually by the single systems, previous work [13, 15] has shown the bisector width to be controlled by the diffusion speed of the reactant. Therefore, it is clear that three distinct bisector types must be formed by this system. It is only because the diffusion speeds of the primary reactants are so similar that the result is almost identical to that constructed by a single-reactant system.

In reality, the rate of the diffusion (reaction) of copper chloride is marginally faster, meaning that the Voronoi cells will be slightly larger than those of ferric chloride. Even these slight differences are emphasised by Fig. 2.8b, where drops are arranged in a regular way. The ideal Voronoi diagram in this case would be a regular array of square cells. However, the bisectors between the dark and light drops (and thus the entire boundaries of the dark Voronoi cells) are noticeably curved.

A Voronoi diagram where sources grow at different rates from the same initiation time is called a multiplicatively weighted Voronoi diagram. This is another generalisation of a planar Voronoi diagram and has specific uses in path planning. However, the chemical system we describe computes a special case of this generalised Voronoi diagram called the multiplicatively weighted crystal growth Voronoi

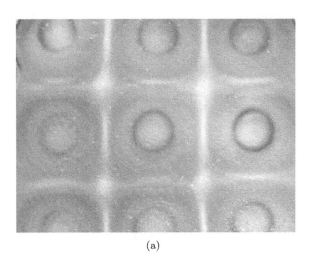

(a)

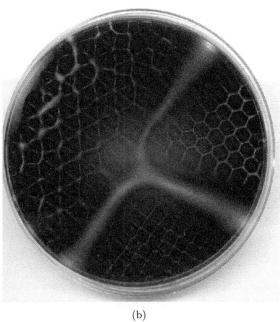

(b)

Figure 2.6: Voronoi diagrams computed on chemical reaction-diffusion processors at low substrate concentrations: (a) reaction of potassium hydroxide (8 M) on 20 mg/ml copper(II) chloride/agar gel substrate (drop size 5 mm), (b) Voronoi-diagram construction in the Prussian blue processor. The reactant drops (ferric chloride 300 mg/ml) were applied in deliberate patterns. The image was obtained by scanning the underside of the Petri dish (9 cm in diameter).

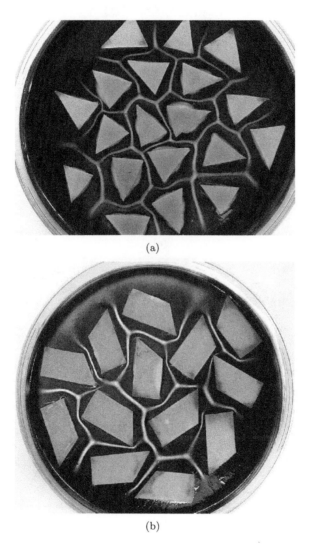

(a)

(b)

Figure 2.7: Voronoi diagrams of planar geometric shapes: triangles (a) and tetragons (b) constructed on the palladium reaction-diffusion processor.

(a)

(b)

Figure 2.8: Mixed cell Voronoi diagram (a) computed when ferric ions (dark) and cupric ions (light) were reacted on a potassium ferrocyanide agar gel substrate (2.5 mg/ml). Drop size 5 mm. (b) The drops are arranged in a regular manner, which emphasises the difference in the speed of diffusion/reaction rates between cupric and ferric ions.

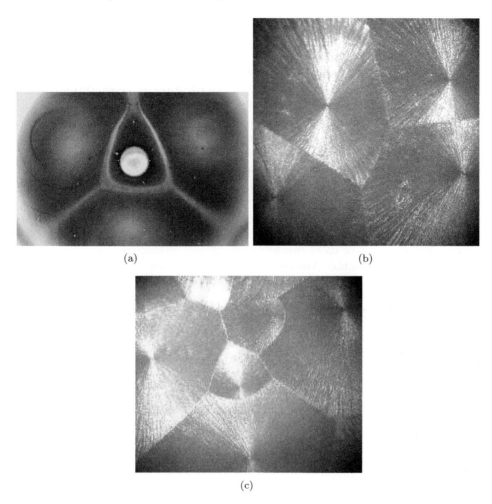

(a)   (b)

(c)

Figure 2.9: (a) A multiplicatively weighted crystal growth Voronoi diagram computed using a constructive chemical processor. (b) Planar Voronoi diagram constructed in a crystallisation process where all growing crystals have equal weights. (c) Weighted Voronoi diagram naturally constructed in a crystallisation process where growing crystals have different weights.

diagram [151]. This describes the partition of the plane into crystals with different growth speeds and was proposed by Schaudt and Drysdale [242].

The multiplicatively weighted crystal growth Voronoi diagram is different from the multiplicatively weighted Voronoi diagram because a crystal cannot grow into an area already occupied by another crystal. A crystal growing with high speed thus grows around slow-growing crystals. Thus, in a crystal-growth diagram the Voronoi region is always connected whereas in the ordinary diagram a region may be disconnected as generators with large weights grow beyond the regions of generators with small weights. Hence, in a multiplicatively weighted crystal growth Voronoi diagram the distance between two points at a given time should be measured by the length of the shortest path that avoids crystal regions generated by that time. For this reason the exact computation of this Voronoi diagram is very hard and only a few approximate algorithms exist [242]. However, this type of generalised Voronoi diagram is seemingly computed naturally in the constructive chemical processors (where outer electrolytes have differing diffusion rates). To emphasise this point Fig. 2.9a shows a system with a very fast diffusing reagent (potassium hydroxide) and a much slower diffusing reagent (potassium iodide) reacted on a gel containing palladium chloride. The difference in bisector and Voronoi cell construction is obvious when compared to the planar case.

It is worth noting here that other types of generalised Voronoi diagrams could be constructed using this simple methodology. If drops of reagent that diffused/reacted at the same speed were placed onto the gels at different times during the reaction process then an additively weighted Voronoi diagram would be constructed. If this was repeated with two or more reagents that diffused or reacted at different rates, then a compoundly weighted Voronoi diagram would result.

Figure 2.9a and c show two Voronoi diagrams constructed by growing crystals. In the first case the crystals were growing at the same speed (Fig. 2.9b) and a planar Voronoi diagram results. In the second example some of the crystal regions grow at different rates (or start growing at different times) and a multiplicatively weighted crystal growth Voronoi diagram (or other generalised Voronoi diagram) results. Thus, these diagrams are computed naturally in crystallisation processes — however, these processes cannot be easily controlled in terms of encoding data, unlike the chemical reaction-diffusion processors.

2.6   When computations go wrong!

Serendipity is a great friend of scientists — they just might not appreciate it at the time!

In an attempt to improve the colour intensity of the mixed processor the concentration of potassium ferrocyanide was increased to 10 mg/ml. The reaction proceeded as normal — however, triangular sections appeared in the growing circular fronts of the copper system. This wasn't supposed to happen! The computation was effectively ruined — but a new class of pattern-forming reactions was discovered.

It turned out that the copper system was subject to an instability which eventually stopped the normal precipitation reaction. Figure 2.10a (see also Fig. 9.14a in colour insert) shows a mixed cell Voronoi diagram which is computed as expected at low substrate concentration. Figure 2.10b (see also Fig. 9.14b in colour insert) on the other hand shows the marked effect on the construction of the Voronoi diagram when the substrate concentration was raised.

The next stage was to investigate the full concentration range for this system. What is interesting is that a new class of Voronoi diagrams termed secondary Voronoi diagrams were computed at high substrate concentrations. These and other facets of the unstable systems will be discussed at more length in the next section.

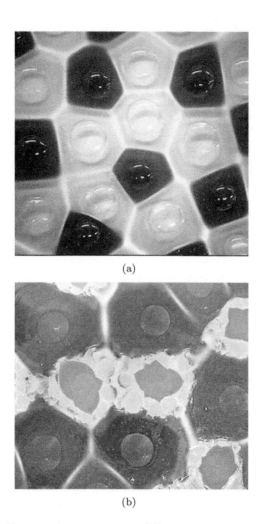

(a)

(b)

Figure 2.10: (a) A Voronoi diagram successfully computed in a constructive chemical processor. (b) The result of the same experiment where the substrate concentration has been increased and the formation of copper ferrocyanide has become unstable.

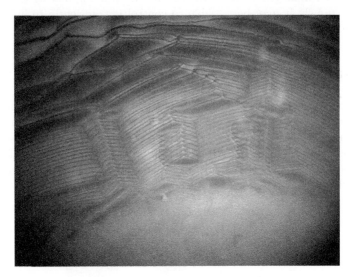

Figure 2.11: Waves formed in the reaction between copper chloride (80 mg/ml) and potassium hydroxide (8 M). Direction of diffusion is approximately from bottom to top. Image size 2 mm by 1.7 mm.

2.7 Unstable processors

2.7.1 Copper hydroxide processor

Primary Voronoi diagrams were constructed in the reaction between the outer electrolyte potassium hydroxide and the substrate copper(II) chloride loaded gel provided the substrate concentration was in the range 2.5–20 mg/ml (see Fig. 2.6). At concentrations above 20 mg/ml the travelling circular wave fronts become unstable and spontaneously split. For a full treatment of the primary front splitting mechanism and a description of extended pattern formations in this non-linear region see [120, 121].

When the substrate-reagent concentration is increased in the copper hydroxide reaction-diffusion system, a Voronoi diagram is not constructed because a chemical instability means that diffusive fronts entrap themselves in a series of growing boundaries surrounding the sites where the initial reactant drops were placed; see Fig. 2.11.

The phenomenon is represented in our discrete models. Let us look back to the 'one reagent and one substrate' cellular-automaton model. We could modify the rule as follows:

$$x^{t+1} = \begin{cases} \beta, \text{ if } x^t = \bullet \text{ and } 1 \le \sigma(x)^t \le \theta \\ \alpha, \text{ if } x^t = \beta \text{ and } 1 \le \sigma(x)^t \le \theta \\ x^t, \text{ otherwise} \end{cases} \qquad (2.4)$$

and let $1 \le \theta \le 8$. For each value of θ we generate a Voronoi diagram. A spectrum of the diagrams generated is shown in Fig. 2.12.

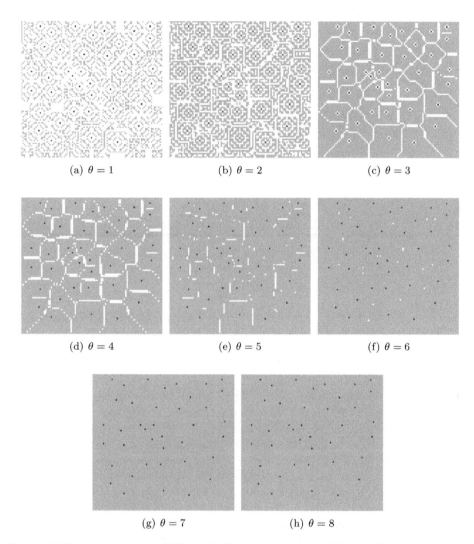

(a) $\theta = 1$ (b) $\theta = 2$ (c) $\theta = 3$

(d) $\theta = 4$ (e) $\theta = 5$ (f) $\theta = 6$

(g) $\theta = 7$ (h) $\theta = 8$

Figure 2.12: A spectrum of Voronoi diagrams generated in a cellular-automaton model of a reaction-diffusion medium (with two reactants); each cell of the automaton is governed by the rule (2.4). Precipitate is shown as grey and the Voronoi edges as white.

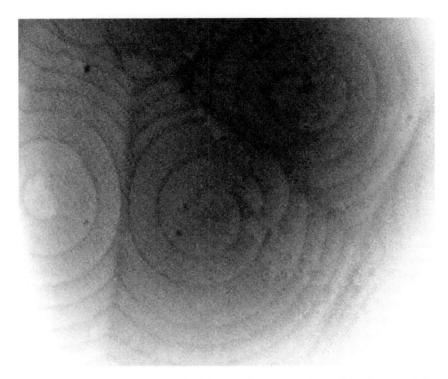

Figure 2.13: Complex spiral and double-spiral patterns formed in the unstable copper hydroxide processor. The formation of Voronoi cells that enclose the complex structures is also evident.

The parametric transition $\theta = 8 \to \theta = 1$ may be analogous to an increase in the reagent concentration in experimental processors. We can also see that a Voronoi diagram constructed for $\theta = 3$ is better, in that it gives more 'continuous' bisectors, than a Voronoi diagram computed for $\theta = 4$ in the previous automata models. However, in the system with $\theta = 3$ 'false' bisectors are generated around the initial 'drops' of the reagent, analogous to the real-life experiments.

If the concentration of copper chloride in the gel is increased still further, then complex structures are formed beneath the application of the potassium hydroxide solution. We highlighted these structures and some possible explanations for their formation in the introduction. Figure 2.13 shows spiral and double-spiral waves of this type. What should be obvious is that each individual structure occupies its own cell. The boundaries of these cells correspond to a Voronoi diagram of the original points of instability. The three-dimensional nature of the structures should also be clear from Fig. 2.13.

An explanation of the formation of the original 'secondary' Voronoi cells was given earlier. The formation of the additional spiral structures within the Voronoi cells in this particular reaction seems to be due to the permeability of the solid boundary to potassium hydroxide. When the original reactive front splits and eventually forms Voronoi cells a physical boundary to diffusion is established (so the reaction might stop at this stage). Thus, the three-dimensional 'Voronoi cells' enclose an area of gel containing unreacted substrate. At some point the reaction starts again in the 'Voronoi cells' as the potassium hydroxide overcomes the barrier to diffusion. This may be due to imperfections in the secondary material or due to some semipermeable nature.

2.7.2 Copper ferrocyanide processor

As mentioned in the case of the copper ferrocyanide system at substrate concentrations at or above 7.5 mg/ml, the circular precipitation fronts are subject to spontaneous splitting [79]. This is because when the speed of diffusion is reduced the reactive region can become saturated with product, and the diffusion front becomes unstable to inhomogeneities in the system and also to the transport of substrate to the reactive region. The 'substrate-leaching' mechanism plays a role in this phase of the reaction — as near the critical saturation point the chemical affinity for the substrate is negligible but the concentration gradient formed in the earlier stages of the reaction will continue to force the transport of substrate towards the diffusion front or reactive region, exacerbating the unstable nature of the advancing front.

We observe that when the original circular wave fronts split the two regressing edges formed from this point follow counter-rotatory orbits until they annihilate either with each other or with other regressing edges. The result is that the precipitation or diffusion front ceases to advance. Thus, in these systems as in the copper hydroxide processor discussed previously a complete Voronoi diagram is not computed. So, the primary controllable pattern formation is lost and the system has entered a highly non-linear region.

If only a small proportion of the fronts become unstable (split) this may result in the formation of a partial Voronoi diagram; see Fig. 2.14. However, if all the original fronts are subject to spontaneous splitting (circular fronts can split many times) a true Voronoi diagram is never formed (Fig. 2.15). Whilst a Voronoi diagram is not computed the sphere of influence of distinct sources is obvious where the extended pattern-forming structures interact. Where these structures approach they annihilate at depleted precipitate zones, indicating that the underlying mechanisms observed for the stable systems continue to operate.

The primary wave splitting is analogous to that observed in the copper hydroxide system [120]. However, the major difference between the two systems is that no new reaction front is formed in the precipitate-free region — instead, the reaction is confined behind a boundary and secondary (see Fig. 2.15) and tertiary pattern-forming structures predominate, see Fig. 2.16. In reality, the boundary is very slightly permeable allowing the formation of tertiary structures many hours or even days after the reaction's initiation.

In the unstable system, once primary wave fronts are annihilated a physical instability in the system forms secondary structures analogous to viscous fingering. This is where fluid-composed tips advance into a more viscous medium by the pressure exerted at the interface of the two fluids [50]. Although the Saffman–Taylor instability, which describes this phenomenon, favours the growth of branches on all length scales the smoothing effect of surface tension leads to a characteristic length

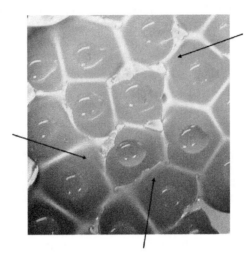

Figure 2.14: The effects on Voronoi-diagram construction of raising the concentration of the potassium ferrocyanide in the reaction to 7.5 mg/ml. At this substrate concentration only a limited number of the original wave fronts became unstable. The arrows mark additional 'ghost' bisectors formed at points where the original bisectors would have been formed if all the wave fronts had remained stable. This provides evidence that the mechanism of bisector formation acts over a long distance and not just in the vicinity of the advancing wave fronts.

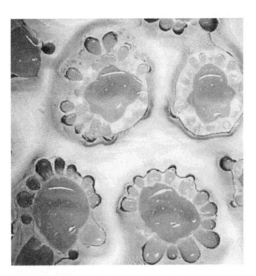

Figure 2.15: Extended pattern formation observed at potassium ferrocyanide concentrations above 7.5 mg/ml and below 20 mg/ml. All wave fronts become unstable and annihilate, whereby the reaction enters a highly non-linear region and secondary (in evidence) and tertiary pattern-forming structures predominate.

Figure 2.16: Tertiary wave structures emanating from a solid boundary in the unstable copper ferrocyanide processor (image size 2.5 mm by 1.9 mm).

scale for the reaction. In this reaction, where there is a relatively low pressure at the interface but a relatively high surface tension, the length scale is large and the formation of fat low branching fingers predominates (see Fig. 2.15).

Tertiary wave structures are also apparent in this system at concentrations between 7.5 mg/ml and 40 mg/ml. For further treatment of this phenomenon see [79]. Figure 2.17 shows bisectors formed when tertiary structures emanating from differing sources approach. Although these depleted regions do not directly translate to a Voronoi diagram, they again exhibit the sphere of influence of separate reactant sources. The depleted zones in the unstable systems also lend weight to the proposition that the mechanism of bisector formation (Voronoi-diagram formation) in these chemical processors is via substrate competition.

2.8 Secondary Voronoi diagrams

At higher concentrations (above 40 mg/ml potassium ferrocyanide) there are no secondary or tertiary structures and pattern formation becomes internalised, i.e. beneath the initial application of the reactant solution. At low substrate concentrations the outer electrolyte diffuses quickly through the entire thickness of the gel and a circular diffusion front is initiated travelling outwards from the edge of the reactant source. However, at even higher substrate concentrations the system is increasingly unstable and despite the thin layer of gel (~ 1 mm) the front becomes increasingly susceptible to the same splitting as it diffuses through the gel.

In this highly unstable copper ferrocyanide system spheroid-like structures are observed to evolve from a point of high precipitate concentration. These spheroid-like structures appear to have a relatively low concentration of precipitate at the centre (ignoring the original point of high precipitation) with increasing precipitate concentration towards the edges. What is actually observed is the difference in precipitate concentration throughout the thickness of the gel (see Fig. 2.18c).

It is postulated that the concentration of product exceeds a critical threshold causing physical precipitation at a number of distinct points; this can be seen

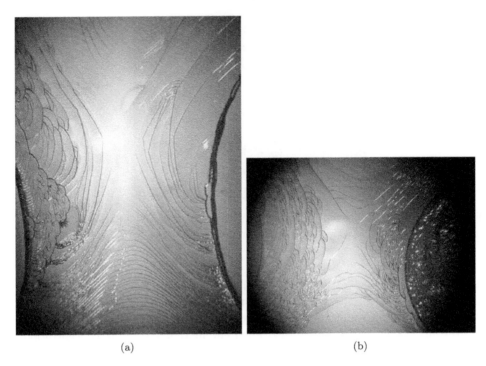

(a)   (b)

Figure 2.17: (a) Formation of bisectors between two tertiary wave-forming structures emanating from distinct sources. (b) A precipitate-free region is only formed
where the sources are directly opposed; however, the effect of this boundary region
is obvious over a large area.

clearly in Fig. 2.18a. These point inhomogeneities in the system cause the uniform
front to split as it penetrates the gel and reaction zones are formed. Thus, the
precipitation does not begin over the whole surface of the gel and at some points
(due to inhomogeneities) it does not start at all. As time evolves the diffusion and
reaction fronts sweep through the gel and these points expand into empty cones,
enclosing an area free of precipitate. As the reaction front proceeds the empty
cones (disc-shaped holes) in the front also grow. The circular margins of the holes
will be referred to as regressing edges of active (precipitating) borders. At points
where the regressing edges of the fronts collide the precipitation reaction stops. As
the horizontal speed of the regressing edges is the same for each cone, the locus of
their collision points constructs a Voronoi diagram of the original points where the
precipitation did not start.

So, where the expanding circular waves of two or more cones collide they annihilate to form a bisector which is relatively high in precipitate. It is these high-
precipitate areas that correspond to the bisectors of the Voronoi diagram (see
Fig. 2.18a and b, and Fig. 9.15 in colour insert). It is apparent that the division of the space beneath the drop into Voronoi cells is in pseudo three dimensions
(see Fig. 2.18b, the schematic in Figs. 2.18c and 2.19).

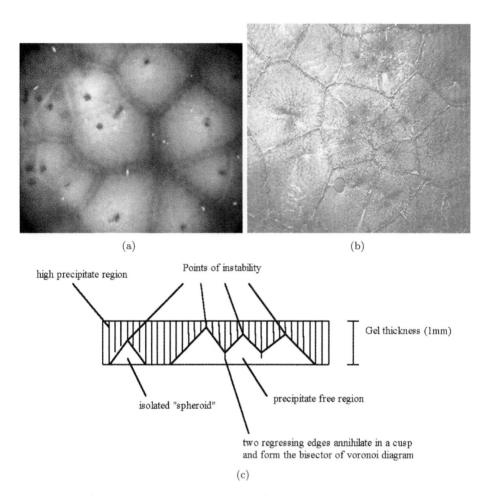

Figure 2.18: An example of the structures formed in the copper ferrocyanide gel reactor at potassium ferrocyanide concentrations between 20 mg/ml and 40 mg/ml. (a) Shows that in the unstable system Voronoi diagrams are formed spontaneously when the reactant is applied to the gel substrate. The dark points at the Voronoi cells' centres correspond to the original points of instability. (b) Shows the reverse side of a reacted gel sheet: the high precipitate concentration at the annihilation points of the fronts (bisector) can be observed (image size 5 mm). (c) Shows a schematic cross section of a gel reactor detailing how Voronoi diagrams are formed in unstable chemical systems.

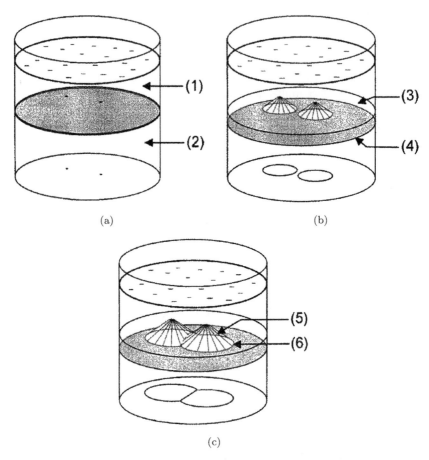

Figure 2.19: Schematic representation of Voronoi-diagram formation through the 'regressing edge' mechanism: (a) the precipitation does not start at some points of the gel surface, (b) these points expand into precipitate-free cones as the front progresses, (c) the regressing edges meet on the line that is an equal distance from the tips of the empty cones. (1) Outer electrolyte, (2) agarose gel containing the inner electrolyte, (3) precipitate, (4) reaction front (active border represented by a dotted surface), (5) passive border, (6) regressing edge. The contours of the patterns are drawn as a projection underneath the vessel.

It is interesting to note that the Voronoi diagram in this case has bisectors of high precipitate concentration, which is the opposite to the stable case but predicted by earlier models of reaction-diffusion systems for computing Voronoi diagrams.

If some of the cones start to develop at a later stage, the resulting polygon will not be of the ordinary Voronoi type, since the belated cones will be able to expand only in a smaller area. The tessellation generated corresponds to a generalisation of a Voronoi diagram, more specifically to a type known as an additively weighted Voronoi diagram. This type of Voronoi diagram differs from the ordinary Voronoi diagram in that when fronts initiated at different times collide they no longer form straight-line bisectors but instead meet at hyperbolic segments.

2.9 Controllability of secondary Voronoi diagrams

So, Voronoi diagrams are also formed in unstable systems. However, these Voronoi diagrams form spontaneously from imperfections in the gel or precipitation points and therefore the resultant Voronoi diagram cannot be controlled. Thus, they have a limited use as chemical processors in their current form.

In an attempt to control the formation of secondary Voronoi diagrams we ran them in a 'Liesegang' setup where the inner electrolyte ($AgNO_3$ or $K_3[Fe(CN)_6]$) was homogenised in agarose gel and the outer electrolyte was layered onto the gel.

At high inner electrolyte concentrations we obtained spontaneous Voronoi-diagram formation (Fig. 2.20) (0.88 M $AgNO_3$ or 0.15 M $K_3[Fe(CN)_6]$) similar to those described above for the case of potassium ferrocyanide.

At low inner electrolyte concentration (below 0.3 M $AgNO_3$ or 0.03 M $K_3[Fe(CN)$ a homogeneous precipitate forms irrespective of the gel's condition. However, when the concentrations were 0.64 M $AgNO_3$ or 0.09 M $K_3[Fe(CN)_6]$ Voronoi diagrams could be formed in a controlled way by marking the gel's surface with a glass needle (Fig. 2.21, see also Fig. 9.16 in colour insert). At these concentrations no natural Voronoi cells would ever be formed and if the gel was not artificially marked a homogeneous precipitate would be formed. An additively weighted Voronoi diagram could be constructed by marking the surface before and during the reaction process.

This system of constructing Voronoi diagrams has some inherent advantages over the stable systems discussed earlier in the chapter. In these reactions the construction of the Voronoi diagram is faster. Also, the Voronoi diagram of a set of points not circles is constructed; thus, the computation is more accurate. As the source of the Voronoi cells are points then the minimum area of a Voronoi cell can be vastly reduced in the unstable systems, thus increasing the complexity of problems that could realistically be calculated.

It is worth noting that for both the stable and unstable Voronoi processors the greater the complexity of the problem, i.e. the greater the number and concentration of points, then the faster the computation. This obviously has to take into account that there is a minimum source separation for each system, which sets the absolute limit of the complexity of any problem which can currently be solved by these processors.

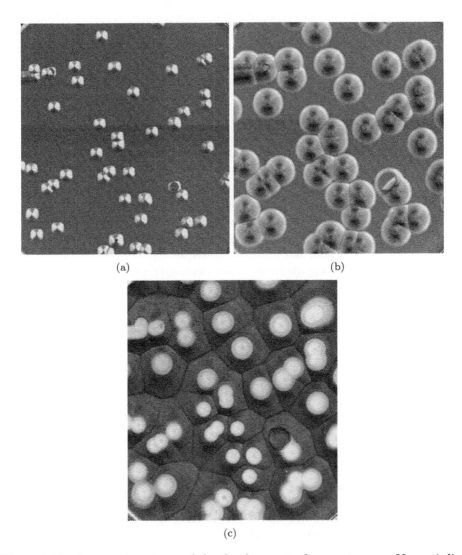

(a)   (b)

(c)

Figure 2.20: Consecutive stages of the development of a spontaneous Voronoi diagram: (a) 35 sec, (b) 138 sec after the initiation of the reaction and (c) at the end of the reaction.

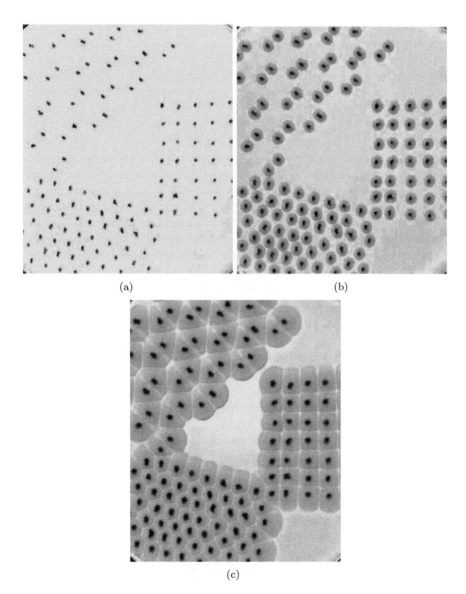

(a)

(b)

(c)

Figure 2.21: Consecutive stages of development of a controlled Voronoi diagram, representing triangular, rectangular and hexagonal arrangements. Elapsed time was (a) 43 sec, (b) 198 sec and (c) 878 sec after the initiation of the reaction.

2.10   Skeleton of planar shape

What is a skeleton of an image and why is it so important? Mathematically, a skeleton of a planar contour is a set of the centres of bi-tangent circles lying inside the contour [67]; a planar connected set is uniquely determined by its skeleton and its convex hull [67]. A skeleton gives us a stable, unique and invariant representation of a shape, which also possesses a rich local support [62, 204]. The skeleton accurately describes most geometric and topological features of an image. Therefore, the skeleton is an effective shape descriptor [215, 170]. This makes the skeleton invaluable for instance in shape recognition, image re-construction, high-level computer vision [122, 87, 101, 166, 172] and morphing operations [186].

How have said skeletons been computed? The first ever algorithm — the grass-fire algorithm — was already 'natural': to set the contour on fire, let the fire spread and quench points where the advancing fire fronts collide; this quenching represents the skeleton of the original contour [54, 55, 67]. So far, the basic variants of skeletonisation, as summarised in [204], include (i) simulation of a grass fire (a good idea to do in massively parallel processors), (ii) analytical construction of a medial axis (too complicated), topological thinning (not accurate and counterintuitive, as indicated in [204]) and a geometric transform [216] and (iii) constructing a skeleton via a distance map or a distance transform [237].

Over ten years ago three physics- and chemistry-based algorithms for skeleton construction were published; they are a reaction-diffusion algorithm [2], an excitable chemical medium based algorithm (based on the Belousov–Zhabotinsky reaction) [229] and an electrostatic field algorithm [1].

The electrostatic field approach to the construction of a skeleton relies on the following facts: (i) the electrostatic field distribution on a surface of a conducting object is proportional to the local curvature of the object and (ii) the field lines are normal to the object's boundary (and also bisect the angles at the corners) [1]. Therefore, the electrostatic field lines represent the segments of a skeleton. Algorithmic implementation of the technique is interesting, possibly not too efficient; however, its potential lies in the real-life fabrication of a skeleton-computing electrostatic computer (we are not aware whether it was ever done). The classical grass-fire transformation algorithm and the unconventional electrostatic algorithm tell us that sites of a given space, corresponding to a skeleton, change their physical characteristics; however, they do not give an obvious way to detect these sites.

The excitable chemical medium approach is very intuitive and clear. A contour is projected onto a light-sensitive Belousov–Zhabotinsky medium to excite the medium. Waves of excitation, recognisable by the colour of reagents, travel inside, as well as outside, this data contour. The sites of collision of the waves represent the points of the skeleton. Unfortunately, as with the grass-fire algorithm, there is no way to 'naturally' detect the sites of the waves' collision in the Belousov–Zhabotinsky medium.

The reaction-diffusion approach rectifies a problem of the detection of the computed skeleton. It is assumed that the reagents diffuse from the data sites, and interact to form a skeleton [2, 4, 7].

2.11 Chemical processors for skeleton computation

The methodology for the production of such processors, including an extended range of systems of this type, can be found detailed in [18]. For the purposes of these experiments reaction-diffusion processors based on palladium chloride and Prussian blue were utilised.

Masks of the images to skeletonise were prepared from absorbent materials such as filter paper. Other methods of applying reagents to specific areas of gel reactors are being investigated, such as the use of ink-jet printing technology, although the current methodology is facile and proves highly effective for the transfer of simple images such as those displayed in Figs. 2.22 and 2.34.

The masks were uniformly soaked with a saturated solution of potassium iodide (or ferric chloride), blotted with additional filter paper to remove the excess and placed onto the surface of the gel. The reaction should be instantaneously initiated at all edges of the mask and diffuse outwards at a uniform rate. This process can be seen at various stages displayed in Fig. 2.22 from the original mask (contour) through reaction stages to the completed reaction showing the skeleton formed within the original data contour.

2.12 Mechanics of skeletonisation

As the potassium iodide (clear solution) diffuses into the gel it reacts with the palladium chloride to form iodo-palladium species. The transition is from yellow (at a $PdCl_2$ concentration of 0.2%) to a dark-brown colouration (iodo-palladium species or PdI_2 formed). If the chemical system is analogous to a mixture of potassium iodide and palladium chloride, then potassium chloroiodopalladite ($K_2PdCl_2I_2$) would form, which could further react to produce potassium iodopalladite (K_2PdI_4). These latter two compounds might be expected to be red-brown, in line with the chloride analogue. Further speciation to produce black palladium iodide is also likely.

The mechanism for the formation of bisectors between the opposing fronts emanating from the contour is identical to that discussed for the formation of Voronoi diagrams in the previous section. Thus, sites of the substrate with very small or absent concentrations of iodo-palladium species represent a skeleton of the data shape. As discussed previously, the bisector's width is proportional to the distance between two points. Thus, in this case the width of the computed skeleton corresponds to the average separation of two points on the contour. This is shown in Fig. 2.22, which gives a space–time dynamic of the construction of a skeleton using a palladium processor.

The mechanism of concentration inhibition can be illustrated in discrete models. Here we construct two cellular-automaton models; these toy models simply aim to show the basics of reaction-diffusion skeleton construction — references to more sophisticated methods can be found in [7]. Neither of the models could be seen as a tool for modelling or simulation of the reaction-diffusion pattern-formation mechanics but rather as an illustrative imitation.

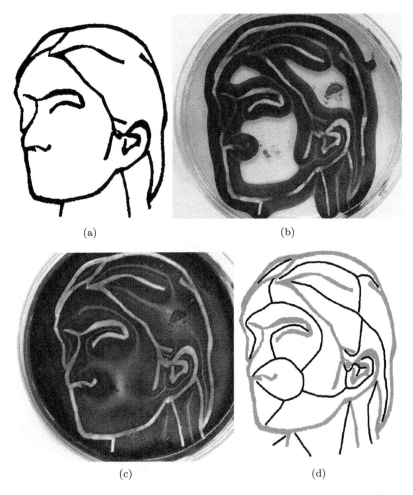

(a)   (b)

(c)   (d)

Figure 2.22: Formation of a skeleton of a simple contour: (a) data contour, (b) snapshot of the early stages of skeleton computation, (c) skeleton computed using a palladium processor, (d) a sketch of the skeleton; the initial contour is shown in light grey.

Results of our experiments can be represented by a configuration of black and white pixels; therefore, it is reasonable to try to simulate the medium by use of a binary cellular automaton. A two-state model can be described as follows. Every cell has two possible states: 0 (empty, uncoloured, no precipitate) and 1 (coloured phase, iodo-palladium species). The cell calculates its next state by the following rule. If the cell is in state 1 it remains in this state forever. If the cell is in state 0 and between one and four of its neighbours are in state 1, then the cell takes the state 1; otherwise, the cell remains in the state 0:

$$x^{t+1} = \begin{cases} 1, & \text{if } (x^t = 1) \vee (x^t = 0) \wedge (\sigma^t(x) \in [1,4]) \\ 0, & \text{if } (x^t = 0) \wedge (\sigma^t(x) \notin [1,4]) \end{cases}$$

where $\sigma^t(x) = \sum_{y \in u(x)} y^t$ and $u(x)$ is the cell x's neighbourhood, consisting of eight neighbours.

The condition $\sigma^t(x) \notin [1,4]$ of the transition $0 \to 0$ reflects our assumption that an above-threshold concentration of reagent, naively expressed in the model by 'concentration' of precipitate, prevents the formation of precipitate in the empty cells.

An example of the space–time dynamic of the two-state cellular-automaton model is provided in Fig. 2.23. As one can see a sketch of a skeleton is given; however, segments of the skeleton are incomplete, disconnected and visibly composed of separate pixels. The constructed skeleton possibly looks so unsatisfactory because we did not consider that the concentrations of some reagents and products are higher at the wave front than at the wave tail. We tried to incorporate this phenomenon into the next model.

A three-state cellular-automaton model of the palladium processor assumes that every cell takes three states: 0 (empty, uncoloured, no precipitate), 1 (coloured, precipitate) and 2 (reagent, wave front). The cell state transition rule is as follows:

$$x^{t+1} = \begin{cases} 0, & \text{if } (x^t = 0) \wedge (\sigma^t(x) = 0) \vee (x^t = 2) \wedge (\sigma^t(x) > 4) \\ 1, & \text{if } (x^t = 2) \wedge (\sigma^t(x) \le 4) \\ 2, & \text{if } (x^t = 0) \wedge ((\sigma^t(x) \in [1,4])) \end{cases}$$

where $\sigma^t(x) = |\{y \in u(x) : y^t = 2\}|$.

An example of the skeleton computation in the three-state model is shown in Fig. 2.24. Whilst the model does not re-construct the skeleton perfectly, the representation is much improved. The three-state model is similar to a cellular-automaton model which may be constructed to model an excitable medium, where the cell state 0 corresponds to a resting cell state, the cell state 2 is analogous to an excited state and the cell state 1 is analogous to the refractory state. However, in contrast to conventional models (i) a cell does not unconditionally switch to the refractory state from the excited state and (ii) a cell does not return to the resting state from the refractory state.

For some applications it is reasonable to prune a skeleton, i.e. delete tiny, and thus unimportant, branches. Can this be done in-flight in our reaction-diffusion processor? Yes, usually, a segment of a skeleton to prune is determined by its adjacency. An adjacency is a number of other skeleton-segment-generating sites

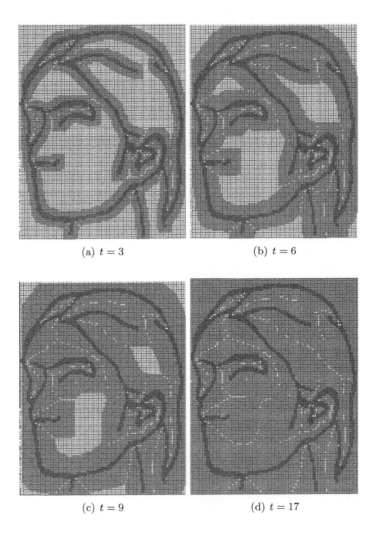

(a) $t = 3$ (b) $t = 6$

(c) $t = 9$ (d) $t = 17$

Figure 2.23: Space–time dynamic of a two-state cellular-automaton model of the reaction-diffusion processor for skeleton computation. White pixels represent sites without a precipitate, grey pixels are sites with the precipitate. The initial contour is artificially indicated on each configuration.

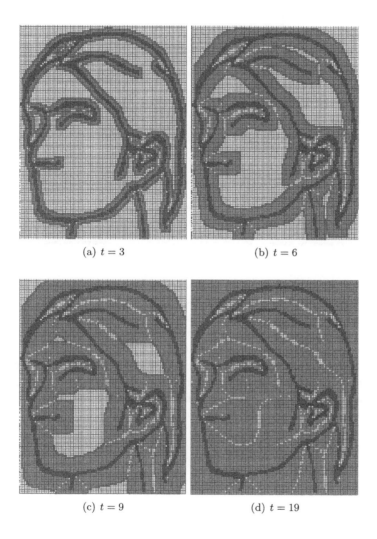

(a) $t = 3$ (b) $t = 6$

(c) $t = 9$ (d) $t = 19$

Figure 2.24: Space–time dynamic of the three-state cellular-automaton model of the reaction-diffusion processor for skeleton computation. White pixels represent sites without precipitate, grey pixels are sites with the precipitate and the dark-grey (red) pixels are sites of the (excited) wave fronts. The initial contour is artificially indicated on each configuration.

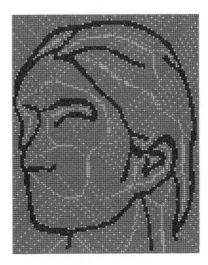

Figure 2.25: A skeleton with abundant (unnecessary) segments constructed in the three-state cellular-automaton model, where the 'threshold' of cell-state change was changed from 4 to 3.

lying between two sites that generate the segment in question [203]. The higher the adjacency, the more the segment is responsible for the description of the topological features of the data image and thus the less chance of this segment being pruned. In our reaction-diffusion processor the width of any particular segment of a skeleton is proportional to the distance between the parts inducing the segment. When the parts are too close no segment is formed. So, pruning is performed 'automatically'.

The number 4, which is a sum of the neighbours of a cell in the reagent state, is a 'magical' threshold that actually determines the sensitivity of the skeleton-constructing processor. This is the analogue of the amount of reagent necessary to produce a precipitate, or an inverse of the amount of product that inhibits precipitation. When we shift the threshold downwards and make it equal to 3 then the structure of the constructed skeleton changes drastically. This is shown in Fig. 2.25, where the skeleton is constructed in the three-state model and unnecessary

$$\begin{matrix} 0 & 0 & 2 \\ 0 & 0 & 2 \\ 0 & 2 & 2 \end{matrix}$$

skeleton sites are generated in cells with neighbourhoods of the form

2.13 Computing skeletons of geometric shapes

Figure 2.26 shows the computation of the skeletons of a number of geometric shapes in a Prussian blue processor. Due to the previously discussed relationship between the separation distance of two points and the bisector width the resulting skeletons all describe two-dimensional shapes. In the case of a square, regular pentagon and regular heptagon the resulting skeletons approximated to a four-pointed, five-pointed and seven-pointed star, respectively. The central region in each case should equate to an area of minimal precipitation, which it does. However, if a perfect

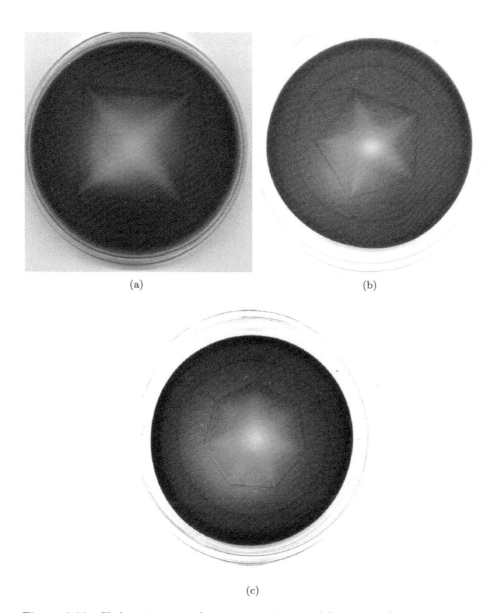

(a)

(b)

(c)

Figure 2.26: Skeletonisation of geometric shapes: (a) square, (b) pentagon and (c) heptagon. The experiments were carried out by soaking masks in ferric chloride and placing them on gels of potassium ferrocyanide. The images of the Petri dishes were collected using a flat-bed scanner.

construct was performed the original shape should be re-constructed in the centre as a precipitate-free/depleted region. This is because in the case of the skeleton-isation of a circular contour the result is not a precipitate-free point but rather a precipitate-free circular region whose size is proportional to the size of the original contour.

Although in the case of the skeletonisation of the square something resembling a precipitate-depleted square is re-constructed at the centre of the four-pointed star, for the cases of the pentagon and heptagon the construct at the centre is close to circular. This is due mainly to the differential speed of precipitation waves emanating from different points of planar boundaries. Thus, there is a smoothing effect and the chemical processor, although initially mapping the angular features of the geometric shape as shown by the construction of an 'n'-pointed star, eventually treats the geometric shape with an increasing number of sides as circular (as the precipitation reaction proceeds).

2.14   Multitasking in chemical processors

Multitasking in a parallel processor implies that the processor implements computations over two separate sets of data and produces two separate results, while both the sets of data and results share the same local memories of the elementary processing units. In an ideal case the tasks do not interfere with each other.

To date, most unconventional specialised (i.e. not performing a universal logical computation) processors carry out only one task at a time because the data, a program of computation and the results are represented by local disturbances of the processor's chemical or physical state.

The implementation of multitasking looks particularly difficult in wave-based computers because the computation is realised via the spreading and interaction of diffusive or phase waves. For example, an excitable chemical medium such as the thin-layer Belousov–Zhabotinsky reaction can only construct one path in one maze during a given experiment [269], navigate only one robot at a time [21] or detect an optimal path along only one tree-like graph at a time [232]; see more examples in [7].

The only way to carry out several computational tasks simultaneously in a reaction-diffusion processor is to spatially segregate them. For example, a palladium processor can construct a Voronoi diagram of a planar set and a skeleton of a planar shape if the planar set and the shape are projected onto separate parts of the reaction medium such that diffusive waves do not interfere; however, this is not multitasking in the truest sense.

Are these reaction-diffusion computers really so bad? Not at all. They are single-task only because there are a small number of reagents constituting the over-all system; by increasing the number of reactants is it possible to increase the number of tasks solved?

Let us check the idea on chemical processors that construct Voronoi diagrams. Given two separate sets of planar points, $\mathbf{P}_1$ and $\mathbf{P}_2$, a chemical processor must construct two 'independent' Voronoi diagrams $VD(\mathbf{P}_1)$ and $VD(\mathbf{P}_2)$, as shown in Fig. 2.27.

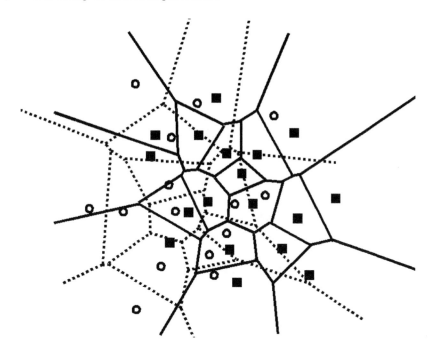

Figure 2.27: A schematic drawing of two super-imposed planar Voronoi diagrams. The first set of planar points, to be separated by the edges of a Voronoi diagram, are shown by solid rectangles; the second set by circles. The edges of the Voronoi diagram pertaining to the first data set are shown by solid lines and the second data set by dotted lines.

How can this be done? Earlier, we considered a very simple model of Voronoi-diagram formation; to adapt it to construct two diagrams you just need to employ one more reactant and one more substrate.

2.14.1   Multitasking in automaton model

Assume that R_1 and R_2 are reactants which do not interact with each other. We represent the elements of the planar set $\mathbf{P}_1$ by drops of the reactant R_1 and the elements of $\mathbf{P}_2$ by drops of R_2. The reactants diffuse and form the precipitate W when they react with the substrate S: $R_1 + S_1 \rightarrow W_1$ and $R_2 + S_2 \rightarrow W_2$. In conditions when diffusive fronts of R_1 (R_2) do not interact with diffusive fronts of R_2 (R_1) but are competing for the substrate or are repelled by the fronts of R_1 (R_2), we have the edges of a Voronoi diagram represented by sites without the precipitates W_1 and W_2.

In this case the bisectors will be uncoloured but the Voronoi cells will be colour coded by the colours of the precipitates W_1 and W_2. The results of the cellular-automaton simulation are shown in Fig. 2.28.

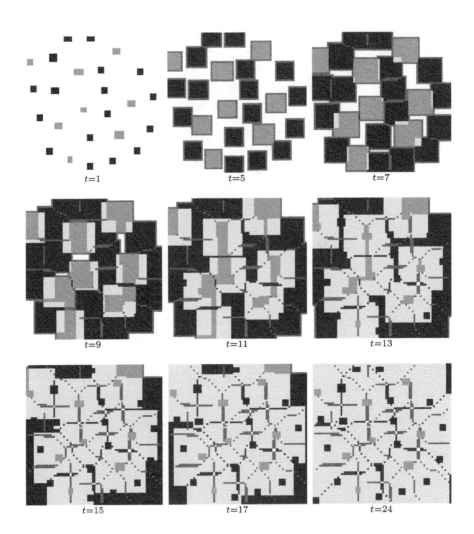

Figure 2.28: A series of configurations of a two-dimensional cellular automaton that simulates the reaction-diffusion construction of two Voronoi diagrams. The initial drops of two separate reagents are shown in brown (the reagent R_1) and green (the reagent R_2). The Voronoi diagrams are shown in cyan — VD($\mathbf{P}_1$) and magenta — VD($\mathbf{P}_2$). See also Fig. 9.17 in colour insert.

2.14.2 Coupling ferric ferrocyanide processor with ferrous ferricyanide processor

To construct an experimental prototype of a multitasking processor we combined the ferric ferrocyanide (Prussian blue) processor, described earlier, with a newly devised ferrous ferricyanide processor. The ferrous ferricyanide processor was constructed using the ferrous ion–ferricyanide couple; the blue-coloured product formed in the reaction is ferrous ferricyanide ($Fe_3(FeCN_6)_2$).

Potassium ferricyanide 2.5mg/ml[2] was mixed with a 1.5% gel of agar. The same methodology was used as to produce thin gel films of the mixture. In this case drops of a ferrous ion solution (300 mg/ml iron(II) sulphate hepta hydrate, 99.5+%, BDH Chemical Ltd.) were applied to initiate the process.

2.14.3 Findings of multitasking

It should be noted that when both ferric and ferrous ions are used in the single parallel processors (reacted on single-substrate gels) the products in both cases are blue. However, when ferric ions are used in the multitasking processor the product is green. This is because the predominant colour of the gel is yellow (due to the ferricyanide) and therefore the combination of the two colours blue (colour of product) and yellow (colour of unreacted gel) gives the product an apparently green colouration. This serves as a good initial indication that the ferric ions do not react with the ferricyanide in the gel. Obviously, the ferrous ions react with the ferricyanide to give a blue product and the ferrocyanide gel is relatively uncoloured, so the product in this case appears blue.

A planar Voronoi diagram VD($\mathbf{P}$) is approximated in a chemical processor with one reactant, drops of which represent elements of the planar set $\mathbf{P}$, when reacted on a mixed substrate gel.

This is demonstrated in Fig. 2.29, where a sequence of images of a two-dimensional chemical processor with the single-reactant ferrous ions (Fig. 2.29, left-hand column) and ferric ions (Fig. 2.29, central column) and double-reactant substrate (potassium ferrocyanide and potassium ferricyanide) gel is shown.

The result is apparently identical to the result obtained where the single reactant was reacted with its respective substrate on a single-substrate processor.

A superposition of two planar Voronoi diagrams VD($\mathbf{P}_1$)∪VD($\mathbf{P}_2$) is expected in a chemical processor with two reactants R_1 and R_2, drops of which represent points of $\mathbf{P}_1$ and $\mathbf{P}_2$, respectively.

Triple subdivision: in a system with two reactants R_1 and R_2 and two substrates S_1 and S_2 three plane subdivisions are formed. The first is a Voronoi diagram formed because of the interaction of reagent fronts R_1 and R_2 (in this case wave fronts do not feel the difference!); the second and third are Voronoi diagrams formed where wave fronts only interacted with the wave fronts of the same reactants after the initial fronts had overlapped.

In Fig. 2.29 (right-hand column) you can see a pattern of ferric ions interlaced with ferrous ions reacting on a double-substrate processor. The reaction is similar

---

[2]99+%, Fisher Scientific UK Ltd., Loughborough, Leicestershire

Figure 2.29: Snapshots of plane subdivision in the two-tasking chemical processor. Right-hand column: the progress of the reaction of ferric ions on a mixed substrate (potassium ferrocyanide and potassium ferricyanide) forming a single Voronoi diagram. Central column: the reaction of ferrous ions on a mixed substrate gel processor forming a single Voronoi diagram. Left-hand column: the reaction of ferric and ferrous ions on a mixed substrate in the process of calculating three Voronoi diagrams.

in the initial stages in that circular uniform wave fronts are initiated at the edges of each reactant drop and move outwards according to the speed of diffusion. However, when the wave fronts approach and the reactant sources are different the wave fronts cross (Figs. 2.30 and 2.31 show this process in progress) — which is in stark contrast to the case where the reactant sources are the same and the wave fronts annihilate.

The mechanics of bisector formation in the double parallel processor are not fully understood. However, the additional Voronoi diagrams formed by wave fronts that have overlapped and then annihilated with wave fronts of the same reactant source appear to be governed by a substrate-competition mechanism. However, these bisectors are very diffuse when compared to the primary bisectors (see Fig. 2.31). There are two reasons for this, the first being the fact that a continuous precipitate layer of the opposite couple will be super-imposed over the bisector. The second

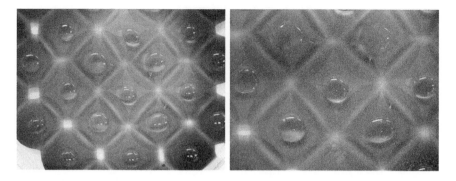

Figure 2.30: Close-up images of the formation of additional Voronoi diagrams by overlapping wave fronts in the double parallel processor.

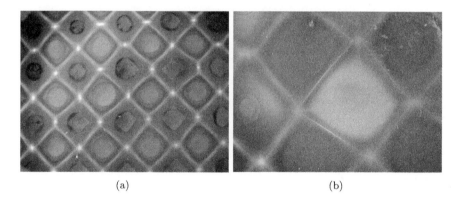

(a)   (b)

Figure 2.31: The intricate fine structure of the primary bisectors in a constructive parallel processor. Also evident are fainter secondary bisectors pertaining to the additional Voronoi diagrams calculated. (a) Shows a reaction nearing completion — the image demonstrates that the diffusing fronts even penetrate the original drops of the opposite reactant source. (b) Shows a completed computation which has subsequently dried, emphasising the permanent nature of the results of the computations with this class of processor.

reason is that until the fronts have actually crossed like wave fronts are not directly opposed (in our experimental setup).

The primary Voronoi diagram would appear on first inspection to be identical to the Voronoi diagrams formed in the single-reactant, single-substrate (or double-substrate) systems. However, Fig. 2.30 shows an image of a completed computation which shows that the primary bisector structure is quite different from the normal type. It has a clear sharp bisector that separates the wider diffuse bisector into two halves.

When discussing the formation of Voronoi diagrams we put forward a mechanism

Figure 2.32: A multitasking parallel chemical processor near to completing the computation of three Voronoi diagrams. The primary bisectors evident in the single processors are very clear, but in this example two additional sets of bisectors are formed so that the plane of the Petri dish is approximately split into 1-cm squares.

whereby the advancing fronts competed for substrate. We also postulated that this force was physical in nature and thus independent of chemical affinity. Therefore, both substrates will diffuse from this separator towards the closest front and into the precipitate-forming region where one (reactive) will react to form precipitate and the other (non-reactive) will remain. Thus, as they approach this separator both the approaching fronts diffuse into an area on either side of it, which is depleted in their respective substrates, and the result is a bisector as in the single system.

At this point as the wave fronts cross the reaction effectively re-sets and the fronts are then opposed to those of the same reactant (in this experimental setup). Thus, each new front exerts a force on the area of gel in advance of it. Thus, when these fronts collide the result is the formation of thin diffuse secondary bisectors as shown in Figs. 2.30 and 2.31. These secondary Voronoi diagrams pertaining to the original locations of identical reactant sources are also permanent, although slightly less easy to observe for the reasons described above (see Fig. 2.32, and Fig. 9.18 in colour insert).

Thus, the result of combining these two chemical couples to form a multitasking parallel processor in chemical media is the formation of three separate Voronoi diagrams (Fig. 2.33).

Figure 2.33: Super-imposed Voronoi diagrams constructed in the two-tasking chemical processor. Two planar sets $\mathbf{P}_1$ and $\mathbf{P}_2$ are represented by drops of two different reagents R_1 (patterned disc) and R_2 (blank disc). A primary Voronoi diagram $\mathrm{VD}(\mathbf{P}_1 \cup \mathbf{P}_2)$ ($\mathbf{P}_1$ and $\mathbf{P}_2$ are not distinguished in this diagram) is shown by solid black lines. Segments of two secondary diagrams $\mathrm{VD}(\mathbf{P}_1)$ and $\mathrm{VD}(\mathbf{P}_2)$ are shown by yellow and red dotted lines, respectively.

2.15 Conclusion

A number of chemical systems exhibit Voronoi diagrams as the output state for distinct sets of parameters. Below a critical substrate concentration trigger fronts have to be induced via the addition of the outer electrolyte at specific points. Trigger fronts propagating from these points interact with fronts initiated at other points and annihilate in bisectors of low precipitate, which correspond to the bisectors of a primary Voronoi diagram.

The same systems were capable of computing a number of different types of generalised Voronoi diagram. These included the computation of a Voronoi diagram where the source of trigger fronts were geometric shapes cut from adsorbent materials and soaked in reactant rather than drops of reagent. A number of weighted diagrams could also be computed if reagents with different diffusion/reaction rates were utilised, or reactants were placed onto the gel at different times during the evolution of the diagram.

In certain systems when the substrate concentration was increased diverse instabilities of the formation process occurred. Increasing the substrate concentration above a lower critical value results in an instability of the trigger fronts leading to imperfect or no Voronoi diagrams. A further increase of the substrate concentration above an upper critical value destabilises the chemical system as a whole such that the application of the outer electrolyte results in spontaneous splitting of the front. The interaction of a number of these expanding circular wave fronts results in the formation of a secondary Voronoi diagram. In this case the bisectors of the Voronoi diagram were constructed from a high precipitate concentration.

From the theoretical point of view, the formation of Voronoi diagrams has been investigated on the basis of cellular-automata models and a reaction-diffusion system of the activator–inhibitor type. Both approaches illustrate the formation of Voronoi diagrams whereby the bisectors are formed due to the mutual interaction of trigger fronts. In the case of the cellular-automata model a scenario has been discussed in which the variation of a control parameter reveals an appropriate parameter interval for the generation of Voronoi diagrams. Either side of this parameter interval the Voronoi diagrams are imperfectly formed or not formed at all.

We detailed how to fabricate a real-life reaction-diffusion chemical processor which constructed a skeleton of a planar shape. Actually, the same processor could be used to construct a Voronoi diagram of several shapes as demonstrated in Fig. 2.34.

In contrast to any existing or virtual massively parallel processor the chemical processor does not artificially discretise the data and results. Instead, an almost 'continuous' skeleton is produced. Clearly, our technique, because it is based on the same principles of spatial computation, inherited most of the disadvantages of Blum's 'classical' grass-fire skeletonisation. Thus, for example, it does not take the concavities of a contour into account; therefore, one could only refer to a unique representation up to some degree of accuracy, see e.g. [116]. There is, however, a slight improvement as well: the chemical skeletonisation is not as sensitive to noise as a 'classical' one; small perturbations of a data shape do not cause changes in the computed skeleton.

OK, I've got the skeleton, how could I classify the skeleton using reaction-diffusion processors? This could be a subject of further studies. For instance, one could project a skeleton onto an excitable chemical medium, e.g. using a light-sensitive modification of a Belousov–Zhabotinsky medium. At several sites of the medium, corresponding to certain points of the skeleton, generators of excitation will form. Some of the generators will be destroyed in competition with other generators. A configuration of the generators that remained 'alive' could reflect, possibly not uniquely, a topology of the skeleton. Some ideas for future investigations can be found in, for example, [7]; also, it is worth mentioning the only existing chemical processor for very simple pattern recognition, designed in [159].

By the end of the chapter we discussed the possibilities of solving two tasks in the same domain (chemical reactor) at the same time. The 'abstract' design of a multitasking parallel chemical processor was as follows. Two separate substrates S_1 and S_2 are mixed with a gel; two sets of planar points $\mathbf{P}_1$ and $\mathbf{P}_2$ are represented as drops of two separate reactants R_1 and R_2. The reactants R_1 and R_2 diffuse and interact with the substrate independently of each other, i.e. R_1 interacts with S_1 and R_2 with S_2. The two separate Voronoi diagrams VD($\mathbf{P}_1$) and VD($\mathbf{P}_2$) would therefore be constructed. However, experimental findings did not support our expectations in full.

Figure 2.34: A structure developed in the palladium processor demonstrates that both skeleton (internal Voronoi diagram) and Voronoi diagram (external skeleton) are constructed in the processor.

The original concept of an ideal multitasking parallel chemical processor, where unlike reaction fronts cross unhindered to form two distinct Voronoi diagrams pertaining to the spatial positions of the like reactant sources, proves to be unobtainable. The actual result is rather a pair of linked chemical processors — where they are inextricably linked due to their existence in the same physical space. Obviously, the multitasking chemical processor solves both tasks somehow, maybe not necessarily as we would have envisaged, but offers a solution in the form of approximated Voronoi diagrams. The accuracy of approximation is a particular field and this will be discussed elsewhere. The direct results of two linked chemical processors even in this simple chemical system is an increase in computational ability and structural complexity.

Indeed, you could speculate firstly that universally no truly unconventional — physics- or chemistry-based or bio-inspired — parallel processes (and thus processors) exist *per se*. That where these processes (processors) can be thought of as parallel, i.e. universal forces such as gravity have a negligible or not fully measurable effect on the computational outcome, they may not be combined in the same physical space without becoming inextricably linked due to the action of these universal forces. This linking of apparently parallel processes and thus processors is a natural route to increasing performance in biological, physical and chemical computing devices.

How many Voronoi diagrams can be constructed in the reaction-diffusion processor? It is hard to find two chemical reactant couples which have limited cross-reactivity; we cannot be sure of the cross-reactivity of the systems reported although for the purposes of multitasking it proved negligible. Therefore, to get three independent systems would be very difficult but not impossible at least theoretically. So, theoretically the number of tasks implemented simultaneously in a real-life reaction-diffusion processor could be unlimited.

Chapter 3

Logical circuits in chemical media

Are reaction-diffusion-based processors constructed in chemical media capable of universal computation (i.e. the implementation of a functionally complete system of logical operations, e.g. negation and conjunction, or equivalence) or at least the realisation of some logical gates? Yes, and this is demonstrated via the implementation of logical gates [127, 128, 283, 53] and the simulation of finite-automata networks [130, 131, 72, 314] in chemical and biochemical systems.

There are three basic ways to implement a logical operation in chemical systems: integral-dynamic based; geometry based, involving the interaction of excitation impulses or diffusive wave fronts in a geometrically constrained medium; and collision based, exploiting the phenomenology of interactions between travelling localisations.

The integral-dynamics-based approach is to exploit chemical kinetics in mass-coupled networks of chemical reactors [127, 128, 72, 314]. Thus, for example, Blittersdorf *et al.* [53] built Boolean logical gates from a network of three chemical reactors, where two of the reactors represent input variables and the third reactor is an output. Binary states of the reactors equate to the level of acidity in the reagent solutions. The computation is performed by setting up pH levels in the input reactors, adjusting the flow rate between the reactors and measuring pH in the output reactor.

In a similar way an exclusive OR gate was realised in a controlled enzymatic dynamic by Zauner and Conrad [72, 314]; this can be seen as a combination of mass-coupled computation and macro-molecular computing, as e.g. realisation of XOR operation in self-assembling of DNA molecules [169].

The integral-dynamics-based approach does not exploit in full the space–time dynamics and the particulars of local interactions between disturbances and patterns in reaction-diffusion systems. Therefore, we do not discuss this further but will provide examples of logical gates in geometrically constrained precipitating chemical systems and in uniform excitable chemical systems exhibiting mobile self-localisations. An overview of early results concerning logical gates in geometrically constrained Belousov–Zhabotinsky systems can be found in [7], so in this chapter we discuss only novel results obtained in recent years.

3.1   Logical gates in precipitating medium

Our approach exploits the fact that in a reaction-diffusion processor two diffusive waves of the same reactant initiated in separate locations form a distinctive pattern when they interact (collide) with each other. The methodology for the production of such processors is originated in [282, 18] and it gets a sophisticated development in recent papers [19, 14, 80].

A reaction-diffusion logical gate based on a palladium chloride processor is designed as follows. A gel (2% agar or agarose by weight) containing palladium chloride (0.2% by weight) is prepared and allowed to set in Petri dishes or on acetate sheets to a thickness of 1 mm. Geometric structures, including T-shaped chambers and coupled gates, are cut from the gel film. The gel film forms an active substrate. The values of logical variables are represented by planar substrates soaked in a reactant solution of potassium iodide (saturated at 20°C), at the input chambers (for example the chambers x and y in Fig. 3.1a).

As the uncoloured solution of potassium iodide diffuses into the palladium chloride gel (yellow) it reacts with the palladium chloride to form iodo-palladium species. Sites where the reaction occurs change from yellow (at a $PdCl_2$ concentration of 0.2%) to dark brown (iodo-palladium species formed). It is assumed that the gel film is homogeneous and uniform and thus wave fronts originating from separate sources travel with the same speed and therefore have the same distance to the point of collision. Sites of the substrate where two or more diffusive wave fronts meet (collide) remain uncoloured; a bisector is formed separating the reactant sources. A more detailed treatment of the chemistry of this and other similar systems is available in previous publications [282, 80, 19, 14].

As discussed previously, the mechanism is thought to be based on substrate competition between the advancing wave fronts. The reactant fronts possess a chemical affinity for the substrate that is maximal at the initiation point of the reaction but that decreases with time. The speed of diffusion also decreases as the reaction proceeds. The physical presence of the diffusing wave front exerts an influence over the gel directly in advance of it. A combination of both chemical and physical mechanisms causes substrate to move from the gel towards the diffusing front. Where the speed of diffusion is high this effect is minimal; however, when the speed of diffusion drops appreciably a dynamic concentration gradient of the substrate in advance of the wave front is formed. The upshot of this is that palladium chloride would be leached from the gel a small (but increasing — proportional to the reaction time and thus the distance separating the reactant sources) distance in front of a travelling wave front. Therefore, the unreacted KI eventually diffuses into a zone of the agar gel near the collision point that is depleted in the substrate, resulting in the formation of substantially less PdI_2; the gel remains uncoloured. Thus, the bisector forms a permanent record of the substrate's concentration gradient just prior to the collision of the fronts. Evidence for this mechanism comes from the bisector width, which is proportional to the distance separating the reactant sources. If the sources are very close, then no bisector or a diffuse bisector is formed; if the sources are just close (a few mm apart), then sharp thin bisectors are formed; if the sources are an appreciable distance apart, wide and more diffuse bisectors are formed (with an obvious concentration gradient). It is this natural tendency to form bisectors of

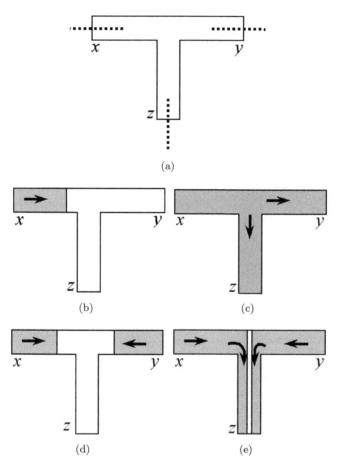

Figure 3.1: Schematic demonstration of the XOR gate. (a) The gate's architecture; sites at which the state of the reactor is measured are shown by the dotted lines. (b) A reactant is added to one of the inputs, $x = T$, the reactant then diffuses along the chambers of the gate and (c) reaches an output, $z = T$. (d) A reactant is added to both inputs simultaneously, $x = y = T$, two wave fronts are initiated and move towards one another, where they interact to form an uncoloured bisector (e), $z = F$.

different thicknesses proportional to the distance separating reactant sources which in the first instance will be exploited to form logical gates in precipitating chemical media.

3.1.1   xor gate

A T-shaped system of chambers comprises the gate in our experimental system: two horizontal chambers, x and y chambers, act as inputs and a vertical chamber, z chamber, as an output. We measure the colour of the input and output chambers in the sites shown in Fig. 3.1a. When we apply reactant at one of the horizontal chambers, e.g. the x chamber, the reactant diffuses and spans along the other two chambers, the y and z chambers. All chambers become filled with the coloured precipitate (Fig. 3.1b and c). If the two input chambers are initiated via addition of the reactant (Fig. 3.1d) then the advancing wave fronts of the diffusing reactant interact at the intersection of the horizontal junctions and the vertical junction (Fig. 3.1e). Thus, an uncoloured strip (bisector) is formed between the two horizontal input chambers; this extends along the vertical section of the gate and corresponds to a 'no-precipitate' value at the measurement point of the output z chamber. The following mapping is implemented in the gate (P stands for precipitate, coloured sites and U stands for no precipitate, uncoloured sites):

x chamber	y chamber	z chamber
U	U	U
U	P	P
P	U	P
P	P	U

Assume that a coloured (precipitate) site represents T (TRUTH) and an uncoloured site (no precipitate) F (FALSE) logical values, then we find that the gate realises the function XOR: x XOR $y = (\neg x \wedge y) \vee (x \wedge \neg y)$. A negation is trivially realised as $\mathbf{T}$ XOR $x = \neg x$, where $\mathbf{T}$ is a constant truth.

Experiments with real chemical gates based on the reaction of a palladium chloride gel and a potassium iodide reactant solution are shown in Fig. 3.2. In this experimental gate the colour intensity at given points gives a measure of the output variables, as it directly correlates with the presence/absence or relative concentration of the primary product (this is shown in detail in Fig. 3.3).

It is apparent that the experimental findings are slightly different from the theory, although it can be seen that the XOR gate is implemented using this experimental setup. In the chemical system a distinct bisector is formed between the two directly opposed fronts emanating from the horizontal chambers x and y; however, this bisector does not extend far down the vertical output chamber z and thus a change in the measurement point would yield a more effective implementation of the XOR gate. Instead, as the wave fronts diffuse away from the collision point and into the vertical output chamber, the bisector becomes increasingly diffuse. However, it is obvious that there is a marked concentration difference at the output chamber corresponding to the interaction of the wave fronts. Thus, the output variable of F is achieved in this experimental setup.

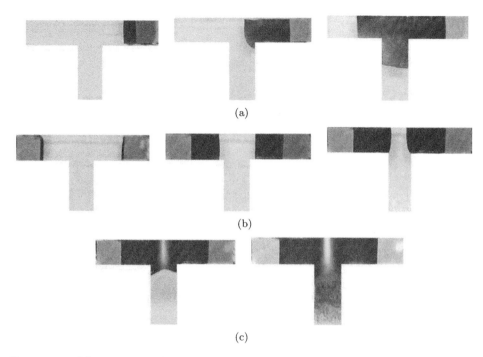

(a)

(b)

(c)

Figure 3.2: Photographs of experimental implementation of the XOR gate. (a) The progression of the reaction-diffusion gate for the inputs $x = F$ and $y = T$. (b) and (c) The progression of the reaction-diffusion gate for the inputs $x = T$ and $y = T$. See also Fig. 9.19 in colour insert.

As the mechanism of bisector formation is based on substrate competition between the advancing fronts, it can be seen how refining the experimental setup would yield a more effective implementation of the XOR gate. If the initial conditions were devised whereby the output chamber would be completely depleted of substrate at the point of collision between the fronts, then the reaction could not proceed further and the fronts are effectively annihilated — the output chamber would contain no precipitate. As discussed earlier, a special case exists where if two reactant sources are separated by an extremely small distance then no bisector is formed. This affords the possibility of constructing an AND gate from an adjusted experimental setup, where a length just below this critical distance separates the two input variables.

The palladium-based reaction discussed here belongs to a larger class of reactions that can all be utilised in the same way to construct XOR and other logical gates. These include the reaction of potassium ferrocyanide gels with a range of transition-metal salts [80] and the reaction of copper(II) salts with sodium hydroxide [120]. It is probable that under the right conditions and carried out in geometrically constrained media that this implementation would extend to a wide range of chemical reactions of this type. However, some of the systems mentioned above have important differences when compared to the palladium chloride system.

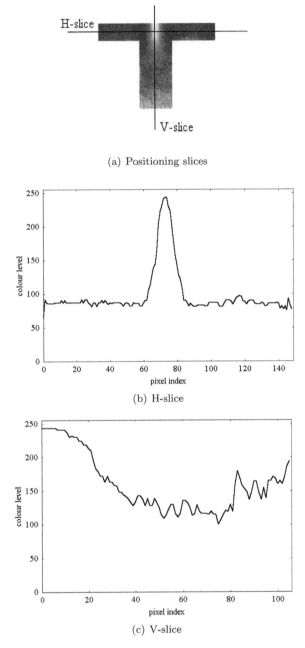

(a) Positioning slices

(b) H-slice

(c) V-slice

Figure 3.3: An example of the value-T representation as a concentration of precipitate. The diagram shows one-pixel horizontal (b) and vertical (c), see scheme in (a), slices of the 8-bit-per-pixel image of the experimental realisation of the XOR gate (Fig. 3.2b).

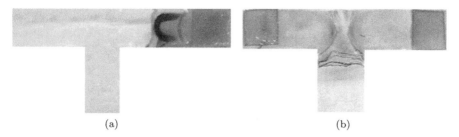

(a)   (b)

Figure 3.4: Implementation of AND gate in an unstable chemical reaction-diffusion system: the reaction of copper chloride and potassium hydroxide. The dark colouration is due to the secondary formation of copper oxide, which occurs when the reaction becomes confined behind a solid boundary. (a) The gate in the input state $x = F$ and $y = T$, reactant is added to one input chamber only. (b) The gate in the input state $x = T$ and $y = T$, reactant is placed at both input chambers. The increased stability of the system resulting in precipitate at the vertical output channel is apparent.

At certain substrate/reactant concentrations (specific to each reaction) the diffusive fronts are subject to spontaneous splitting caused by a chemical instability and the fronts cease to progress. Therefore, it is apparent that changing either concentration in the right range will result in a negation (any input results in an output of F) rather than an XOR gate.

3.1.2   Towards and gate

The mechanism of substrate competition between the fronts may allow an unstable negation gate to be converted into a true AND gate. In an unstable system any input variable results in an output of F as the precipitation never reaches the output channel. However, if the experimental conditions are chosen carefully the stability of two otherwise unstable fronts may be maintained only when there is an input variable (reactant solution) in both channels. Thus, only when the x and y chambers are initiated will any precipitate be detected at z. This is because the instability has its basis in exceeding a critical reagent concentration and only where two fronts compete for this same reagent will this concentration not be exceeded. Work to control reactions in this non-linear region of their evolution is complex practically, although theoretically it is possible. Some preliminary results (Fig. 3.4) show a gate of this type (based on the reaction of copper(II) chloride (60 mg/ml) gel and potassium hydroxide (8 M)) working in some capacity. The single input variable at y does not maintain stability and does not even reach the intersection point and the opening of the output chamber. When there are input variables at both x and y they maintain stability until near the point of contact, and subsequently precipitation can be seen to extend into the vertical output chamber z. The reproducibility of this gate implementation is low, and it is included merely to serve as an indicator of future areas for development in using this class of non-linear reactions for universal computation.

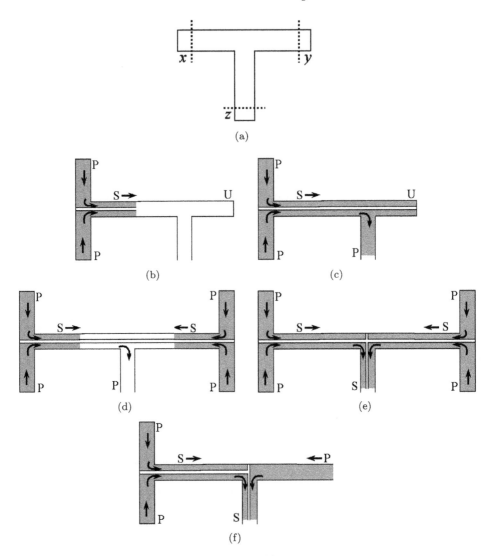

Figure 3.5: Three-valued gate. (a) Dotted lines show sites where measurements are taken. (b) and (c) Two snapshots of the implementation of $S \circ U = P$. (d) and (e) Implementation of the composition $S \circ S = S$. (f) Final configuration of a gate which performed computation $S \circ P = S$.

3.1.3   Three-valued compositions

If we arrange the measurement sites across the chambers (Fig. 3.5a) then an un-coloured strip, call it S, can be detected. In the previous section we constructed a binary composition, which happens to represent a logical XOR gate. What will be the structure of the algebraic system $A = \langle U, P, S, \circ \rangle$? We already know the partial structure of the composition (we can put S instead of U as a result of the composition $P \circ P$, because now we can sense a difference between the uncoloured chamber and an uncoloured strip; symbols denoted as ? show compositions which are yet to be defined):

$\circ$	U	P	S
U	U	P	?
P	P	S	?
S	?	?	?

Let us re-construct the rest of the table. How does an uncoloured site U interact with the strip S? It is not sensible to generate a strip *per se*, so we consider the state S as an input to the gate from another gate (Fig. 3.5b and c). First, the area with an uncoloured strip is generated via collision of two diffusive wave fronts. The strip moves along the horizontal chamber as shown in Fig. 3.5b. When the growing structure gets to the branching point the upper part continues to spread, while the bottom part releases the reactant which diffuses into the vertical (output) branch; thus, it becomes coloured (Fig. 3.5c). Thus, we implemented the composition $S \circ U = U \circ S = P$. Consider two uncoloured strips that move towards one another (Fig. 3.5d); then, the upper wave fronts interact and form a short segment of uncoloured sites and a long vertical segment of uncoloured sites (Fig. 3.5e); thus, we have an implementation of $S \circ S = S$.

The composition $S \circ P = P \circ S = S$ is realised in a similar manner (Fig. 3.5f). So, the completed table looks as follows:

$\circ$	U	P	S
U	U	P	P
P	P	S	S
S	P	S	S

The table gives the structure of the algebra A. The algebra A is commutative but not associative (e.g. $S \circ (P \circ U) = S$ and $(S \circ P) \circ U = P$). It does not have minimal generators, neither units nor zeros. There are just two idempotents, U and S.

We are unaware of any known logics that might have the composition $\circ$ as one of its operators. To construct a three-valued logical system we must represent logical values by the states U, P and S. Three assignments of logical values may look sensible (however not truly satisfactory); $*$ is a third logical value, usually interpreted as nonsense:

$E_1(S = T, P = F, U = *)$

$\circ$	T	F	$*$
T	T	T	F
F	T	T	F
$*$	F	F	$*$

$E_2(S = T, U = F, P = *)$

$\circ$	T	F	$*$
T	T	$*$	T
F	$*$	F	$*$
$*$	T	$*$	T

$E_3(U = T, P = F, S =$

$\circ$	T	F	$*$
T	T	F	F
F	F	$*$	$*$
$*$	F	$*$	$*$

An operator represented by the table E_1 gives a constant T at the set $\{T, F\}$. The tables E_2 and E_3 could in principle be considered as tables of conjunction operators $\wedge_2$ and $\wedge_3$. The composition $* \wedge_2 * = T$ makes the gate $\wedge_2$ useless. The $\wedge_3$ gate may be applied for certain types of non-classical reasoning if a proper interpretation of $F \wedge_3 F = *$ and $* \wedge_3 * = T$ can be found. We leave this for future studies.

Figure 3.6: A chemical implementation of a three-valued logic $S \circ S = S$ gate (see scheme in Fig. 3.5d and e) in the palladium chloride reaction-diffusion processor. The analysis of the results is complicated but the main point is that it shows that combinations of gates can be achieved in these real chemical systems. Some leaching of the product is apparent in the original gates due to the high relative concentrations of the reactant — however, a bisector (S) is still apparent in these original gates. A secondary bisector (S) can also be observed in the third gate — the width and intensity of this bisector should give a measure of the original input states of the first and the second gates.

A chemical implementation of a three-valued composition is shown in Fig. 3.6 and equates to the composition shown in Fig. 3.5e. The gates must be implemented on a much smaller scale to allow time for completion of the reaction before syneresis of the gel affects the results. Clearly, the fact that the uncoloured strip does not extend to the output variable in this form gives some differences between experimental and theoretical results. The real systems prove to have an in-built smoothing mechanism. However, this said the output from each single gate corresponds directly to the concentration of products — which is given by the relative composition of reactants. Therefore, the output from each type of gate is distinct and its exerted influence on the next gate must be distinct depending on the configuration of those gates. For example in Fig. 3.5e the reactant concentration at the input of the second gate will be double what it would have been if the single-input gates were utilised in the first stage; so effectively the width of the uncoloured strip S equals half the width of the precipitate representation of P. Therefore, in this case the strip S formed at the second junction will be approximately twice as wide as the case where single input variable gates formed the inputs to the same gate.

Eventually, reactant depletion will control the output variables from these combined systems — and could be used to differentiate between distinct input histories. For example, the combination of various two-input gates will eventually produce a precipitate (output — albeit mediated by collisions) where single-input gates have become totally depleted in substrate and thus produce no output.

3.2   Collision-based computing in excitable media

All of the experimental prototypes of reaction-diffusion processors so far implemented are based on exploiting the interaction of wave fronts in a geometrically constrained chemical medium, i.e. the computation is based on a stationary architecture of the medium's inhomogeneities. Constrained by stationary wires and gates, reaction-diffusion chemical universal processors pose little computational novelty and have no dynamical re-configuration ability because they simply imitate architectures of silicon computing devices.

To appreciate in full the massive parallelism of thin-layer chemical media and to free the chemical processors from the limitations of fixed computing architectures, we adopt an unconventional paradigm of dynamical, architecture-less, or collision-based, computing. The paradigm originates from the computational universality of the Game of Life [51], conservative logic and the billiard-ball model [103] and their cellular-automaton implementations [171]. A collision-based computation employs mobile compact patterns; in our particular case they are self-localised excitations in an active non-linear medium. The localisations travel in space and perform computation (implement logical gates) when they collide with each other. There are no predetermined stationary wires — a trajectory of the travelling pattern is a momentary wire — thus, almost any part of the medium's space can be used as a wire. Truth values of logical variables are given by either absence or presence of a localisation or by various types of localisations. The state of the art in collision-based computing is presented in [8].

Solitons, defects in tubulin micro-tubules, excitons in Scheibe aggregates and breathers in polymer chains are most frequently considered candidates for a role as information carriers in nature-inspired collision-based computers; see the overview in [7]. It is experimentally difficult to reproduce all these artefacts in natural systems; therefore, the existence of mobile localisations in an experiment-friendly reaction-diffusion medium would open new horizons for fabrication of collision-based computers. Until recently, we have had very little if any information about the interaction of mobile localisations in two- or three-dimensional reaction-diffusion media. However, Schenk et al. [244] and Bode et al. [58] demonstrated the existence and rich interaction of quasi-particles (dissipative solitons) in a three-component reaction-diffusion system. The basis for collision-based universality of reaction-diffusion chemical media was finally laid when Sendiña-Nadal *et al.* [250] experimentally proved the existence of localised excitations — travelling wave fragments which behave like quasi-particles — in a photosensitive subexcitable Belousov–Zhabotinsky (BZ) medium.

In the present section we aim to computationally demonstrate how logical circuits can be fabricated in a subexcitable BZ medium via collisions between travelling wave fragments. While implementing collision-based logical operations is relatively straightforward, more attention should be paid to the control of signal propagation in the homogeneous medium. For example, to realise a (non-conservative) analogue of the Fredkin–Toffoli–Margolus billiard-ball model of interaction logic [103, 171], we must somehow 'fabricate' a reflector to control information-quanta trajectories. It has been demonstrated widely that by applying light of varying intensity we can control the excitation dynamics in the BZ medium [47, 218, 117, 100], wave velocity [243] and pattern formation [299]. Of particular interest is experimental evidence of light-induced back-propagating waves, wave-front splitting and phase shifting [308]; we can also manipulate the medium's excitability by varying the intensity of the medium's illumination [64]. Based on these facts, we show how to

control signal-wave fragments by varying the geometric configuration of excitatory
and inhibitory segments of impurity reflectors.

3.2.1 Model of subexcitable medium

We based our model on a two-variable Oregonator equation [97, 285] adapted to a
light-sensitive BZ reaction with applied illumination [47, 152]:

$$\frac{\partial u}{\partial t} = \frac{1}{\epsilon}\left(u - u^2 - (fv + \phi)\frac{u - q}{u + q}\right) + D_u \nabla^2 u,$$

$$\frac{\partial v}{\partial t} = u - v,$$

where variables u and v represent local concentrations of bromous acid $HBrO_2$
and the oxidised form of the catalyst ruthenium Ru(III), ϵ sets up a ratio of time
scales of the variables u and v, q is a scaling parameter depending on reaction
rates, f is a stoichiometric coefficient and ϕ is a light-induced bromide-production
rate proportional to the intensity of illumination (an excitability parameter — a
moderate intensity of light will facilitate the excitation process, a higher intensity
will produce excessive quantities of bromide, which suppresses the reaction). We
assumed that the catalyst is immobilised in a thin layer of gel; therefore, there is no
diffusion term for v. To integrate the system we used a Euler method with five-node
Laplacian operator, time step $\Delta t = 10^{-3}$ and grid-point spacing $\Delta x = 0.15$, with
the following parameters: $\phi = \phi_0 + A/2$, $A = 0.0011109$, $\phi_0 = 0.0766$, $\epsilon = 0.03$,
$f = 1.4$ and $q = 0.002$. When adjusting parameters of the model we took into
account that a decrease in ϵ results in unbounded growth of excitation activity,
while by increasing f we may roughly control the outcomes of wave collision [234].

The chosen parameters correspond to a region of 'higher excitability of the
subexcitability regime' outlined in [250] (see also how to adjust f and q in [227])
that supports propagation of sustained wave fragments (Fig. 3.7a). These wave
fragments are used as quanta of information in our design of CB logical circuits.
The waves were initiated by locally disturbing initial concentrations of species,
e.g. 10 grid sites in a chain are given value $u = 1.0$ each; this generated two or
more localised wave fragments, similar to the counter-propagating waves induced
by temporary illumination in experiments [308]. The travelling wave fragments
keep their shape for around 4×10^3–10^4 steps of simulation (4–10 time units), then
decrease in size and vanish. The wave's lifetime is sufficient however to implement
logical gates; this also allows us not to worry about 'garbage collection' in the
computational medium (the 'garbage', by-products of the collision unused in the
subsequent steps of the computation, will just disappear, annihilate, by itself!).

3.2.2 Implementation of basic operations

We model signals by travelling wave fragments [250, 47]: a sustainable propagating
wave fragment (Fig. 3.7a) represents the TRUE value of a logical variable corresponding to the wave's trajectory (momentarily a wire). To demonstrate that a
physical system is logically universal it is enough to implement negation and conjunction or disjunction in the spatio-temporal dynamics of the system. To realise

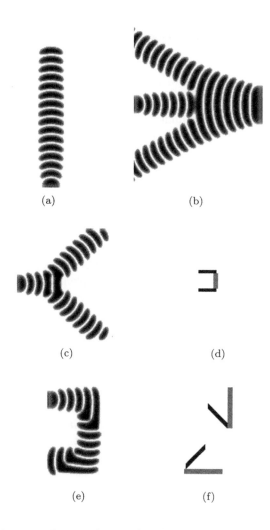

(a)   (b)

(c)   (d)

(e)   (f)

Figure 3.7: Basic operations with signals. Overlay of images taken every 0.5 time units. Exciting domains of impurities are shown in black, inhibiting domains of impurities are shown in grey. (a) Wave fragment travelling north. (b) Signal branching without impurities: a wave fragment travelling east splits into two wave fragments (travelling south-east and north-east) when it collides with a smaller wave fragment travelling west. (c) Signal branching with impurity: a wave fragment travelling west is split by an impurity (d) into two waves travelling north-west and south-west. (e) Signal routing (U-turn) with impurities: wave fragments travelling east are routed north and then west by two impurities (f). An impurity reflector consists of inhibitory (grey) and excitatory (black) chains of grid sites.

(a)   (b)

Figure 3.8: Two wave fragments undergo angled collision and implement inter-action gate $\langle x, y \rangle \rightarrow \langle x\overline{y}, xy, \overline{x}y \rangle$. (a) In this example $x = 1$ and $y = 1$, both wave fragments are present initially. Overlay of images taken every 0.5 time units. (b) Scheme of the gate. In the upper left and bottom left hand corners of (a) we see the domains of wave generation, two echo wave fragments are also generated, they travel outwards away from the gate area and thus do not interfere with the computation.

a fully functional logical circuit, we must also know how to operate input and out-put signals in the system's dynamics, namely to implement signal branching and routing; delay can be realised via appropriate routing.

We can branch a signal using two techniques. Firstly, we can collide a smaller auxiliary wave into a wave fragment representing the signal; the signal wave will split into two signals (these daughter waves shrink slightly down to a stable size and then travel with a constant shape a further 4×10^3 time steps of the simulation) and the auxiliary wave is annihilated in the collision (Fig. 3.7b). Secondly, we can temporarily and locally apply illumination impurities on a signal wave's trajectory to change the properties of the medium at this point and thus cause the signal to split (Fig. 3.7c and d). We must mention that it was already demonstrated in [308] that a wave front influenced by strong illumination (inhibitory segments of the impurity) splits and its ends do not form spirals, as is the typical situation for excitable media. A control impurity, or reflector, consists of a few segments of sites where the illumination level is slightly above or below the overall illumination level of the medium. Combining excitatory and inhibitory segments, we can precisely control the wave's trajectory, e.g. realise the U-turn of a signal (Fig. 3.7e and f).

A billiard-ball model like an interaction gate [103, 171] has two inputs — x and y, and four outputs — $x\overline{y}$ (ball x moves undisturbed in the absence of ball y), $\overline{x}y$ (ball y moves undisturbed in the absence of ball x) and twice xy (balls x and y change their trajectories when they collide with each other). We were unable to make wave fragments implement an exact billiard-ball gate, because the interacting waves either fuse or one of the waves annihilates as a result of the collision with another wave.

However, we have implemented a BZ (non-conservative) version of an interaction gate with two inputs and three outputs, i.e. just one xy output instead of two. This BZ collision gate is shown in Fig. 3.8.

The rich dynamics of the BZ medium allow us also to implement complicated logical operations in a single-interaction event. An example of a composite gate with three inputs and six outputs is shown in Fig. 3.9. As we see in Fig. 3.9,

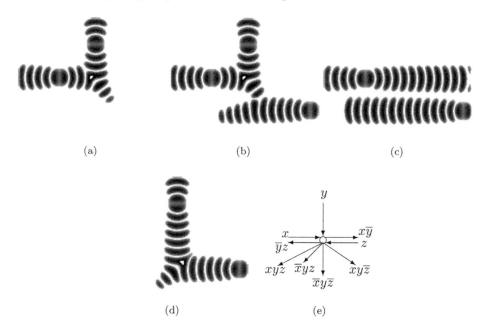

(a)   (b)   (c)

(d)   (e)

Figure 3.9: Implementation of $\langle x, y, z \rangle \rightarrow \langle x\overline{y}, \overline{y}z, xyz, \overline{x}yz, \overline{x}y\overline{z}, xy\overline{z} \rangle$ interaction gate. Overlay of images of wave fragments taken every 0.5 time units. The following combinations of the input configuration are shown: (a) $x = 1$, $y = 1$, $z = 0$, north–south wave collides with east–west wave. (b) $x = 1$, $y = 1$, $z = 1$, north–south wave collides with east–west wave, and with west–east wave. (c) $x = 1$, $y = 0$, $z = 1$, west–east and east–west wave fragments pass near each other without interaction. (d) $x = 0$, $y = 1$, $z = 1$, north–south wave collides with east–west wave. (e) Scheme of the gate.

some outputs, e.g. $\overline{x}yz$, are represented by gradually vanishing wave fragments. The situation can be dealt with by either using a very compact architecture of the logical gates or by installing temporary amplifiers made from excitatory fragments of illumination impurities.

3.3 Laboratory prototype of collision-based computer

As discussed previously the formation and control (modulation) of small BZ fragments has previously been undertaken in a modulated light-sensitive BZ reaction [250]. This medium was considered to be subexcitable and the direction and growth of small wave fragments could be controlled by carefully controlling a projected light intensity.

From a computational point of view these results are interesting as they provide a method for implementing collision-based logic gates and the theoretical work discussed in previous sections and in [9] have shown some examples of how gates based on this system can be constructed. However, from an experimental point of view it poses some problems; whilst these are not insurmountable it is a fact that a high degree of control and auxiliary equipment is required even to control the modulation of a single wave fragment, and this will be further complicated if two or more fragments are to be collided. There are two other foreseen difficulties with this methodology, namely does the modulation of light which has to be so carefully controlled to maintain the fragment in a semi-stable state correspond to a 'pseudo-wired' architecture? Also, if a collision was implemented in this system what light modulation would be applied (and at what point) to the resulting fragment and would this therefore compromise the computational results? If no modulation was applied, presumably the resulting fragment would expand and dominate the reactor space or conversely die. There is one major advantage that this system would possess over the one described in the following sections and that is some degree of inherent controllability. However, some possible limitations of this system have been highlighted above.

We devised an alternative system whereby small semi-stable wave fragments were readily produced in a BZ-type reaction over a long period and readily underwent collisions with other fragments, often multiple sequences of collisions. Thus, the experimental system we will describe serves as a proof of principle prototype for assessing collision-based computations. The control aspect is not considered in detail as we do not aim to provide an architecture-less system capable of directed computation — however, obviously there is some degree of control factored into the experimental design allowing fragments to exist *per se* but never to dominate the reactor space. Our aim is just to attribute computational behaviour to the dynamics and interactions of wave fragments in a truly architecture-less BZ medium.

3.3.1   Experimental design of collision-based processor

The reactor used during the experiments is based on a ferroin-catalysed analogue of the BZ reaction. A 2.5 cm by 2 cm piece of photographic paper (Kodak Ultra ISO 400) was cut and coated with 0.25 ml of a solution of 1, 10-phenanthroline iron(II) sulphate complex (ferroin 0.025 M, Sigma Chemical Co.). Any excess was removed and the evenly coated film left to dry in ambient conditions for 10 min.

The piece of photographic paper was then added to a Petri dish and 2 ml of a BZ stock solution added to the top surface. The stock solution was prepared as follows and was based on a recipe in [98]. An acidic bromate stock solution incorporating potassium bromate ($KBrO_3$) and sulphuric acid ($[BrO_3^-] = 0.5$ M and $[H^+] = 0.59$ M) was prepared (solution A) and stock solutions of malonic acid $[CH_2(CO_2H)_2] = 0.5$ M (solution B) and sodium bromide $[Br^-] = 0.97$ M (solution C) were prepared. The following quantities of the stock solutions were mixed: 3.5 ml solution A, 1.75 ml solution B and 0.6 ml solution C. The solution was stirred until it turned from brown to colourless prior to use in the experiments.

The piece of photographic paper was immersed in the BZ solution for ~ 5 min, in which time the back surface of the photographic paper had turned from a dark-brown to a deep-orange colour. At this stage the photographic paper was placed between two clean glass microscope slides. Some additional BZ stock solution was introduced between the slides using a pipette and the assembly was sealed using a clip taken from a gel-electrophoresis assembly (Atto Corporation).

The whole assembly was then viewed under a binocular microscope (Prior-James Swift) at a low magnification of ×10. Pictures were recorded using a Fujifilm 2600z digital camera (zoom lens $f = 6$–18 mm equivalent to $f = 38$–114 mm on a 35 mm camera) using the built-in macro function with a resolution of 640×480 pixels and using a transmissive lighting method. No filters were used.

3.3.2 Phenomenology of wave fragments

Approximately the same wave fragments could be observed from either side of the reactor, indicating that they were travelling in the bulk material (emulsion) of the photographic paper. The effect on the chemistry of the BZ reaction of using photographic paper is likely to be complex when compared to the use of conventional support substrates such as agarose or silica gels and is outside the scope of this current discussion. The photographic paper will initially contain inhibitors to the reaction in the form of silver halides (chloride or bromide ions) — there will be some conversion to an activator species (silver) with light exposure (experiments carried out in ambient light), although this reaction is limited and reversible under acidic conditions.

The fact that wave fragments and not the classical target waves and spirals are observed suggests that the chemical environment is inhibitory — the reaction is 'subexcitable'. The time for small travelling wave fragments to form varied, but was later than approximately 15 min and the wave formations could be observed for over 2 hours continuously. The average velocity of the waves was ∼ 1 mm/min, although this increased at the point just prior to collision. It was also highly dependent on the curvature of the fragments. The number of waves or wave fragments on average within the reactor space increased with reaction time as expected; however, at no stage did the wave fragments dominate the reactor space as would be observed for most conventional BZ systems especially with no external modulation.

Wave fragments observed ranged from 0.4 mm in length (such as the fragment termed the 'BZ bullet' — see Fig. 3.10, and also Fig. 9.20 in colour insert) to several millimetres; however, large fragments travelling in the medium were short lived and subject to spontaneous splitting thereby forming two or more smaller travelling fragments. Small travelling fragments were also subject to extinction prior to any collision, e.g. see Fig. 3.10, where the straight fragment on the left-hand side disappears prior to any collision with the convex fragment. Experimental observations suggest that this effect is more likely (but not solely) if the fragments are approaching a collision point.

Observations also indicate that waves travelling on the same trajectories as previous wave fragments (within a certain time interval) are more likely to undergo extinction or spontaneous splitting. This effect can also be seen in the wake of collisions where a daughter fragment will diminish if it travels back over the path of one of the parent fragments. This is strongly indicative that the BZ medium run in this setup has a long refractory period. The reaction has been carried out in this format on many occasions and apart from a high proportion and large variety of travelling wave fragments it is difficult to assign any typical behaviour. The examples covered in this chapter aim to give a representative sample of the types of collisions between these fragments and some computational significance. However,

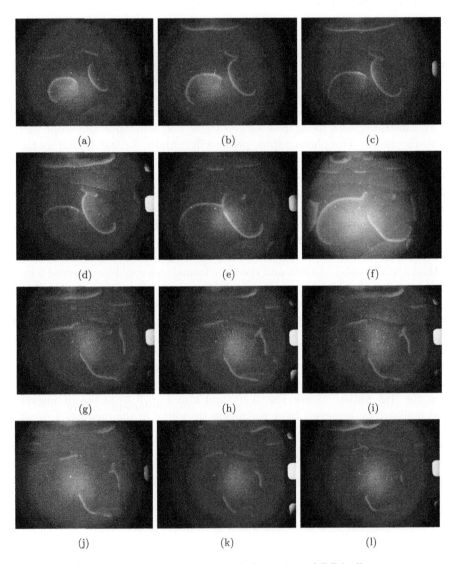

Figure 3.10: Spatio-temporal dynamics of BZ bullets.

due to the high volume of information collected it is not possible to cover all aspects of the behaviour.

Therefore, the following section will set out some of the noteworthy features of the reaction and some typical and atypical collision types, some of which are covered in the computational schemes but others that are not.

Perhaps most noteworthy is that as mentioned a number of small submillimetre wave fragments were stable (did not expand or diminish appreciably in size) over distances of many millimetres and could therefore be observed undergoing collisions. If a specific example of the 'BZ bullet' (Fig. 3.10) is taken its size was stable at between 0.3 and 0.4 mm and it was stable for many minutes travelling a distance of 0.4 cm, at which point it underwent collision with a larger fragment. The 'BZ bullet' was originally formed from a collision between a straight fragment travelling on a north-east to south-west trajectory and a convex fragment expanding along its north-west axis; the leading tips of both fragments meet and a small fragment ('BZ bullet') results travelling on a north-west to south-east trajectory.

The type of wave fragments observed varied greatly, with the dominant type being a convex-type wave comprising two counter-rotating spiral tips — waves of this type were the only fragments observed to grow significantly in the reactor space. Obviously, a large number of the more unusually shaped fragments, e.g. non-symmetrical, concave and small stable fragments, in the reactor space were the result of previous collisions or spontaneous splitting of larger fragments. A typical collision involving these convex fragments when expanding on directly opposite trajectories (convex to convex collision) results in two small daughter fragments travelling in opposite directions at 90° angles to the original trajectories of the parent fragments (see Fig. 3.13). Figure 3.10 shows an alternative scenario where fragments of this type collide when not on opposite trajectories, resulting in larger fused fragments whose trajectory is predominantly dictated by the surviving spiral tip.

Another type of possible interaction is if a convex wave approaches an obstacle, i.e. an impurity or the edge of the reactor space. In the example shown in Fig. 3.11 (see also Fig. 9.21 in colour insert) a convex wave fragment travelling from south to north (Fig. 3.11a and b) that was the daughter fragment of a collision between two convex waves travelling on east to west and west to east trajectories collides with the reactor's boundary. The result is that two daughter fragments are formed which move on east to west and west to east trajectories along the edge of the reactor (Fig. 3.11c–e). Thus in experimental systems logical operations may be realised not only via collisions with other waves but also by interaction with temporary or permanent obstacles (impurities) as predicted by theoretical models.

Another scenario is when the fragments expand and collide when moving in the same direction resulting in a small daughter fragment travelling in the opposite direction to the original parent fragments and a larger fused concave fragment moving in the same direction as the parents. If they collide whilst moving on opposite diagonals (e.g. north-west to south-east and south-west to north-east) and are size matched, then no daughter fragments may be produced.

Another typical collision is that of a convex fragment and a concave fragment (result of a previous collision); this convex to concave collision (if at the point of maximum curvature and the fragments are size matched) results in two 'v'-shaped

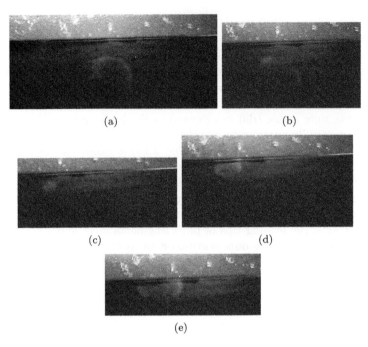

(a)   (b)

(c)   (d)

(e)

Figure 3.11: Wave-fragment's collision with the reactor's boundary.

daughter fragments which undergo 'self-annihilation'. This term 'self-annihilation' is used where fragments formed due to collisions are very short lived as they contain integral parts which continue to move on opposite trajectories or have resultant parts which are directly opposed to other parts of a larger fragment (a form of self-collision). It must be noted that size-matched collisions are not the norm in this case and the concave fragments tend not to be symmetrical; thus, convex to concave head-on collisions in this case typically result in one travelling daughter fragment and one 'v'-shaped short-lived fragment.

There were also a number of linear fragments that typically either split or diminished in size as they travelled in the reactor space. Typical collisions observed for this type of fragment were usually with convex fragments. However, on occasion two-size matched straight fragments will collide and annihilate completely.

Figure 3.14 shows a collision of a linear fragment with a larger concave fragment, resulting in two daughter fragments moving in the same direction as the larger parent fragment but moving on opposite trajectories away from the point of collision. This is in very good agreement with the theoretical results described for the collision of a large and a small fragment with the daughter fragments travelling on approximately south-east and north-east trajectories from a collision of parent fragments travelling east and west, respectively. If the larger fragment had also been straight, the distance between the daughter fragments would have remained constant. The experimental observations are consistent with a typical concave fragment being formed via a convex–convex collision and the retained counter-rotating spiral tips altering the trajectory of the daughter fragments when involved in the

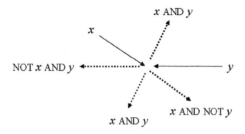

Figure 3.12: Gate $\langle x, y \rangle \rightarrow \langle x$ AND y, x AND NOT y, NOT x AND $y \rangle$.

subsequent collision.

Figure 3.16 shows a straight fragment colliding almost end on to a convex fragment — the result is a fusing and self-annihilation of one spiral tip, which cause a shift in the trajectory of the larger parent fragment as the remaining spiral tip dominates the final trajectory of the daughter fragment.

As mentioned, at some stages during the reaction fragments may spontaneously split — the most marked occasion is when a convex fragment splits into two daughter fragments moving in opposite directions to each other. This is almost identical to the case where two convex fragments collide convex face on (Fig. 3.13). It is assumed that this spontaneous splitting is due to the depletion of the activator species in the vicinity of the travelling front. This can occur when a fragment travels along a previously travelled trajectory within or close to the refractory period. It may also occur when two fragments converge, suggesting competition between the two fronts for the activator species. In this reactor the inhibitory nature of the gelatin–silver halide emulsion must be considered a factor in producing a subexcitable environment. However, another remote possibility is that local impurities in the reactor may act as a temporary barrier to diffusion. This possibility is dealt with and expanded in the next section. Perhaps more likely is that diffusion processes are extremely limited in the reactor setup described *per se*.

All of the aforementioned collisions and observed behaviours can be assigned some computational significance (see the following sections). The experimental results fit well with a theoretical framework for collision-based computing discussed in [9].

3.3.3   Examples of experimental collision gates

To describe interaction gates realised in collisions of wave fragments we adopt the formalism of [5]: $\langle a_1, \ldots, a_k \rangle \rightarrow \langle b_1, \ldots, b_m \rangle$ as a transformation of a set of logical values of k input trajectories to a set of logical values of m output trajectories.

The gate $\langle x, y \rangle \rightarrow \langle x$ AND y, x AND NOT y, NOT x AND $y \rangle$ (Fig. 3.12) is the most common gate implemented in almost any non-linear medium with mobile self-localisations.

When two convex wave fragments approach each other (Fig. 3.13a–c), they collide (Fig. 3.13d) and the convex faces of the wave fragments merge and annihilate (Fig. 3.13e and f; see wave-interaction scheme in Fig. 3.13i). However, the respec-

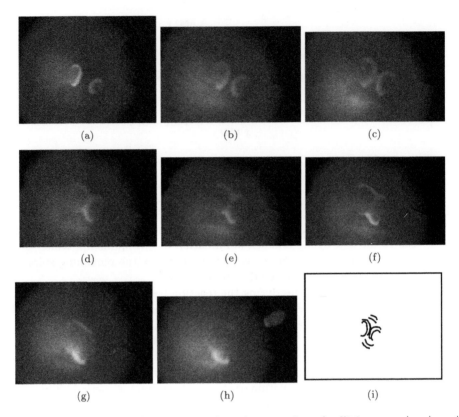

Figure 3.13: Snapshots of experimental implementation of collision gate $\langle x, y \rangle \rightarrow \langle x$ AND y, x AND NOT y, NOT x AND $y \rangle$ (see Fig. 3.12) in BZ medium, x = TRUTH, y = TRUTH. (a)–(h) Snapshots of excitation dynamics, (i) scheme of wave-fragments collision. See also Fig. 9.22 in colour insert.

tive spiral tips of the fragments form new wave fragments; these daughter fragments are initially concave (Fig. 3.13f and g), then become convex (Fig. 3.13h) and continue travelling along new trajectories. When one of the wave fragments is not present, another fragment will travel along its original trajectory. So, assuming that the presence of a wave fragment along trajectory x is representative of logical TRUTH and the absence of the fragment FALSE, then the unchanged trajectory of each fragment represents operation x AND NOT y and NOT x AND y, respectively. Trajectories of daughter wave fragments represent operation x AND y (Fig. 3.12). This is analogous to the famous Fredkin–Toffoli interaction gate [103]; however, the BZ gate just simulates but does not 'truly' support conservativeness.

The gate has $\langle x, y \rangle \rightarrow \langle x$ AND y, x AND NOT y, NOT x AND $y \rangle$: two output trajectories which represent values of the same logical variable x AND y, so the gate can be used in signal splitting.

There is one more scenario of wave-fragment interaction — they just merge into one wave fragment when they collide with each other (Fig. 3.14a–e); see the scheme

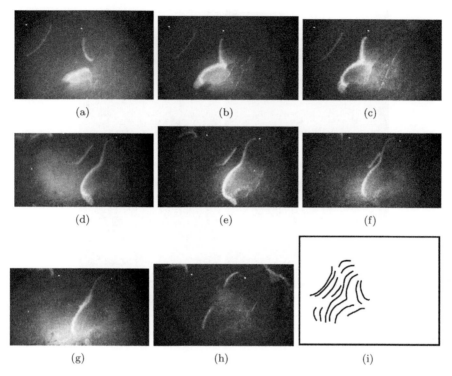

Figure 3.14: Snapshots of experimental implementation of multiple collision between wave fragments. The dynamics realised gate $\langle x, y \rangle \rightarrow \langle x$ AND y, NOT x AND y, x AND NOT y, NOT x OR NOT $y \rangle$, shown in Fig. 3.15. See also Fig. 9.23 in colour insert.

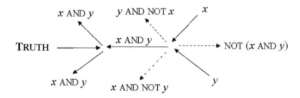

Figure 3.15: Gate $\langle x, y \rangle \rightarrow \langle x$ AND y, NOT x AND y, x AND NOT y, NOT x OR NOT $y \rangle$.

of collision in Fig. 3.14i.

In this case the colliding wave fragments implement gate $\langle x, y \rangle \rightarrow \langle x$ AND y, x AND NOT y, NOT x AND $y \rangle$ with two input and three output trajectories (Fig. 3.15, the output trajectory labelled x AND y in the far right-hand part of the gate).

A wave fragment can be split into two independent wave fragments when a smaller wave fragment collides with it (Fig. 3.14f and h). Thus, assuming that a smaller wave fragment (as that travelling south-west in Fig. 3.14) represents a

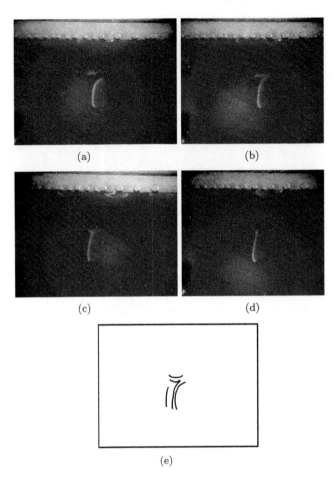

Figure 3.16: Experimental implementation of signal's reflection. Trajectory of a wave fragment representing a logical variable x is changed by colliding a control wave fragment with the wave fragment representing x. (a)–(d) Snapshots of excitation dynamics, (e) scheme of wave-fragment collisions.

constant TRUTH, and the bigger wave fragment, x, we get a splitting of a signal x into two signals x. Via the combination, merging and splitting of wave fragments it is apparent that almost any kind of logical circuit can be implemented.

An example of cascaded gates is shown in Fig. 3.15: $\langle x, y \rangle \rightarrow \langle x$ AND y, NOT x AND y, x AND NOT y, NOT x OR NOT $y \rangle$: the gate has three input trajectories, for x and y, and constant truth, and five output trajectories: two x AND y, one NOT x OR NOT y, one x AND NOT y and one NOT x AND y. Space–time dynamics of the gate for $x =$ TRUTH, $y =$ TRUTH is shown in Fig. 3.14.

Trajectories of wave fragments are changed during collisions, so that we can implement reflection (deviation of the signal trajectory) of one signal by colliding another signal into it, as show in Fig. 3.16. Sites of the collision correspond to

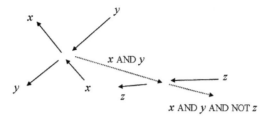

Figure 3.17: A computational scheme derived from dynamics of wave fragments shown in Fig. 3.10.

positions of 'momentary' reflectors.

In certain conditions two colliding wave fragments do not change their velocity vectors but continue to travel along their original trajectories and generate travelling daughter wave fragments, as shown in Fig. 3.10, which may undergo subsequent collisions with other wave fragments. An example of a possible computational scheme derived from dynamics of wave fragments in Fig. 3.10 is shown in Fig. 3.17.

There is also an option to incorporate stationary reflectors of signals in the BZ medium — the reflectors are represented by impurities. As discussed earlier, waves are subject to spontaneous splitting in the devised reactor; there are many possible reasons outlined and one includes the presence of impurities in the reactor. Presently, if these impurities do act as a barrier to diffusion it is a natural and uncontrollable phenomenon. However, we discussed cases whereby spontaneous splitting of certain parent fragments resulted in daughter fragments identical to those produced in collision-based gates. Therefore, as mentioned it may be possible to incorporate impurities into the BZ reactor that would cause this splitting or bring about some deviation in the trajectory (reflection) of the travelling wave fragments. Ideally, the positioning of such impurities could be controlled during the lifetime of the reaction in order to bring about directed behaviour. This may be via the application of external fields and ideally would be a reversible and dynamic process. Another option would be to use high local light intensities or a combination of locally applied light gradients — then rather than manipulating the fragment's motion continuously it may be possible to control it transiently to induce certain types of collisions or act as a reflector, etc.

3.4 Hexagonal reaction-diffusion automaton

So far, the collision-based universality of chemical systems was demonstrated in simple cellular-automaton models of excitable media, in numerical simulation or in experiments with the Belousov–Zhabotinsky (BZ) system. There is however a variety of chemical systems which could be potentially used to implement dynamical computation [79]; some of them involve complicated chemical reactions with activator and inhibitor species.

In the present section we aim to fulfil a double objective. First, to give an example of collision-based computing in reaction-diffusion system with inhibitor and activator species — which differs from the 'classical' model of an excitable medium. Second, to provide an example of localisation-based computing in hexagonal cellular automata [226] — so far we are aware that computational universality of hexagonal automata was proved by embedding a Fredkin gate [187] but have no evidence that hexagonal automata support gliders in their 'classical' excitable-medium model, as those described for orthogonal lattices in [5]. In our studies we employ the beehive cellular automaton rule, discovered by Wuensche [306, 305], which exhibits glider dynamics, and allows for a reaction-diffusion interpretation. Using the particulars of glider collision we construct basic logical gates and signal-routing operations sufficient to demonstrate computational dynamical universality of the hexagonal cellular automata. The theoretical results discussed in this section will be used in future for the experimental implementation of collision-based computing devices in 'non-standard' chemical reaction-diffusion systems.

3.4.1   Hexagonal cellular automaton: beehive rule

We can convert the rule-transition table presented in [306, 305] to more compact matrix form $\mathbf{M} = (m_{ij})$, where $0 \leq i \leq j \leq 6$, $0 \leq i + j \leq 6$, and $m_{ij} \in \{0, 1, 2\}$:

$$M = \begin{cases} 0 & 1 & 2 & 1 & 2 & 0 & 0 \\ 0 & 2 & 2 & 2 & 1 & 1 \\ 0 & 0 & 2 & 2 & 0 \\ 0 & 2 & 2 & 0 \\ 0 & 0 & 2 \\ 2 & 0 \\ 0 \end{cases}.$$

Every cell x of the hexagonal lattice updates its state in discrete time t as follows: $x^t = m_{\sigma_2^t(x), \sigma_1^t(x)}$, where $\sigma_z^t(x) = |\{y^t = z : y \in \mathbf{U}(x)\}|$; $\mathbf{U}$ is a hexagonal neighbourhood of x. Cell x is not included in its neighbourhood; therefore, the state transitions are independent of states of the cell x itself.

Starting its evolution in a random initial configuration (Fig. 3.18a) the automaton exhibits mobile localised patterns — gliders — which dominate the lattice at the concluding phase of development (Fig. 3.18b–d). The gliders either leave the lattice due to absorbing boundary conditions or continue travelling undisturbed, along non-intersecting trajectories, if boundaries are periodic. A glider is composed of one cell in state 1, which is a head of the mobile localisation, and a tail of four cells in state 2, as shown below in an example of a glider travelling west:

$$\begin{matrix} 0 & 0 & 0 & 0 & \\ & 0 & 2 & 2 & \\ 0 & 1 & 0 & 0 & 0. \\ & 0 & 2 & 2 & \\ 0 & 0 & 0 & 0 & \end{matrix}$$

Detailed analysis of glider dynamics is provided in [306, 305].

3.4.2   Reaction-diffusion interpretation

The glider's structure — active head and following tail — indicates a possibility of reaction-diffusion interpretation of cell state transition rules. Assume that the

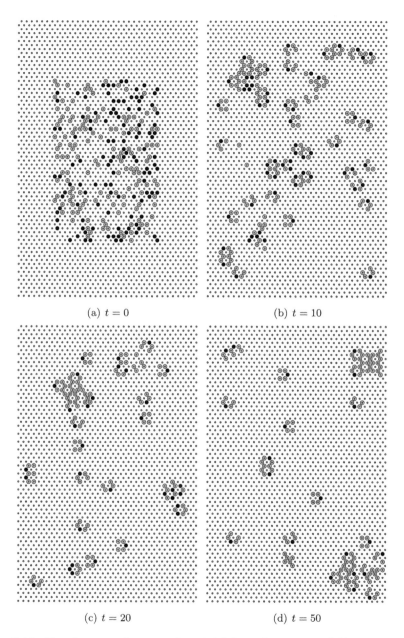

(a) $t = 0$

(b) $t = 10$

(c) $t = 20$

(d) $t = 50$

Figure 3.18: Development from random initial configuration. Cells in state 1 are shown as ●, state 2 as ⊚ and state 3 as ○.

automaton simulates a chemical medium with three reagents E (equivalent of cell state 1), I (equivalent of state 2) and S (equivalent to state 0). A cell takes state z when all six neighbours are in the state z only for $z = 0$; therefore, reagent S can be earmarked as a substrate.

State 1 is at the very front of the propagating patterns, so that E is an activator or an excitation state. It was demonstrated in [306] that changing of values of entries m_{03}, m_{14} and m_{15} does not affect formation and propagation of gliders; for simplicity we can take them equal to 1. So, this leaves us with only one condition of cell activation — one neighbour is in state 1 and the others are in state 0. Presence of even one neighbour in state 2 prevents action. Therefore, we can say that reagent I is an inhibitor of the activation reaction.

Both reagents E and I are diffusive/react with substrate in 'pure' solutions: $m_{05} = 2$ and $m_{50} = 1$; however, at higher concentrations they self-inhibit the reaction: $m_{06} = m_{60} = 0$.

To access reactions producing reagent I we can again simplify the transition matrix M, based on [306], and assign value 0 to 'wildcard' entries m_{04}, m_{13} and m_{23}. Entry $m_{02} = 2$ tells us that reagent I can be produced in concentration-dependent inhibition of reagent E.

Reagent I is involved in reaction with E (entry $m_{42} = 2$) with formation of I, so that reagent E plays a role of catalyst for I. Remaining entries of the matrix M determine that transition to state 2 happens when $1 \leq \sigma_2 \leq 3$, $1 \leq \sigma_1 \leq 4$ and $1 \leq \sigma_0 \leq 4$ but not for $(\sigma_2 = 2, \sigma_1 = 1)$.

So, the final state-transition matrix will be as follows:

$$
M^{\mathrm{RD}} = \left\{
\begin{matrix}
0 & 1 & 2 & 0 & 0 & 0 & 0 \\
0 & 2 & 0 & 0 & 0 & 0 & \\
0 & 0 & 2 & 0 & 0 & & \\
0 & 2 & 2 & 0 & & & \\
0 & 0 & 2 & & & & \\
2 & 0 & & & & & \\
0 & & & & & &
\end{matrix}
\right\}.
$$

The automaton with cell-transition states determined by M^{RD} simulate a reaction-diffusion system of three reagents with activator E, inhibitor I and substrate S, the dynamics of which is governed by the quasi-chemical reactions below, where $1 \leq \gamma_I \leq 3$, $1 \leq \gamma_E \leq 4$ and $1 \leq \gamma_S \leq 4$ but not $(\gamma_I = 2, \gamma_E = 1)$; and, $\beta_E > 3$, $\beta_I \neq 5$:

$$E + 5S \rightarrow E,$$
$$I + 5S \rightarrow I,$$
$$2E + 4S \rightarrow I,$$
$$4I + 4E \rightarrow I,$$
$$\beta_E E \rightarrow S,$$
$$\beta_I I \rightarrow S,$$
$$\gamma_I I + \gamma_E E + \gamma_S S \rightarrow I.$$

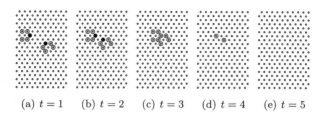

(a) $t = 1$ (b) $t = 2$ (c) $t = 3$ (d) $t = 4$ (e) $t = 5$

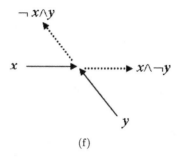

(f)

Figure 3.19: Glider heading east (represents x) collides with glider heading north-west (represents y); both gliders annihilate as a result of collision. Collision-dependent trajectories are shown by dotted lines.

In the system the inhibitor I is produced by activator E, and both reagents E and I degrade at certain concentrations.

3.4.3   Glider-interaction operations

The simplest gate $\langle x, y \rangle \rightarrow \langle x \wedge \neg y, \neg x \wedge y \rangle$ is implemented when two gliders collide and annihilate as a result of the collision (Fig. 3.19a–e). The undisturbed trajectory of the glider representing the value of x is interpreted as $x \wedge \neg y$, and glider y as $\neg x \wedge y$ (Fig. 3.19f).

To generate constant TRUTH signals we can use generators of gliders — glider guns. So far, no stationary glider guns are found in the studied reaction-diffusion automaton; however, several types of mobile guns were discovered and classified in [306]. An example of a mobile gun generating three streams of gliders is shown in Fig. 3.20. The automaton exhibits glider guns only when $M_{03} = 1$, which means a higher degree of non-linearity of chemical reactions, underlying the automaton rules. This means that activator E reacts with substrate S when the number of E molecules equals one or three, and two molecules of activator produce inhibitor I.

However, logical universality gives us just a hypothetical chance to implement real computation architectures in a reaction-diffusion automaton: a few more operations — at least reflection and multiplication — are needed to feel comfortable about the computational potential of the automaton.

For certain initial positions of gliders one glider is reflected (i.e. inverts its velocity vector) when it collides with another glider. Thus, in Fig. 3.21a–j we can see that when the glider heading east collides with the glider heading north-

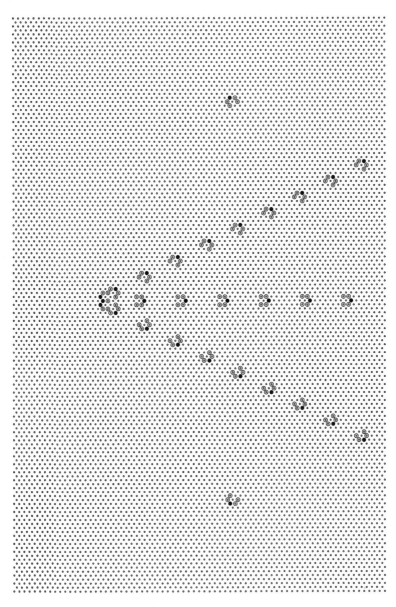

Figure 3.20: Mobile glider gun travels west and emits three streams of gliders in north-east, east and south-east directions.

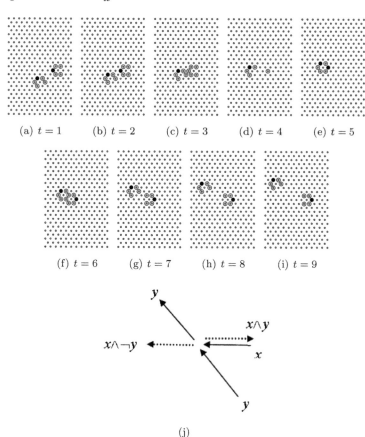

(a) $t = 1$ (b) $t = 2$ (c) $t = 3$ (d) $t = 4$ (e) $t = 5$

(f) $t = 6$ (g) $t = 7$ (h) $t = 8$ (i) $t = 9$

(j)

Figure 3.21: Reflection. Glider travelling north-east (y) acts as mobile reflector for glider travelling east (x). Collision-dependent trajectories are shown by dotted lines.

east, the former continues travelling along its initial trajectory while the latter travels to its original position. The glider acting as a mobile reflector continues travelling towards the originally specified direction. Interpreting the presence of a reflector glider as the TRUTH value of y and another glider as x we construct the following gate (Fig. 3.21k): $\langle x, y \rangle \rightarrow \langle x \wedge y, y, x \wedge \neg y \rangle$. The phenomenon can be used to implement proper routing of mobile signals by colliding mobile reflectors into them. A delay can be realised by employing several mobile reflectors which shuffle the signal between them for a certain period of time.

There are several versions of signal multiplication implementable in the automaton. In descriptions of binary collisions leading to multiplication we will assign one glider to be signal x and another glider to be multiplicator m. A collision of signal x (glider heading north-west) with the multiplicator (glider travelling east) shown in Fig. 3.22 leads to the destruction of the multiplicator, and the formation of four copies of x running north-west, south-west, east and west.

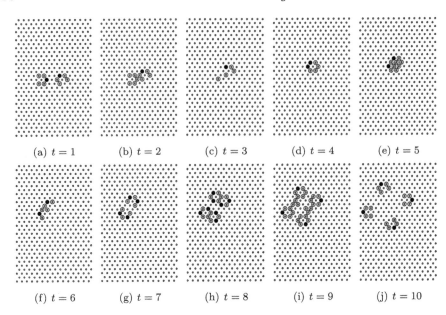

| (a) $t = 1$ | (b) $t = 2$ | (c) $t = 3$ | (d) $t = 4$ | (e) $t = 5$ |

| (f) $t = 6$ | (g) $t = 7$ | (h) $t = 8$ | (i) $t = 9$ | (j) $t = 10$ |

Figure 3.22: Multiplication with destruction of multiplicator.

For certain conditions (Fig. 3.23a) of the collisions the multiplicator continues travelling almost undisturbed (Fig. 3.23g) while the signal x is multiplied to four signals (Fig. 3.23e–i).

The two previous examples show that we can precisely tune the signals' trajectories by using disposable and re-usable multiplicators.

3.5   Conclusion

The chapter highlighted novel topics of universal computation in reaction-diffusion systems: implementations of logical gates in precipitating chemical systems (geometrically constrained, or structure-based computing), and dynamical architecture-less computing in excitable chemical systems.

The chapter uncovers new findings in structure-based computing in reaction-diffusion media, where the results of computation are given as a stationary spatial distribution of precipitate concentration. The palladium chemical processor, described in the chapter, is a disposable device; it could not be re-used as is the case with gates operating in a conventional computer. This is a disadvantage of the palladium processor, particularly when we compare it with excitable chemical media. However, the stationary spatial configuration of the processor could be interpreted as evolved hardware [180]; so, our results could be the first step in designing evolving chemical processors.

We discussed a laboratory prototype of a reaction-diffusion XOR gate. The XOR gate is employed in almost any field of computer science: from real-life circuits, self-testing systems, error-correcting codes, cryptographic systems to image processing and arithmetic logic unit design to biologically realistic algorithms of learning.

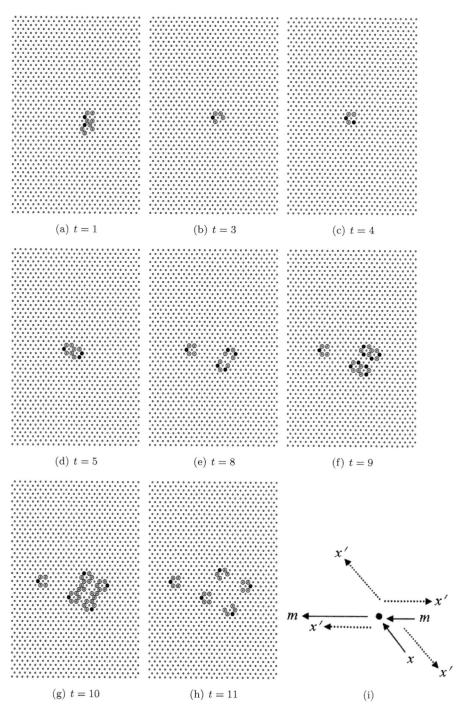

(a) $t = 1$ (b) $t = 3$ (c) $t = 4$

(d) $t = 5$ (e) $t = 8$ (f) $t = 9$

(g) $t = 10$ (h) $t = 11$ (i)

Figure 3.23: Multiplication without destruction of multiplicator. Glider m (multiplicator) travelling west multiplies glider x travelling north-west. Four copies x' of signal x travel west, east, north-west and south-east. Multiplicator m continues travelling along original trajectory. Collision-dependent trajectories are shown by dotted lines.

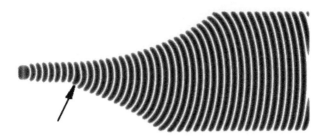

Figure 3.24: Small perturbation of a travelling wave fragment — a decrease of light intensity in a local domain of five by nine grid sites (domain of perturbation is indicated by arrow) — changes geometry of the wave fragment and thus causes uncontrollable growth of the wave front.

So, ideally, our design could be applied in fabrication of all-wet unconventional computers and reaction-diffusion information-processing units.

Also we proved, from a common-sense point of view, that even non-excitable chemical systems can form a base for designing general-purpose computers — the palladium processor, which does logical computation in the chapter, was previously applied to some problems of computational geometry, image processing and path planning [19, 14, 80].

In computational models of subexcitable chemical systems we demonstrated the existence, analysed the phenomenology and designed logical gates via the collision of travelling wave fragments. The subexcitable medium realises an architecture-less computation because there are no predetermined stationary wires, a trajectory of the travelling wave fragment is a momentary wire: almost any part of the medium's space can be used as a wire at some stage of the computation. To realise a Boolean logical gate we represent a logical TRUTH by the presence of a wave fragment and logical FALSE by the absence of the wave fragment. When two or more wave fragments collide, they may fuse, annihilate, generate new wave fragments or at least change their trajectories or velocity vectors. Thus, Boolean variables represented by the fragments are changed and the computation is implemented.

We have shown that the studied medium is highly sensitive to local perturbations and even a tiny local change in illumination may lead to drastic modification of the medium's dynamics (see example in Fig. 3.24). Thus, the computation occurs at the edge of instability.

The Oregonator model utilised in computational studies proved to be a satisfactory fast-prototyping tool for qualitative design of collision-based circuits for further experimental realisations of subexcitable chemical processors.

We provided experimental evidence of collision-based computation in a BZ medium with immobilised ferroin catalyst. We demonstrated that under certain conditions compact wave fragments develop in the medium. The wave fragments travel in the medium and implement logical gates when they collide with each other. We have shown how to embed non-trivial logical circuits in an experimental BZ medium. A spectrum of logical gates realised in the experimental system are similar to those implemented in numerical models of soliton-based computation in a bulk medium (see [8]); the key point is that we provided experimental verification of theoretical results.

The experimental setup described in the chapter is particularly encouraging because the reaction can continue for over two hours; the chemical reactor never becomes overloaded with wave fragments because conditions do not favour wave growth and the reactor resides in a reduced steady state with only minimal wave fragments populating the space. This gives us an opportunity to employ temporal cascading of complicated logical circuits.

We employed the beehive hexagonal cellular automaton [306] to design a discrete model of a 'non-trivial' chemical reaction-diffusion system. The system comprises three species — substrate, activator and inhibitor. Reactions between the activator and the substrate are concentration sensitive and highly non-linear; at a certain concentration of the activator the inhibitor is produced. The system exhibits compact travelling patterns — gliders — in its space–time dynamics. We constructed the basic logical gates based on details of particular glider collisions. We also demonstrated how signals — quanta of information represented by gliders — can be routed by colliding them with other control gliders. We provided an example of a compact pattern generator — a glider gun — which is essential for implementing negation. Therefore, we demonstrated that the reaction-diffusion hexagonal cellular automaton is logically universal, allows the embedding of logical circuits and can potentially implement meaningful computation operations. The chemical interpretation of the cell state transition rules could make this model a computational prototype for further designs of laboratory prototypes of reaction-diffusion dynamical or collision-based processors.

Chapter 4

Reaction-diffusion controllers for robots

In this chapter we present our pioneering results concerning the development of experimental reaction-diffusion processors for use as robotic controllers. Several computational models and laboratory implementations of chemical on-board robot controllers are discussed. They include Belousov–Zhabotinsky processors for mobile robot guidance, precipitating and excitable chemical media processors for robot collision avoidance and navigation, and a Belousov–Zhabotinsky chemical processor for controlling a robotic hand.

4.1 Robot taxis controlled by a Belousov–Zhabotinsky medium

We considered it a natural progression to have a chemical processor on-board of a mobile robot, thus allowing for dynamical interaction between the robot, its environment and the chemical processor. The feasibility of this approach is discussed in the present section, where we develop an experimental prototype of an on-board chemical processor, which exploits the excitation dynamics in the Belousov–Zhabotinsky reaction. We offer an experimental setup of a mobile robot with an on-board chemical controller and discuss robot-guiding techniques.

4.1.1 Guiding robots using non-linear media

Our main objective was to prove that the Belousov–Zhabotinsky (BZ) [312] reaction could be used to navigate a mobile robot in a real-life environment. Our approach involved the use of an on-board thin-layer BZ reactor and the aim was to demonstrate that by intermittent stimulation of the reactor one could implement a sensible guidance of the mobile robot. Ideally, the stimulation should come from and describe some feature of the robot's environment.

Preliminary algorithmic techniques for the chemical navigation of robots have been discussed in our early publications [7, 16, 17, 22]. In this work we describe a novel wet-ware implementation, which allows the direction towards a source of stimulation to be extracted from the spatio-temporal dynamics of the chemical controller.

As discussed, the BZ reaction exhibits a property known as excitability. An excitable system has a steady state and is stable to small perturbations; however, if the perturbations exceed a critical threshold then the system responds with an excitation event. In the case of a thin-layer architecture this results in a circular wave travelling from the source of stimulation. Parts of the wave front annihilate when they reach the boundaries of the chemical reactor (excitation waves do not reflect); however, other parts propagate towards the centre of the reactor. We will demonstrate that the asymmetry of an excitation wave front can be detected using an optical interface and thus normals to the wave front can be extracted from snapshots of the BZ medium. Using these we can construct a vector indicating the position of stimulation (initiation point of the circular wave) relative to the reactor's centre. This vector is then communicated to the mobile robot's motor controllers in order to execute certain types of motion.

4.1.2   Designing on-board chemical controller

We prepared a thin-layer BZ reaction using a recipe adapted from [98]: an acidic bromate stock solution incorporating potassium bromate and sulphuric acid ($[BrO_3^-]$ = 0.5 M and $[H^+]$ = 0.59 M) (solution A); a solution of malonic acid (solution B) ($[CH_2(CO_2H)_2]$ = 0.5 M); and a solution of sodium bromide (solution C) ($[Br^-]$ = 0.97 M). Ferroin (1, 10-phenanthroline iron(II) sulphate, 0.025 M) was used as a catalyst and as a visual indicator of the excitation activity in the BZ medium. To prepare a thin layer of the BZ medium we mixed solutions A (7 ml), B (3.5 ml) and C (1.2 ml) and finally when the solution had become colourless ferroin (1 ml) was added and the mixture was transferred to a Petri dish (layer thickness 1 mm).

Excitation waves in the BZ reaction were first initiated using a silver colloid solution; in preliminary experiments this was added to the reaction by a human operator.

The chemical controller was placed on board an off the shelf wheeled mobile robot (Fig. 4.1a). The robot is about 23 cm in diameter and able to turn on the spot; wheel motors are controlled by a Motorola 68332 on-board processor. The robot features a horizontal platform, where the Petri dish (9 cm in diameter) is fixed, and a stand with a digital camera Logitech QuickCam (in 120 × 160 pixels resolution mode), to record excitation dynamics. Robot controller and camera were connected to a PC via serial port RS-232 and USB 1.0, respectively (see scheme in Fig. 4.1b).

Because vibrations affect the wave dynamics in the liquid-layer BZ reaction we tried to make the movements of the robot as smooth as possible; thus, in our experiments the robot moved with a speed of ~ 1 cm/sec. We found that rotational movements were particularly disruptive and caused the spread of the excitation from the initial wave fronts at speeds in excess of those seen purely for a diffusion-driven process. The wave fronts also became diffuse/blurred and elliptical rather than circular. Therefore, to minimise this effect, we made the robot rotate with a very low speed of around 1 deg/sec. Obviously, this problem can be easily overcome by immobilising the catalyst for the BZ reaction in a gel and bathing the gel in

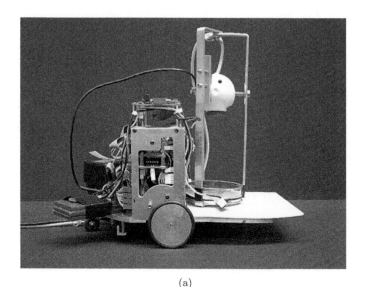

(a)

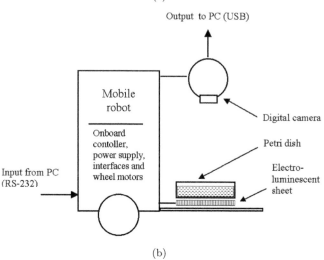

(b)

Figure 4.1: Scheme (a) and photograph (b) of mobile robot with on-board chemical reactor.

catalyst-free reagents. However, for the purposes of these preliminary experiments a liquid layer was favoured.

To enhance the images of the excitation dynamics the Petri dish was illuminated from underneath by a flexible electro-luminescent sheet (0.7-mm thick) cut to the shape of the dish. The sheet, powered via the robot's batteries, produces a uniform blue light without associated heat; thus, it has a minimal effect on the physico-chemical processes in the reactor. For the purposes of preliminary experiments we found that it was desirable to perform these in low ambient light environments to gain the maximum image quality.

Algorithms of analysing the spatial dynamics of the excitation (calculation of the local and global vectors indicating the position of the source of stimulation), and 'PC–robot' and 'camera–PC' interfaces were written in C and compiled with XHM68K (modules residing in the robot's on-board processor) and Borland C (modules residing in the PC).

4.1.3   Experiments on robot taxis

The main goal of our research concerning chemical controllers for robots is to develop a stand-alone on-board controller which can sense and process information from the robot's environment, calculate the current mode of the robot's movement and communicate a set of appropriate commands to the robot's motor controllers [7, 16, 17]. Therefore, we put the following constraints on a possible solution of the problem:

- An algorithmic solution must be local — if the BZ medium is mapped onto a two-dimensional array, e.g. a cellular automaton, then every cell of the array must be able to calculate its own local vector from the configurations of its neighbourhood.

- Only spatial data are allowed — a vector towards the position of a stimulation source may be extracted from just one snapshot of the medium's excitation dynamics.

- A robot-control unit (including subsystems based in the PC) is not allowed to remember local or global vectors calculated at the previous stages (that is, we tried to reduce the hardware role mainly to interfacing and decoding).

Detecting single source of stimulation

To prove that a vector towards a source of stimulation is extractable under these three constraints, it is adequate to demonstrate that the asymmetry of excitation waves can be detected from digital snapshots of the medium's dynamics.

As we can see in Fig. 4.2, the shape of the excitation wave front is visually represented by a spatial distribution of red, green and blue components of each pixel's value. The distribution of the blue-component values efficiently characterises all parts of the wave front, including zones corresponding to sharp excitation and zones where the medium's micro-volumes gradually return to the reduced steady state (excitable state) via a refractory state. Therefore, only the blue components

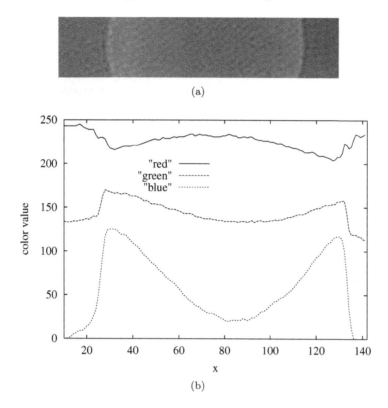

(a)

(b)

Figure 4.2: Demonstration of the asymmetry of wave fronts as detected from snapshots of the BZ medium: (a) a fragment of a snapshot of an expanding target wave; (b) values of the red, green and blue components of pixels along the x-axis cross section averaged along the y axis. The distribution preserves its form but the exact values in each experiment vary depending on the concentration of reactants.

of the pixels' colour were considered in re-constructing the stimulation position. All normals to the wave front, calculated from the gradients of the blue-colour values, are orientated towards a site where the wave originated. Therefore, a simple analysis of the spatial distribution of normals would give the robot an indicative position of the stimulation source.

For every pixel (i, j), from its blue-colour value b_{ij} of a two-dimensional image we calculate a local vector $\mathbf{v}_{ij}$ as follows. For $l, m \in \{-1, 0, 1\}$: if $b_{ij} - b_{i+l\,j+m} > \gamma$ then $(v_i, v_j) = (v_i, v_j) + (l, m)$ and $s + +$; $\mathbf{v}_{ij} = (v_i/s, v_j/s)$. Parameter γ could be selected individually at the beginning of every experiment (depending on the overall level of illumination); 15 and 45 graduations usually gave us the best results (see also Fig. 4.2b).

An example of local vectors calculated from image Fig. 4.3a, snapshot, and Fig. 4.3b, blue component, is shown in Fig. 4.3c. The global vector (shown in bold in Fig. 4.3c) is calculated from a sum of all the local vectors $\mathbf{v} = \frac{1}{qp} \sum_{\substack{1 \le i \le q \\ 1 \le j \le p}} \mathbf{v}_{ij}$

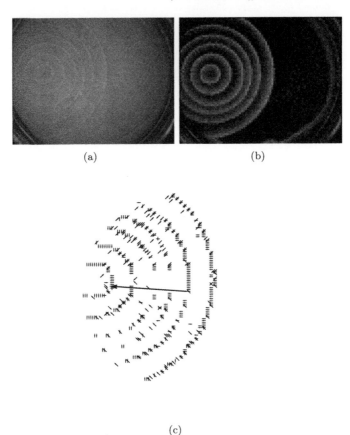

(a)   (b)

(c)

Figure 4.3: Extraction of stimulation position from one snapshot of the BZ medium. (a) Digital photograph (160 × 120 pixels) of the medium, (b) blue component extracted; (c) local vectors are calculated for every site of the medium. The global vector towards the source of stimulation is shown in (c).

($q = 120$ and $p = 160$ in our experiments); this almost perfectly indicates the position of the stimulation source relative to the centre of the Petri dish.

The algorithm does not work if the source of stimulation is represented by an almost perfect circular wave front, all parts of which are inside the processed zone of the reactor — in this case all local vectors will cancel each other out. However, this can be coped with by introducing a breathing receptive field, which changes its size during the experiments, or simply by waiting until some parts of the wave front reach the edges of the reactor and disappear. The dynamically changing receptive field could also reduce the influence of phase waves spontaneously generated at edges of the reactor.

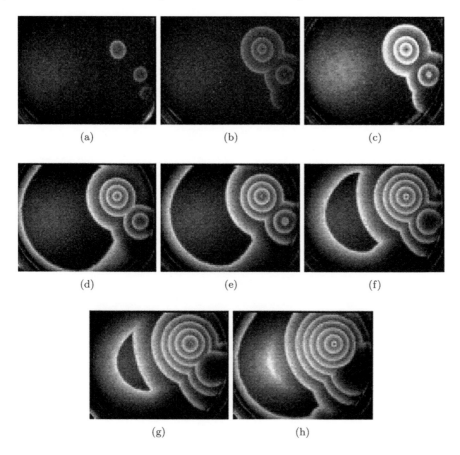

Figure 4.4: Snapshots of BZ medium; intensity of blue component of pixel colour is displayed.

Several sources, decision making and primitive memory

Consider a more 'dynamical' example in Figs. 4.4 and 4.5. Three sources of target waves were initiated at different time intervals. At first the robot detects no stimulation because wave fronts are perfectly closed and circular (Figs. 4.4a and 4.5a). Then, a wave front of an older target wave reaches the edge of the reactor and breaks. The robot detects a vector orientated towards the position of stimulation (Figs. 4.4b and 4.5b). At the same time another (originally second from top in Fig. 4.4a) source of target waves matures (Fig. 4.4c–e) and this causes deviation of the calculated global vector (Fig. 4.5c–e). The second source of target waves was initiated using a smaller quantity of silver colloid than was used for the initiation of the first source; therefore, this second site of stimulation generates just four wave fronts and then disappears (Fig. 4.4f–h) — at this point the global vector rotates back to the first source of stimulation (Fig. 4.4f–h).

In real life, robots often choose between several sources of stimulation. How

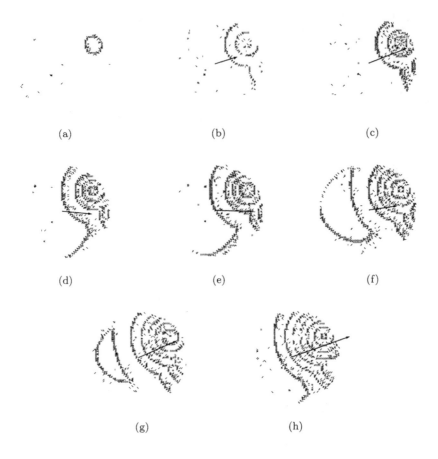

Figure 4.5: Local vector fields and global vectors (shown in bold) extracted from snapshots of BZ medium in Fig. 4.4.

could such a type of decision making be implemented in an on-board chemical controller? Given two sources of point-wise stimulation, the robot chooses the older source. This happens because the older point of stimulation is represented by broken wave fronts, a greater number of wave fronts and more spatially asymmetric wave fronts when compared to the younger sources of stimulation (Fig. 4.6a and b). When the older source of excitation disappears the younger one is chosen as the current target. Therefore, we can say that the on-board chemical controller memorises the temporal order of the stimulations. Moreover, a fine balance could be achieved between the intensity and the time tag of any stimulation to achieve a more precise trajectory of the robot's motion.

Types of motion control

Once stimulated the robot moves in circles because it re-calculates the position of the stimulation source every time irrespective of its previous position and does not

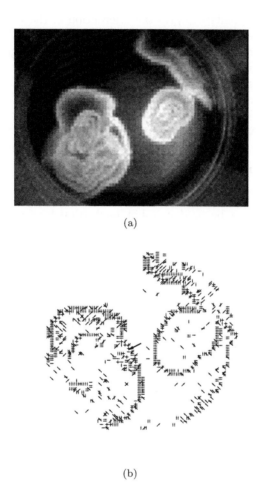

(a)

(b)

Figure 4.6: An example of decision-making procedure in chemical processor: selection of strongest generator of target waves, which represents a 'strongest' source of stimulation. (a) Blue component of BZ image, (b) local vector field and global vector (in bold).

remember previously extracted vectors. This circular motion has been previously described in simulation [7, 17] as a typical behavioural mode of mobile excitable lattices with a wide interval of node sensitivity. A relatively straightforward motion could be achieved by stimulating the chemical reactor twice and positioning the points of stimulation at opposite sites of the reactor. Then, in an ideal case, a bisector of these two points will give the orientation of the velocity vector. Also, a circular motion can be prevented by reducing the excitability of the chemical controller and thus making it impossible for a generator of target waves to persist. In this case just a few wave fronts will travel across the reactor and disappear. Obviously, the robot must analyse snapshots of the medium after longer delays, e.g. once per 10–15 min.

To slow down or stop the robot we could 'block' the velocity vector with several sources of wave generation, as shown in Fig. 4.7.

4.2 Path planning

In this section we produce an experimental 'proof-of-concept' that reaction-diffusion chemical processors can be used in robot navigation by employing two distinct experimental reaction-diffusion chemical controllers for the task of robot-motion planning.

The collision-free path-planning problem — given an experimental arena with obstacles, calculate a shortest path between two selected sites in such a manner that when travelling along the path a robot does not collide with any obstacles — is one of the basic issues in computational robotics and is particulary important for autonomous robots [276, 137, 286, 197]. To solve the problem one must

- select a connected subset of planar sites that are farthest from the obstacles, i.e. to approximate a planar Voronoi diagram, and then

- calculate the shortest path along the selected subset of planar points.

These two steps of optimal path calculation can be implemented in 'one go' when the paradigm of artificial potential fields is employed. In this model obstacles generate repulsive fields while a destination point generates an attractive field. A robot starts at a given site within the space and moves, guided by the attractive and repulsive field gradients, until it reaches the destination point.

A field approach to a shortest-path computation has been used since the early 19th century; for example, by employing a graph of a resistance network, applying a voltage to the graph and then measuring the resistance or capacitance of the network. However, the main ideas on field-based computing were introduced in an algorithmic form in [294] and then were implemented in cellular-automaton models of reaction-diffusion and excitable media [3]. In order to calculate the shortest path it is sufficient to represent the robot as a positive charge and the destination site as a negative charge. A potential field is then generated at the destination site and this spreads throughout the space whilst remaining at zero at the sites of the obstacles. An optimal path is computed by evaluating the potentials at the sites of the robot's route. This technique has been used, with some modifications, in [102], and has also been employed in [134, 135]. Other examples of successful

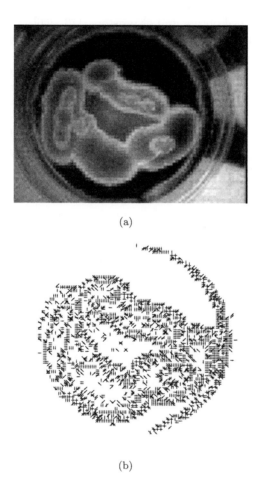

(a)

(b)

Figure 4.7: Stopping robot by 'trapping' global vector between several generators of target waves: (a) blue component of BZ image, (b) local vector field and global vector (in bold).

implementations of the potential field technique include the simulation of a potential field with virtual charges [251], integration of a potential field using simulated annealing techniques [211], evolving artificial potential fields [287], modifications of potential functions and numerical construction of a potential field as an array of discrete distance values [46]. The potential field approach is also useful for massively parallel VLSI implementation of robot path planning, where the space with obstacles is mapped onto a grid of resistors [173, 266].

The analogy of the potential field was also employed in several 'physics-based' techniques. Thus, researchers [302] imitate an artificial potential field as heat transfer, where a distance function is represented by variable thermal conductivity and an optimal path is computed as a flow through the minimal resistances. In [247] the optimal obstacle-free path is dynamically generated via simulation of a diffusion process, and a robot finds a destination site by comparing concentration gradients. It is also possible to represent a destination site as a source of a fluid; if the fluid is viscous then friction coefficients are proportional to the distance from the obstacles [167]. The optimal path can also be calculated from velocity potentials.

The approach presented in this work implicitly employs the potential field paradigm — one field is produced by an experimental reaction-diffusion processor, another is produced by an additional chemical reactor, as described in Sect. 4.2.1 controlling the motion of a pixel bot, or computed using a cellular-automaton model of an excitable chemical medium. For example, a repulsive field is generated in thin-layer chemical processors and an attractive field is developed during the space–time excitation dynamics of a non-linear medium linked to the chemical processors.

4.2.1 Pixbot in a Petri dish

To test the principles of robot navigation we utilised the concept of a pixbot, a pixel-sized robot, travelling on the snapshots of the BZ medium.

Reaction-diffusion chemical processors, for experimenting with pixbot, were prepared using a thin-layer BZ reaction [98] mixture of three solutions:

A: an acidic bromate stock solution incorporating potassium bromate and sulphuric acid ($[BrO_3^-] = 0.5$ M and $[H^+] = 0.59$ M),

B: a solution of malonic acid ($[CH_2(CO_2H)_2] = 0.5$ M),

C: a solution of sodium bromide ($[Br^-] = 0.97$ M).

Ferroin (1, 10-phenanthroline iron(II) sulphate, 0.025 M) was used as a catalyst. The catalyst was also a visual indicator of the excitation activity in the BZ medium. To prepare a thin layer of the BZ medium we mixed solutions A (7 ml), B (3.5 ml) and C (1.2 ml). When the solution had become colourless ferroin (1 ml) was added and the mixture was transferred to a Petri dish (layer thickness 1 mm). Excitation waves in the BZ reaction were initiated using a silver colloid solution. The targets and obstacles were represented via the local excitation waves in two separate BZ reactors.

Space–time dynamics of the excitation in the reaction medium were recorded using a FujiFilm 2600z digital camera (zoom lens $f = 6$–18 mm equivalent to

$f = 38$–114 mm on a 35 mm camera). The images were recorded using the built-in macro function with a resolution of 640×480 pixels and using a transmissive lighting method. No filters were used. The software models were compiled in `lccwin32` with `WinAPI`. As regards the typical computation time, it is related to the speed of the wave front in the BZ medium. The excitation wave front moves with the speed of around 0.2-2 mm/min (dependent on many factors). Essentially, a matter of minutes are needed for the entire experiment if a Petri dish of 9 cm is used.

Principles of navigation

We navigate the robot in reaction-diffusion media using a combination of attractive and repulsive patterns generated by the excitation waves in chemical media. In [13, 15] we demonstrated the feasibility of this approach; however, that was utilising attractive wave fronts implemented in cellular-automaton models and repulsive wave fronts generated in real chemical systems. Ideally, it would be reasonable to have two excitable chemical systems (whose wave fronts can be detected using different visual indicators) physically co-existing in one reactor. In such an ideal case the robot will be attracted by the wave fronts of one system — which represents a target — and repelled by the wave fronts of the other system — which represent obstacles. This physical co-existence does not make an easy experimental option.

Therefore, in the present work we navigate the robot using two separate and isolated (from each other) chemical reactors containing the BZ medium. Obstacles are mapped onto one reactor and targets onto another. We assume that the robot detects the concentration of the chemical species using optical sensors from spatial snapshots of the BZ medium's activity. Analysis of the wave profiles [20] shows that in a RGB colour scheme gradations of blue colour very efficiently characterise the excitation-wave dynamics in the BZ medium.

A cross-sectional profile of the BZ wave front (Fig. 4.8) — showing a steep rise at the wave-front's head and gradual descent at the wave-front's tail — give an indication of how robot-navigation techniques based on repulsion and attraction could be designed. Moving through the medium the robot avoids sites of 'repulsion', i.e. points of the medium where blue-colour gradations exceed a certain threshold. This guarantees that the robot will in all likelihood not cross a repulsive wave front. Thus, a collision with any obstacles is avoided. The relatively gradual descent of the wave-front's tail allows the robot to move along a gradient of colour gradation to reach the target. The excitation wave patterns are discrete, i.e. compared to the situation involving an attractive field the robot cannot move 'smoothly' along the colour gradients. Initially, the robot must somehow 'jump' onto the top of a wave front and only then freely roll down the tail. The 'jump' will be implemented using a form of 'energy accumulation' when modelling the robot's behaviour. The wave-front's head is much shorter than the wave-front's tail; therefore, once it has crossed the wave front, the robot will be trapped inside the target wave because it will not be able to accumulate enough energy to jump back over the tail.

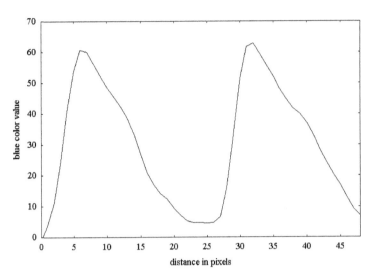

Figure 4.8: Cross section of two excitation waves taken from an image of the BZ medium. The wave profile is represented by the values of the blue-colour component.

The pixbot and its behaviour

To verify the robot-navigation ideas described above we constructed a software model of the 'pixbot' — a pixel-sized robot which moves in discrete steps on images (attractive medium $\mathbf{A}$ and repelling medium $\mathbf{R}$) of the spatial excitation dynamics of the BZ medium. Let a_{ij} and r_{ij} be blue-colour values of pixels (i,j) in images $\mathbf{A}$ and $\mathbf{R}$, and $(x,y)^t$ be the pixbot's coordinates at time step t. At each step of discrete time the pixbot can move to one of eight pixels closest to $(x,y)^t$. The pixbot coordinates are calculated as follows: $p^t = (x,y)^t$:

$$p^{t+1} = p^t + \vartheta(t, t-m)[f(p^t, g, \mathbf{A}, \mathbf{R})\chi(f(p^t, g, \mathbf{A}, \mathbf{R}))$$
$$+(1 - \chi(f(p^t, g, \mathbf{A}, \mathbf{R})))(\varrho(p^t, \mathbf{R}))]$$
$$+(1 - \vartheta(t, t-m)(\varrho(p^t, \mathbf{R})),$$

where $\chi(z) = 1$ if $z \neq 0$ and $\chi(z) = 0$ otherwise; $\vartheta(t, t-m) = 0$ if $|p^t - p^{t-m}| < k$ and $\vartheta(t, t-m) = 1$ otherwise;

$$f(p^t, g, \mathbf{A}, \mathbf{R}) = |\mathbf{V}|^{-1} \sum_{(u_i, u_j) \in \mathbf{V}} (u_i, u_j),$$

$$\mathbf{V} = \{(v_i, v_j) \in \{-1, 0, 1\} : |a_{p^t} - a_{p^t + (v_i, v_j)}| > g, \, r_{p^t + (v_i, v_j)} < c\}$$

and

$$\varrho(p^t, \mathbf{R}, m) = \mathtt{random}\{(v_i, v_j) \in \{-1, 0, 1\} : r_{p^t + (v_i, v_j)} < c\}.$$

Here g and c are constants depending on the initial concentration of reactants; in most experiments, $2 \leq c \leq 5$, $15 \leq g \leq 30$, $10 \leq m \leq 20$. The function ϑ plays the role of a 'kinetic energy accumulator': if the pixbot spends too much time wandering in the same local domain it is forced to jump randomly, this will allow the pixbot to mount wave fronts. The function $f()$ selects a site neighbouring p^t along the preferable descent of **A** and the minimum values of **B**. If such a site does not exist then a site with no obstacle wave is selected at random.

Because of the significant influence of a stochastic component on the pixbot's behaviour, the pixbot reaches the target by various routes (Fig. 4.9a and b) and it may wander for a while between two wave fronts, generated by the same target (Fig. 4.9). There is no guarantee that the best target (i.e. oldest or strongest) will be chosen because as soon as the first wave front is crossed there is no way back (Fig. 4.9a and c).

An example of the pixbot's collision-free movement towards a target is shown in Fig. 4.10. The trajectory is relatively straightforward when the obstacles are absent (Fig. 4.10a); however, it is characterised by a high density of sites visited by the pixbot when the obstacles are present (Fig. 4.10a and b). The disorder of the trajectories is more visible in domains where the pixbot's path deviates from the straight line between the start point and the target (Fig. 4.10b).

To test the stability of the algorithm in the conditions expected in an on-board chemical reactor, we placed the Petri dish with the BZ medium onto a specially constructed table rotating at $1°$/sec. The centrifugal forces disturb the wave fronts and make their structure more irregular; this does not prevent the pixbot from reaching the target (Fig. 4.11). This demonstrates that the reaction-diffusion approach will work when implemented in real-life conditions.

4.2.2 Navigating real robots: the problem and its solution

The collision-free path-planning problem can be considered to be finding the shortest path between two marked sites on a plane in the presence of obstacles. In the design of reaction-diffusion processors we assume that a robot has a single body, the obstacles are stationary and a complete map of the obstacles is known *a priori*. An image of the robot arena is projected onto a reaction-diffusion chemical medium, and the centres or corner points of the obstacles are represented by local disturbances in the reactant concentrations in the medium. These perturbations generate diffusive gradients and diffusive wave fronts, in the palladium-based chemical processor, or excitation waves, in a BZ chemical processor. The spreading waves build a repulsive field derived from the obstacles. This repulsive field created by the chemical waves is then used by a cellular-automaton model of an excitable medium to construct a tree, rooted in a selected destination site, of all shortest paths from the arena's sites to the destination site.

4.2.3 Computing a repulsive field in the palladium processor

A distance field around obstacles has been approximated in a palladium processor — a reaction-diffusion chemical processor based on the reaction of palladium chloride gels with potassium iodide. The experimental method for the preparation of palladium processors was detailed in Chap. 2.

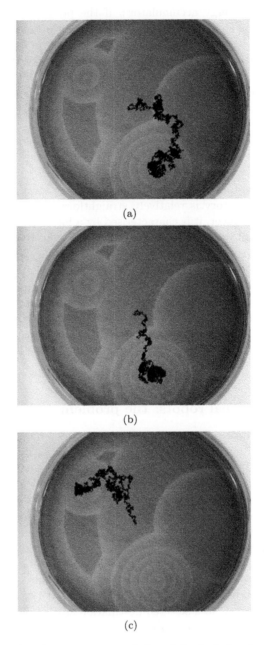

Figure 4.9: The stochastic component of the pixbot's behaviour determines that various routes are used to reach the target (a) and (b), and when two targets exist in the same reactor the selection is random (a), (b) and (c). No obstacles are present in this particular experiment. The trajectories of the pixbot are shown superimposed on images of the BZ medium.

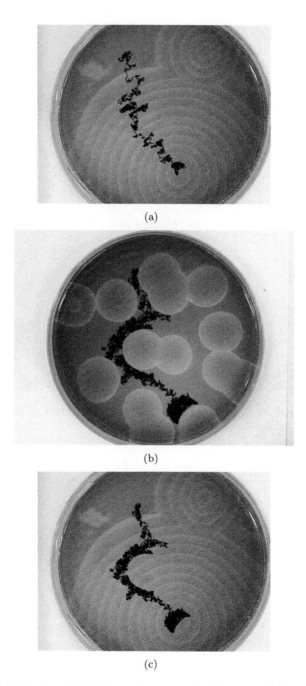

(a)

(b)

(c)

Figure 4.10: Trajectories of pixbot moving towards the target (a) without obstacles and with obstacles (b) and (c). See also Fig. 9.24 in colour insert.

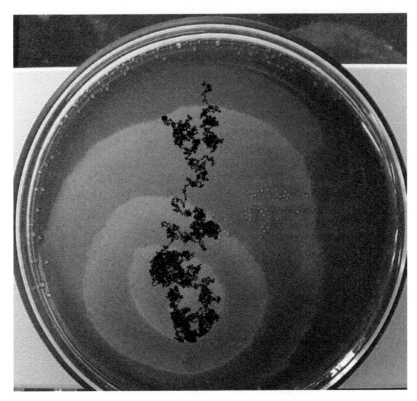

Figure 4.11: Trajectory of pixbot moving towards the target when BZ medium is subjected to centrifugal forces.

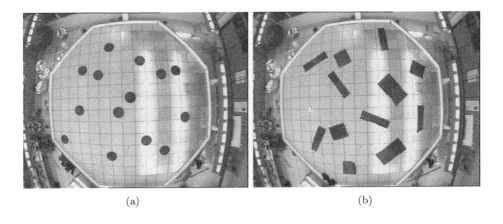

Figure 4.12: Photographs of the experimental robotic arena (approximately 9 m in diameter) show configurations of circular (a) and rectangular (b) obstacles.

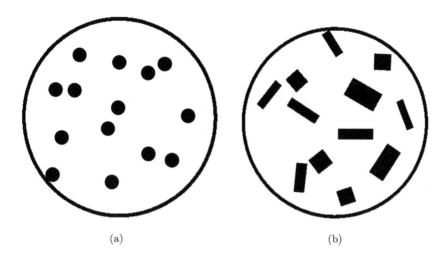

(a)   (b)

Figure 4.13: Models of circular (a) and (b) rectangular obstacles, shown in Fig. 4.12, prepared from a filter paper to use in reaction-diffusion chemical processors. The models are precisely scaled down configurations of obstacles in the real robotic arena (scale 1 m : 1 cm approximately).

Photographs of the experimental arena Fig. 4.12 were scaled down to the size of a Petri dish ($\sim$ 9 cm) and were used as templates to prepare models of the obstacles (Fig. 4.13) from filter paper (any other absorbent material can be used). The filter-paper models of obstacles were soaked with a solution of potassium iodide (saturated at $20°$C), blotted with additional filter paper to remove excess potassium iodide and placed onto the surface of the gel.

The clear solution of potassium iodide diffuses from the edges of the 'obstacles' into the palladium chloride loaded gel where it reacts to form iodo-palladium species. During the development of the palladium processor colour transitions from yellow (palladium chloride gel) to a dark brown are observed, depending on the presence of certain iodo-palladium species in the gel.

At sites where two or more diffusive fronts meet each other almost no precipitate is formed; these sites therefore remain uncoloured — thus, uncoloured sites of the gel represent bisectors of a Voronoi diagram, generated by the obstacles.

Uncoloured or significantly less coloured sites of the gel represent a subset of planar points which are as far from the obstacles as possible. Therefore, they indicate all available routes around the obstacles (Fig. 4.14). The optical density of a given site in the medium is proportional to the precipitate concentration. Thus, taking a grey-colour snapshot of the reaction medium we get a two-dimensional array of pixels, the values of which are inversely proportional to the precipitate concentration.

As we see from an example shown in Fig. 4.15 a repulsive field, represented by the concentration of precipitate, is nearly perfect at sites corresponding to the bisectors of a Voronoi diagram.

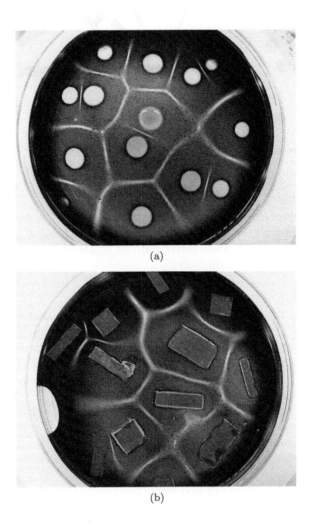

(a)

(b)

Figure 4.14: Photographs of the palladium processor; the processor started its development with a configuration of (a) circular and (b) rectangular obstacles. Uncoloured sites of the processor represent all collision-free routes in the experimental arena.

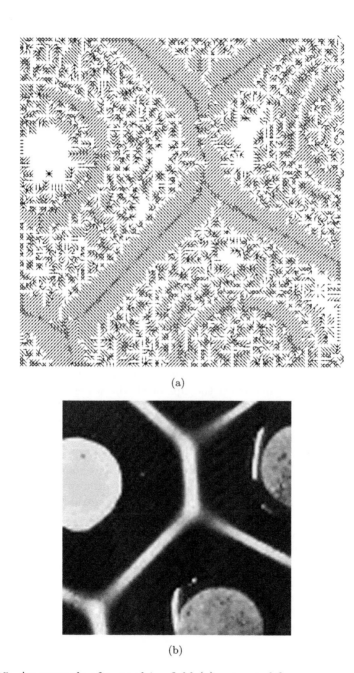

(a)

(b)

Figure 4.15: An example of a repulsive field (a) extracted from a segment (b) of a Voronoi diagram of the discs.

Notice that the palladium processor computes a continuous Voronoi diagram in Euclidean space. Obviously, this is because molecules of the reagents are not aligned in any regular grid but either diffusively move or occupy sites of an amorphous irregular structure or random grid. However, as we will see below, a discrete model with regular topology gives us reasonably good results.

In the next step we project the resultant configuration of the palladium processor back onto the arena and the problem of collision-free motion is then solved: a robot moves along lightest parts of the experimental arena. This does not solve however an optimal, i.e. the shortest-path, problem; this is discussed later. In further sections we represent a grey-level image of a palladium processor in a matrix $\mathbf{R}_P$: the greater the value of a cell of $\mathbf{R}_P$, the further away from an obstacle it is.

4.2.4   Computing a repulsive field in the BZ chemical processor

We prepared a thin-layer implementation of the BZ reaction using a recipe adapted from [98] and used earlier in this chapter.

Excitation waves in the BZ reaction were initiated using silver — the addition of silver at specific sites reversibly removes bromide ions from the sites; bromide ions act as the main inhibitor of autocatalysis in the BZ reaction and therefore the local removal of the ions triggers wave generation at a specific site.

To map the 'obstacle model' (Figs. 4.12 and 4.13) onto the BZ layer we represent centres of the circular obstacles and corner points of the rectangular obstacles by an array of thin silver wires; the wires are positioned parallel to each other and perpendicular to the surface of the BZ medium. To start the evolution of the BZ processor we briefly immersed the tips of all the wires into the BZ mixture. Single circular excitation waves were generated at all sites of the BZ medium that were contacted by the silver wires; these wave fronts then travelled throughout the medium and where the waves collide with each other they are annihilated, thus eventually re-setting the medium (Fig. 4.16).

The chemical medium is assumed to be uniform and homogeneous; therefore, all wave fronts travel the same distance in a fixed period of time. When two or more wave fronts meet they annihilate; therefore, every site of the chemical medium can be 'covered' by a wave front originating from only one site within the medium. Figuratively, excitation wave fronts can be 'used by obstacles' to subdivide the space into cells, where every site inside a cell, generated by a specific obstacle, is closer to the obstacle than to any other obstacle. Every site is covered by a wave front only once because in our experiment initiation of excitation waves by use of a silver wire gave only one circular wave (in almost all cases). Wave fronts recorded at time t of the BZ medium's development represent all those points of the medium that lie at a distance tv from the centres of the obstacles, where v is the wave speed (~ 0.1 mm/sec).

To build a distance field (Fig. 4.17) for a given configuration of obstacles, we project the obstacles onto the BZ medium and extract the positions of the wave fronts at regular intervals: a matrix of the time-lapsed positions of the wave fronts represents a discrete distance field. Thus, 8-bit RGB colour images $\mathbf{S} = \{S^1, \ldots, S^m\}$ of the BZ medium were taken at intervals of 15 sec (see several snapshots in Fig. 4.16).

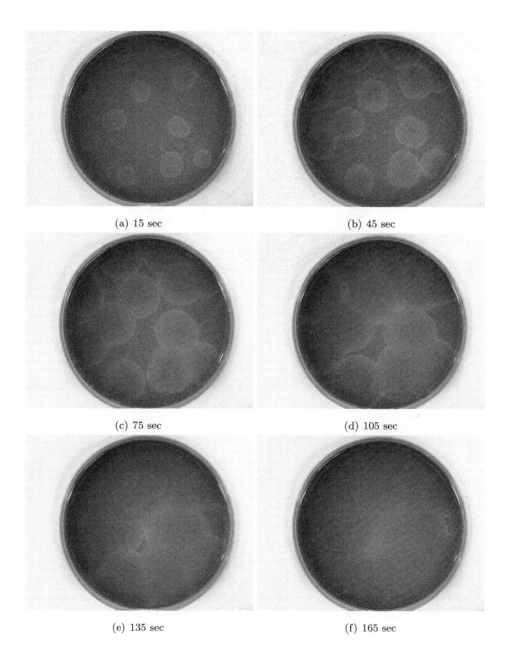

(a) 15 sec

(b) 45 sec

(c) 75 sec

(d) 105 sec

(e) 135 sec

(f) 165 sec

Figure 4.16: Photographs of the BZ medium, which was excited by a template of silver wires, corresponding to the geometry and configuration of the rectangular obstacles shown in Figs. 4.12 and 4.13.

Figure 4.17: Contours of excitation wave fronts made up from super-imposed images from Fig. 4.16

A series $\mathbf{S}$ of the images was transformed to a grey-level matrix $\mathbf{R} = (r_x)_{x \in \mathbf{Z}^2}$ using the following rule:

$$\mathbf{R} = \prod_{t=1}^{m} R^{t-1} \circ S^t,$$

where all elements of R^0 are zeros, and

$$r_x^t = r_x^{t-1} \star s_x^t = \begin{cases} t, \text{ if } s_x^{t-1} = 0 \text{ and } B(s_x^t) > \beta \\ s_x^{t-1}, \text{ otherwise} \end{cases}$$

$B(s_x^t)$ is a blue component of an RGB colour of the pixel x of snapshot S^t of the BZ medium taken at time step t. Again, we considered only the blue component of the colour images because only this component unambiguously represents the position of a wave front at any given time in the reaction's evolution. The value of β may vary from 0 to 50 depending on the light intensity used during the experiments; $\beta = 40$ was used in the experiment discussed in the present study. The matrix $\mathbf{R}_{BZ}$ computed from $\mathbf{S}$ in two experiments, for circular and rectangular obstacles, is shown in Fig. 4.18.

The matrix $\mathbf{R}_{BZ}$ represents a distance scalar field derived from a configuration of obstacles, the greater the value of a cell in $\mathbf{R}_{BZ}$ the further away it is from the closest obstacle.

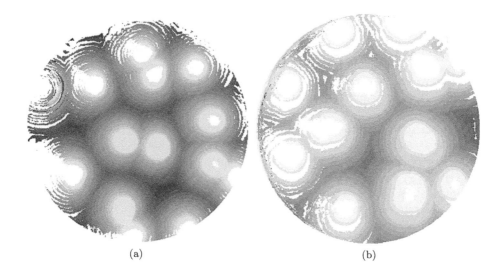

(a)   (b)

Figure 4.18: Distance matrix **R** computed from snapshots of the BZ medium 'excited' by circular (a) and rectangular (b) obstacles. Values of **R** are shown by inverse grey-level pixels.

4.2.5   Computing a tree of the shortest paths

As we have shown in the previous sections, reaction-diffusion chemical processors compute the distance matrix **R**, $\mathbf{R} = \mathbf{R}_P$ or $\mathbf{R} = \mathbf{R}_B Z$. This matrix represents a set of all possible obstacle-free routes in an experimental arena, and thus can be used as a representation of a robots environment. To navigate between two selected sites of an arena the robot must compute a shortest path from the source to a destination site — the matrix **R** is used as a basic data-structure in the computation of the shortest path.

We have designed a model of a discrete excitable medium to assist reaction-diffusion chemical processors in the computation of the shortest path. An excitation wave in a uniform medium travels along the shortest path; therefore, to compute the shortest path we excite the source site, observe how excitation waves spread in the space and record the local 'histories' of the travelling wave fronts. The computation is considered to be finished when the excitation front reaches the destination site. Then, the shortest path is extracted from the 'histories' of the spreading excitation waves. In principle, this technique can be implemented directly in an excitable chemical medium, e.g. in the BZ reaction, as demonstrated in [269, 28] and independently discovered in cellular-automaton models of excitation in [3]; however, for the purposes of this work we will stick to a discrete model to specify certain particular details of the computation.

To simulate excitation wave dynamics we employed a two-dimensional cellular automaton — an array of finite automata, called cells, which take finite numbers of states, and update their states in parallel and in discrete time, each cell calculating its next state depending on the states of its eight closest neighbours [3]. To record the 'histories' of the excitation 'trajectories' we supplied each cell of the automaton with a pointer, which points to the cell's neighbour which excited this cell. The shortest path is extracted from a configuration of pointers, obtained after running an excitation from a source cell to a destination cell.

Every cell x of a cellular automaton updates its state in discrete time t depending on the states of its eight closest neighbouring cells, defined by a rectangular 3×3 cell template $u(x)$. At time step t cell x takes a compound state $\langle x^t, p_x^t, r_x \rangle$, where $x \in \{\bullet, +, -\}$ and $p_x^t \in \{(z_i, z_j)|z_i, z_j \in \{-1, 0, 1\}\} \cup \{\lambda\}$, and $0 \leq r_x \leq 255$.

The component x^t simulates a 'physical' state of the cell x and takes resting ($\bullet$), excited ($+$) and refractory ($-$) states. A resting cell x is excited if at least one of its neighbours is excited and a value r_x of the corresponding cell of the matrix $\mathbf{R}$ exceeds a certain threshold θ. The cell x changes its state from excited to refractory and from refractory to resting unconditionally, i.e. independent of the states of its neighbouring cells.

The component p_x^t is the state of a pointer, which can be seen as an arrow centred at x and looking towards one of eight neighbours of x or nowhere ($p_x^t = \lambda$), i.e. 'disconnected'. A pointer p_x^t at a cell x points to the closest neighbour y of the cell x whose corresponding value r_y (of the matrix $\mathbf{R}$) is maximal over the matrix $\mathbf{R}$ values of the cell x's other neighbours which are excited at time t.

So far, the cell state transition rules are as follows:

$$x^{t+1} = \begin{cases} +, & \text{if } x^t = \bullet \text{ and } \sum_{y \in u(x)} \{y^t = +\} > 0 \text{ and } r_x \geq \theta \\ -, & \text{if } x^t = + \\ \bullet, & \text{otherwise} \end{cases}$$

$$p_x^t = \begin{cases} y \in u(x) : y^{t-1} = + \text{ and } x^t = + \text{ and } r_y = \min\{r_x | z \in u(x) \text{ and } z^{t-1} = +\} \\ p_x^{t-1}, & \text{otherwise} \end{cases}$$

The condition $r_x \geq \theta$, which is undoubtedly optional, restricts a set of excitable cells from the whole cellular-automaton lattice to a sublattice $(G) = \{x | r_x \geq \theta\}$; this does not decrease the time spent on the shortest-path computation — however, it reduces the number of cells excited at each time step.

Let at the beginning of the computation only one cell z of $\mathbf{G}$ be excited, all other cells be resting and all pointers take the state λ; then at moment $n \leq t \leq n^2$ the configuration of the pointers of the cellular automaton represents a minimum spanning tree T rooted at the cell x.

An excitation wave front, originated in cell z, passes over all sites of $\mathbf{G}$, and updates the states of their pointers (Fig. 4.19). A pointer at a cell can look towards only one neighbour, so a directed graph T is acyclic. The domain $\mathbf{G}$ is assumed to be connected. So, the graph T is a spanning tree. All cells update their states in parallel and using the same rules; thus, the excitation front gets to a cell x along a shortest path. Therefore, T is a minimum spanning tree.

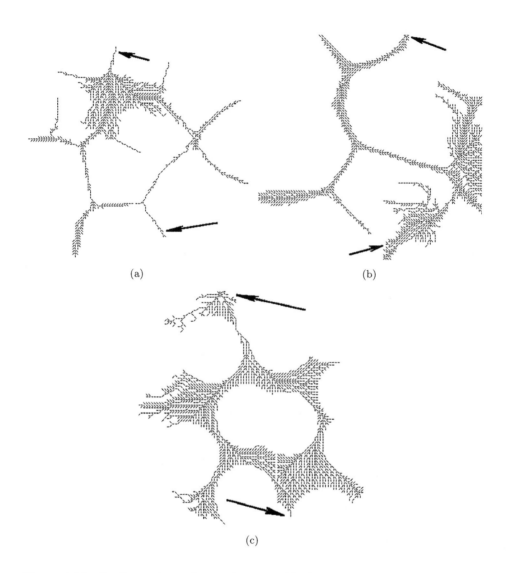

(a)   (b)

(c)

Figure 4.19: Configurations of spanning trees T of pointers computed in cellular-automaton models of excitable medium from distance field matrices $\mathbf{R}_P$, circular obstacles (a) and rectangular obstacles (b), and $\mathbf{R}_{BZ}$, circular obstacles (c). Destination (upper part) and source (lower part) sites of the arena are indicated by arrows.

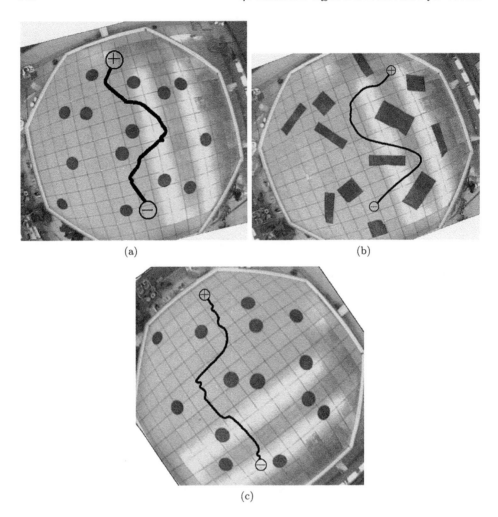

(a)   (b)

(c)

Figure 4.20: Shortest paths mapped onto the robotic arena with obstacles. The path is computed from the experimental thin-layer chemical medium — palladium processor, circular obstacles (a) and rectangular obstacles (b), and the BZ processor, circular obstacles (c) — assisted with two-dimensional cellular-automaton models of an excitable medium.

The tree T represents a shortest path around the obstacles (at a maximum distance from any obstacle) from any site of $\mathbf{G}$ towards the destination site z.

The tree T is rooted at z; therefore, starting at any site of $x \in \mathbf{G}$, and moving along the directed edges of T a robot will inevitably reach the site z. The path is shortest because it is a chain of T. The condition $r_y = \min\{r_z | z \in u(x)$ and $z^{t-1} = +\}$ guarantees that the computed path travels along sites with minimal values of $\mathbf{R}$ cells and thus each site of the path lies at a maximum distance from the obstacles (Fig. 4.20).

To execute real-time navigation a list of vector representations of the computed shortest path is loaded into the memory of the robot's on-board controller. The robot then implements rotations and forward motion determined by the list of vectors.

4.2.6 Complexity and performance

All reaction-diffusion algorithms are space consuming [7], and this is the case for the experimental chemical processor and its cellular-automaton 'accomplice'. In the automaton model a discrete lattice has $n \times n$ sites; each site has m levels of precipitate concentration. Thus, a space complexity of a 'one destination many sources shortest path' spanning tree computation is $O(mn^2)$, where $O(m)$ is essentially a boundary on the cell's local memory. It takes only $O(n)$ steps in the case of the 'normal' excitation rule and $O(n^2)$ steps in the case of the 'conditioned by colour' excitation rule. Extraction of any particular path can take $O(n^2)$ steps. If a Voronoi diagram is computed in a cellular-automaton processor then the automaton algorithm has $O(n^2)$ space complexity; pleasantly, the cell's local memory has $O(1)$ capacity.

Let us consider an experimental palladium processor as a massively parallel processor. How many elementary processing units does this massively parallel processor have? To answer this one should calculate the number of micro-volumes in the chemical medium. The effective diffusion length for the system is 0.075 mm (this is obtained from experimental observations of the bisector width). This gives us approximately 10^6 micro-volumes (in a Petri dish of 9-cm diameter).

Each micro-volume is an elementary processor. Three reagents are involved in the computation. At the beginning of computation each elementary processor contains a certain amount of palladium chloride ($PdCl_2$); this can be considered as a part of the elementary processor's program. Obstacles are represented by a solution of potassium iodide (KI), molecules of which diffuse in the network of elementary processors. So, a molecule of KI can be seen as a data message. The result of the reaction $PdCl_2$ is converted to iodo-palladium species, ultimately to PdI_2; the reaction itself may be seen as a hard-wired part of the local program. Depending on how much $PdCl_2$ was converted to PdI_2 in an elementary processor, the processor takes a colour ranging from yellow to dark brown. So, the concentration of PdI_2 is an output state of the elementary processor. As we found, even 16 gradations of palladium iodide concentration are enough to calculate a decent optimal path from an experimental Voronoi diagram. Therefore, we could say that each elementary processor of the palladium processor has $O(1)$ local memory. An approximate diffusion speed for this particular type of reaction is 0.005 mm/s. Therefore, an elementary processing unit switches its state once per 15 sec. The massively parallel chemical processor is essentially asynchronous; therefore, we could suggest a computation rate of $\sim 7 \times 10^4$ operations per second. The approximated parameters of the palladium processor are similar to those suggested by Agladze *et al.* [28] for a BZ processor. Alternatively, we could see the palladium processor as an aggregate of masses of very simple mobile processors (as is the case with swarm-based computing, see e.g. [7]). Each molecule is a processor, that moves due to diffusion, and physically changes its state when a chemical reaction occurs.

If computation is implemented in a standard Petri dish, ~ 9 cm in diameter, there are 3.39×10^{19} molecules of $PdCl_2$ in the substrate. Imagine that obstacles are represented by drops of KI. Each drop contains 1.45×10^{19} molecules of KI. Therefore, in this representation, the palladium processor has $(3.39 + \frac{1.45}{2}p) \times 10^{19} = (3.39 + 0.725p) \times 10^{19}$ elementary mobile processors, because for one computation event we require one molecule of $PdCl_2$ but two molecules of KI if we assume the simplest reaction $PdCl_2 + 2KI \rightarrow PdI_2$, where p is a number of obstacles represented by drops of potassium iodide.

4.3 Controlling a robotic hand

As already discussed, reaction-diffusion chemical computing is a feasible way to develop unconventional robotics; however, no experimental results about the real-time interaction of robotic devices with reaction-diffusion chemical systems have so far been reported. Results of this type would certainly fill the gap in our understanding of hybrid (wet + hard) ware systems. Therefore, as a first approach we designed and implemented a series of experiments[1] on two-way physical interactions between a spatially distributed excitable chemical medium and a robotic hand: the hand excites the medium, whilst the space–time dynamics of the medium controls the motions of the fingers of the robotic hand [307].

In the next sections we outline the hardware architecture of the sensor and motor parts of the robotic hand, and the design of the reaction-diffusion chemical processor. We also discuss basic modes of the closed-loop system comprising the robotic hand and chemical medium couple.

4.3.1 Interfacing chemical processor with robotic hand

In experiments we used the robotic hand developed in [139, 307]. Each finger of the robotic hand consists of two pivots with springs and wire guides attached (Fig. 4.21a).

In order to realise the function of the spring as a wire guide, we used a spring coil type of stainless steel outer wire as a guide for a nylon inner wire. When a finger does not bend the spring coil type of outer wire keeps it straight. If a finger needs torque, the spring coil type of outer wire bends to realise the shortest route of the inner wire; the elastic wires are connected to servomotors (Fig. 4.21b). The motors are controlled by a micro-processor H8Tiny produced by Hitachi. The micro-processor has eight inputs from photosensors, and outputs linked to 10 servomotors. The micro-processor encodes signals from the sensors to run the motor drives. The finger motors are powered by 5 V, and the microprocessors are powered by 9 V. The hand is mounted above the chemical reactor and glass capillary tubes filled with silver colloid solution are attached to the fingertips to stimulate the chemical medium.

The chemical processor is prepared in a thin layer BZ reaction using a recipe adapted from Field and Winfree [98] and detailed earlier in the chapter.

---

[1]The experiments were implemented by Prof. Hiroshi Yokoi, Tokyo University during his visit to UWE, Bristol

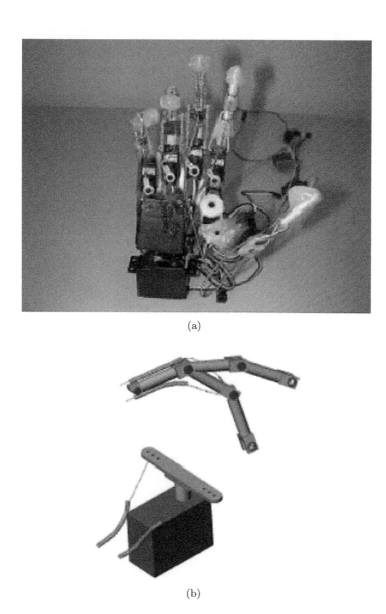

(a)

(b)

Figure 4.21: Hardware architecture of robotic hand. (a) Overall view of the hand.
(b) Actuating fingers via elastic wires (top) and by servomotor (bottom) [307].

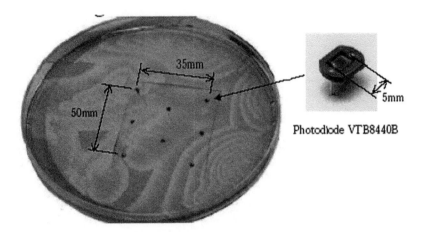

Figure 4.22: Sensory system of the BZ hand. Experimental chemical reactor with photosensors underneath (left), sensors are visible as black sites on the photograph, and a view of the single sensor (right). In this photograph we can also see light-coloured excitation wave fronts [307].

4.3.2   Modes of interaction of BZ medium with robotic hand

As mentioned, to enable the robotic hand to interact with the chemical processor we attached glass capillary tubes filled with a solution of colloidal silver to the fingertips, and adjusted the hand above the Petri dish containing the BZ reaction solution in such a manner that when a finger bends the tube is very slightly immersed in the mixture and a small fairly reproducible quantity of silver colloid is released, thus starting excitation wave dynamics in the chemical medium.

To detect the excitation wave fronts in the chemical reactor we used a photodiode VTB8440B, with a spectral application ranging from 330 nm to 720 nm (Fig. 4.22). When a wave front travels through the medium and across the sensor this is detected and the sensor's output current increases; this increase is detected by the micro-controller and this starts the servomotor causing the finger corresponding to that photosensor to bend. The fingers are programmed to unbend a few seconds after bending.

A typical development in the studied system was as follows: when one or more fingers bend for the first time (due to a solitary circular wave initiated at the reactor's edges and moving towards the centre of the reactor that crosses all the photosensors), colloidal silver from glass capillary tubes, attached to the fingers, is deposited into a local area of the medium and initiates a series of excitation waves (Fig. 4.23). A series of waves is initiated as silver colloid is physically deposited into the system and, depending on the amount (proportional to the time the finger is in contact with the surface), will usually initiate a number of waves at or near the contact point when the system has regained its excitability.

As silver diffuses from the sites of contact multiple excitation sites may be established, leading to complex waves or spiral waves (also caused by the disturbance of the reactor surface). When the excitation waves spread, they stimulate photosen-

Figure 4.23: Colloidal silver on the finger 'nails' excites target waves in the BZ medium [307]. See also Fig. 9.25 in colour insert.

sors, attached to the underneath of the Petri dish, and this causes further bending of the fingers. Let us consider the behaviour of the system when up to three fingers have the ability to move. The positions of the sensors controlling these fingers and the sites where the needles attached to the fingers touch the medium are shown in Fig. 4.24. When only one finger is operational, it excites, in the first instance, a generalised target wave in the medium. The wave fronts of the target wave are detected by the finger's sensor and the finger bends, delivering more activator (silver) to the system and continuing the excitation process or waves. Thus, the

system enters a mode of self-excited oscillatory motion (Fig. 4.25). If two fingers are operational and just one finger bends, then, after some transient period, the robotic hand's fingers act in anti-phase because they reciprocally stimulate each other via the wave fronts of the target waves they excite (Fig. 4.25b).

When both fingers touch the BZ medium at the beginning of the experiment the system starts to exhibit synchronous oscillations (Fig. 4.25c). This happens because the wave fronts of the target waves, initiated by the movement of the fingers, cancel each other out when they collide, and, thus, as in the general situation, each finger excites only itself. When three fingers are operated simultaneously (Fig. 4.23) highly complex behaviours are observed (whatever the starting configuration) — eventually the system may appear to reach an almost synchronised or repetitive motion but this is then subject to major fluctuations.

The wave front interaction scenario of finger-motion control was rectified in

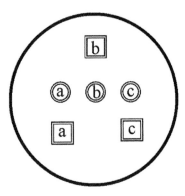

Figure 4.24: Schematic relative positions (circles) of sensors affecting index (a), middle (b) and third (c) digits of the robotic hand, and sites (rectangles) where these fingers stimulate the BZ medium.

a two-dimensional cellular-automaton model of the system 'BZ medium–robotic hand'. The model comprises a $n \times n$ cell lattice, where every cell takes three states: resting, excited and refractory, and updates its states depending on the states of its eight closest neighbours. The scheme from Fig. 4.24 is mapped onto the lattice as follows (Fig. 4.26). The sensors a, b and c are positioned in the cells with integer coordinates $(1, n/2)$, $(n/2, n/2)$ and $(n, n/2)$. The stimulation sites a, b and c (rectangles in Fig. 4.24) have integer coordinates $(1, n-1)$, $(n/2, 1)$ and $(n-1, n-1)$.

The cellular-automaton model mimics processes in the BZ medium at a very basic level. A resting cell becomes excited if at least one of the neighbours is excited or the cell is assigned to its unique finger and the finger bends at this step of simulation. An excited cell takes a refractory state, and the refractory cell in turn takes a resting state unconditionally. Links between 'sensor' cells and 'stimulation' cells are established as follows: each of three 'sensor' cells corresponds to its unique 'stimulation' cell; the resting 'stimulation' cell takes an excited state if one of its eight closest neighbours is excited or its unique 'sensor' cell is excited. Patterns of excitation, mostly target waves initiated by the fingers' 'stimulation' cells excited once at the beginning of the simulation, spread on the lattice and activate 'sensor' cells; this causes excitation of the 'stimulation' cells and the cycle repeats. An example of excitation dynamics and finger-motion patterns are shown in Figs. 4.27 and 4.28, respectively: initially, the glass pipettes of the index and middle digits do not touch the medium while that of the third digit does.

In simulation experiments with various sizes of cellular-automaton lattice ($n = 5, \ldots, 50$) we found that any initial configuration of medium excitation (caused by one or more of three digits) falls either in the fixed circle of oscillation between patterns $\wedge\vee\vee$ (middle and third digits excite the medium) and $\vee\wedge\wedge$ (only index digit excited the medium), or in the fixed points $\vee\vee\wedge$ and $\vee\vee\vee$ (Fig. 4.29). The global transitions of the 'BZ medium–hand' system, as simulated in the cellular automaton, are sensitive to the oddness of the lattice size due to the discrete topology of the excitation dynamics.

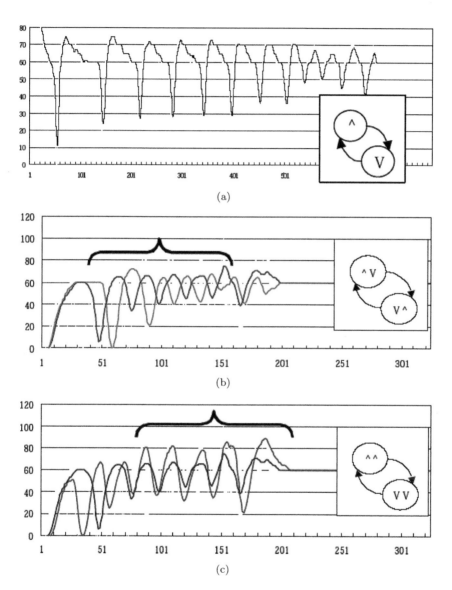

Figure 4.25: Typical modes of 'fingers–BZ medium' system for one (a) and two (b) and (c) fingers. Voltage on the photosensor is shown on the vertical axis; time in seconds is shown on the horizontal axis. The corresponding movements of the fingers are outlined in the insert, where ∧ means up and ∨ down [307].

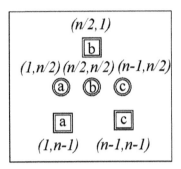

Figure 4.26: Configuration of sensors and stimulation zones (in square brackets) mapped onto the cellular-automaton lattice of $n \times n$ cells, corresponding to a three operational fingers setup of index (a), middle (b) and third (c) digits.

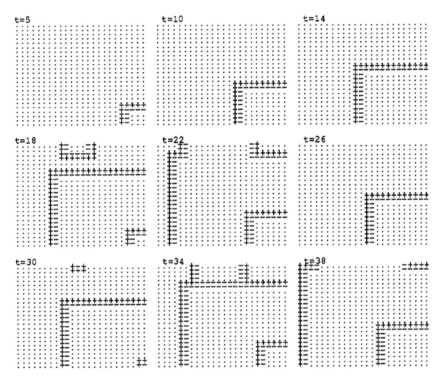

Figure 4.27: Configurations of cellular-automaton model of 25×25 cells of excitation dynamics in BZ medium interacting with a robotic hand; resting site is shown by $\cdot$, excited by $+$, refractory by $-$. This corresponds to finger-movement pattern (Fig. 4.28) with initial configuration when the third digit stimulates the medium. Initial configuration of fingers is $\wedge\wedge\vee$.

∧∧∨ ⟼ ∧∧∧ ⟼ ...13 times ...⟼ ∧∧∧ ⟼ ∧∨∨ ⟼ ∧∧∧ ⟼ ...10 times
...⟼ ∧∧∧ ⟼ ∨∧∧ ⟼ ∧∧∧ ⟼ ∧∧∧ ⟼ ∧∨∨ ⟼ ∧∧∧ ⟼ ...10 times ...⟼
∧∧∧ ⟼ ∨∧∧ ⟼ ...

Figure 4.28: Patterns of finger movements interacting with excitation dynamics
of cellular-automaton model of BZ medium, configurations of which are shown in
Fig. 4.27.

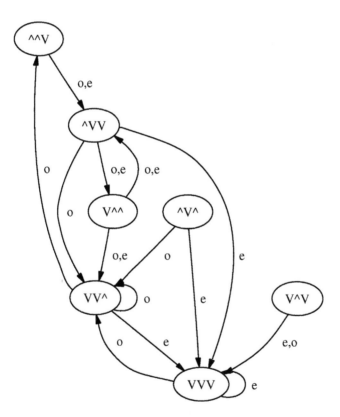

Figure 4.29: Transitions between finger-movement patterns interacting with BZ
medium. Finger up symbolised by ∧, down by ∨. Labels 'e' and 'o' specify that
the transition happens exclusively in cellular-automaton model with even and odd
lattice sizes, respectively.

4.4   Conclusion

In experiments we have shown that it is indeed possible to guide a mobile robot
with an on-board excitable chemical reactor. To stimulate the robot, and to implement its taxis, we locally perturb concentrations of chemical species in the reactor.
Excitation wave fronts propagating in the reactor carry (in their asymmetric wave
fronts) information about the original position of the stimulation source. Therefore, from images of spatio-temporal dynamics of excitation in the chemical reactor
the robot extracts a direction towards the source of stimulation and implements
rotation and forward motion.

The experimental approach, described in the book, has its drawbacks. Let us
discuss how to overcome them.

The robot moves slowly to prevent reaction-diffusion processes in the chemical
controller from being significantly disturbed in the lifetime of the reactor. It is
possible to increase the robot's speed but this will increase vibrations, and thus
will disrupt the geometry of the wave fronts and make calculations of source or
direction of stimulation less accurate. In scenarios of intermittent movement, the
wave fronts will re-focus in a vibration-free environment (when the robot stays still).
This does not follow if the disruption to the wave fronts is large, e.g. more than
four wavelengths (0.5 cm). We would envisage therefore doing the reaction in an
immobilised gel or membrane layer if vibrations (due to high speed of robot motion
or uneven surface of experimental arena) were too disruptive to the measuring
process. Whilst the use of a gel would no doubt improve the accuracy/ease of
vector calculation, it would result in slower overall dynamics and precludes the use
of the motion effect in a positive mode, i.e. the effect of the robot's motion on the
chemical reactor could constitute feedback to the controller.

Contact stimulation of the robot by a human operator is a far from perfect
experimental solution, it was used rather as a proof-of-concept than a final design.
Ideally, we would like to project an image of the surroundings onto the robotic
chemical processor without using silver stimulation. The answer lies in the light-
sensitive BZ system [154], where environment (key points and obstacles) can be
mapped via laser or fibre-optic assemblies onto the BZ processor, and then updated
during the robot's movement. Our preliminary experiments [78] show that a laser
beam (635-nm He/Ne, 1.5 mW) can initiate a wave in a part of the BZ medium
that is at steady state and can also affect already existing waves in previously
stimulated parts of the reaction, i.e. a wave passing a laser beam splits into two
wave fragments.

Another way to implement remote stimulation would be to use a light-sensitive
BZ processor acting as the robot's outer shell (sandwiched between two polymeric
layers with the light-sensitive catalyst immobilised on the inner surface). Fresh BZ
reagents would be actively pumped around the gap between the polymeric layers
in continuous contact with the gel/catalyst layer. Then, we could place a camera
inside this BZ shell, make the shell transparent, utilise filters and thus analyse the
changing excitation patterns (speed of existing waves changed or new wave centres
initiated or destroyed) on the shell, which would be changed according to local
light intensity. Therefore, the robot would not only be able to follow light but also
recognise and avoid shapes or obstacles which change light intensity and directions.

In an ideal situation the chemical reactors/controllers would be linked directly to the motor controllers of the robot. Whilst this is not easy, we envisage that a number of linked chemical reactors maintained in an oscillating or bistable state could be utilised. In this setup one reactor of the linked network would be assigned to a specific controller. In the case of the BZ reaction the oscillation results in a change in potential — therefore, this may be utilised directly via an interface to pass information about the state of the reactor to the controller. If each reactor has a continuous frequency or amplitude of oscillation then the robot assumes a searching motion pattern. However, if the behaviour of the reactors is linked to some control parameter such as temperature, light, etc., then as the robot moves in an environment the output of the reactors may be differentially affected. Thus, for example, if the robot approaches a point of bright illumination the oscillations in the reactors on one side may be suppressed. This change in frequency is fed forward to the controller, which applies a differential to the motion controller either to orientate towards or away from the light source. In the absence of illumination or in even illumination conditions eventually the reactors should return to the steady state. There are many problems associated with a practical implementation of this type, but it would reduce some of the conventional processing steps required in the present arrangement.

So, we have discussed some possible methods of mapping the robot's environment in real time to the chemical controller and also methods of linking the chemical controllers to the motor controllers; however, both approaches still require a large amount of conventional processing. There are now polymer materials [275] which can oscillate or move in other controllable ways when linked to chemical reactions such as the BZ reaction. The residual motion of these materials can then be further controlled by applying external forcing such as light. Thus, it should be possible to construct simple robots purely from such polymers and controlled via chemical reactions and their direct interaction with the robot's environment.

Dissonance of chemical and silicon components is an obvious drawback of our experimental setup. Chemical processors are not efficient when integrated in conventional hardware architectures. Any standard wheeled robot with primitive photosensors will incomparably outperform a chemical-medium-controlled robot in any navigation task. Not only are reaction-diffusion algorithms, when implemented in chemical media, really slow (it may take many minutes for an excitation wave to cross a Petri dish), but also analysis of the medium's space–time dynamics consumes more computational resources compared to any simple program that can guide a standard mobile robot. The state of things will change drastically when we enter the field of unconventional robotics (see e.g. [149]). Wave-based control in disordered media will ideally suit amorphous robotic architectures, like those based on electro-activated [257] and oscillating polymers [275].

In Sect. 4.2 we showed how to approximate sets of collision-free routes in experimental chemical thin-layer reactors — the palladium processor and the BZ processor. These chemical processors 'label' sites of an experimental arena, which are most distant from obstacles, whereby a robot minimises a risk of collision with obstacles if it travels along these 'labelled' routes. The palladium processor is a 'monotonous' system and it does not require any supporting computing devices to approximate a set of collision-free routes. This set is represented by sites of the medium without precipitate and thus uncoloured. The processor is however disposable since we could not restore its initial configuration, and thus it can be used just once. The BZ processor is re-usable, because when all excitation fronts, generated by obstacles, collide and annihilate the whole medium returns to its resting state and it can be re-excited. However, there are no stationary structures — which may represent the processor's memory as precipitate-concentration profiles do in the palladium processor — in the BZ medium; therefore, the medium needs the assistance of an external computing device to store the results of the computation.

To find all collision-free routes was the first step of the reaction-diffusion path planning. The second step was to calculate a shortest collision-free path between two given sites of the experimental arena. We implemented this step using a cellular-automaton model of an excitable (reaction-diffusion) medium. Luckily, this discrete excitable medium calculates not just a single path but a 'one-destination-many-sources' spanning tree of all shortest free paths from every site of an obstacle-free subspace of the arena to one selected site. A robot gets immediate benefits even at the first step of the reaction-diffusion computation, i.e. in order to avoid collision with an obstacle it must stick to the less coloured sites of the palladium processor and must not intersect the steep parts of the excitation wave fronts in the BZ processor.

There are many limitations of the present realisations of thin-layer chemical controllers. Firstly, sites within the chemical processors, corresponding to obstacles in the robotic arena, are perturbed by mechanical application of reagents. Optical excitation is desirable. Secondly, in the present experimental setup neither the reaction-diffusion chemical controllers nor the robot interact with their environments in real time. It is possible to assemble several chemical processors on board the robot and assign different tasks, e.g. taxis, obstacle avoidance and interaction with other robots, to different processors and correspondingly interpret space–time phenomena of wave dynamics. Thirdly, diffusion wave fronts in the palladium processor and excitation wave fronts in the BZ processor travel very slowly; therefore, real-time reactions of obstacle avoidance are feasible only if the robot travels slowly (e.g. 1 mm per minute). To increase the speed of the reaction-diffusion processing, one can either scale down the chemical processor or employ other types of faster wave-based media.

We experimentally demonstrated a non-trivial concurrent interaction between an excitable chemical medium and a robotic sensing and actuating device. We proved that it is indeed possible to achieve some sensible control of a robotic hand using travelling and interacting excitation wave fronts in a liquid-phase BZ system. We discovered several patterns of finger movements: self-excited motion, synchronised motion of fingers and reciprocal motion of fingers. To demonstrate viability of the approach we studied a closed-loop system where the robotic hand excites wave dynamics and these dynamics cause further movement of the hand's fingers. The approach discusses pioneering experiments on robotic hand control by a liquid-phase active chemical medium; therefore, certain problems were omitted from the analyses. Thus, for example, we did not take into account that the fingers disturb the chemical mixture in the dish and the medium's surface; this leads to disruption of the excitation wave fronts and this may cause unnecessary formation of spiral waves, multiple excitation sites due to physical re-location and diffusion of silver or generalised oxidation areas due to a high concentration of the activator species. These factors all result in highly complex hand or finger movements especially where more than two fingers are capable of stimulating the chemical medium. In future experiments we aim to adapt the present version of the chemical controller to enable the system's interaction with its environment and implementation of purposeful operations by the robotic hand.

Chapter 5

Programming reaction-diffusion processors

Despite promising preliminary results in reaction-diffusion computing, the field still remains art rather than science; most reaction-diffusion processors are produced on an *ad hoc* basis without structured top-down approaches, mathematical verification, rigorous methodology or relevance to other domains of advanced computing. There is a need to develop a coherent theoretical foundation of reaction-diffusion computing in chemical media. Particular attention should be paid to issues of programmability, because by making reaction-diffusion processors programmable we will transform them from marginal outcasts and curiosities to plausible competitors of conventional architectures and devices.

5.1 Controllability

Figure 5.1 shows two images, where Fig. 5.1a represents the natural disordered state of a certain type of chemical reaction. The behaviour even under controlled reaction conditions would be unpredictable. However, a natural feature of the reaction is to form expanding cones of precipitate whose circular edges interact to form cell-like structures with straight edges. To exploit this feature of circular interacting waves and the resulting pattern formation, we reduced the concentration of the reagent in the gel.

Now, if the reactant is poured over the gel surface we obtain a uniform precipitate, i.e. no cones are initiated. However, if we repeat the experiment but this time mark the surface of the gel with an array of glass needles prior to adding the reactant solution, then we can induce a travelling wave at each marked point. Where these circular waves collide they form straight-line segments and an ordered and predictable pattern results (Fig. 5.1b). The resulting pattern known as a Voronoi diagram is significant in computational geometry. Thus, via a simple process of control we have guided the natural dynamics of the chemical reaction to realise a useful computation.

Therefore, whilst all chemical reactions can be considered to be chemical processors, it is this ability to interface with and control the reaction to obtain a desired

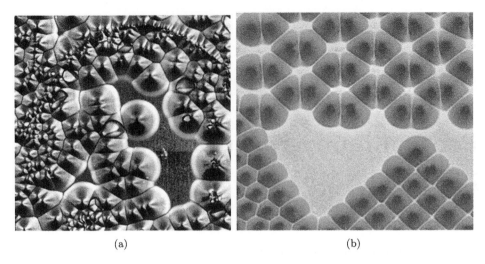

(a)   (b)

Figure 5.1: Controlling chemical processes: (a) disordered natural behaviour of a precipitating chemical reaction, (b) controlled ordered behaviour exhibited by a prototype chemical processor [81].

output state which will be the definition used for the purposes of this book.

5.2 How to program reaction-diffusion computers?

Controllability is an inherent constituent of programmability.

- How do real chemical media respond to changes in physical conditions?

- Are they controllable?

- If yes then what properties of the media can be used most effectively to program these chemical systems?

Despite the fact that the problem of controlling reaction-diffusion media did not receive proper attention until recently, some preliminary although rather mosaic results have become accessible in the last decade. There is no lucid view of the subject and this will be a major future task to build a theoretical and experimental framework of the controllability of chemical media. Below, we provide some details of the findings related to the external control of chemical media.

5.2.1 Electric field

A large amount of the literature related to theoretical and experimental studies concerning the controllability of reaction-diffusion media deals with the application of an electric field.

For example, in a thin-layer Belousov–Zhabotinsky (BZ) reactor stimulated by an electric field the following phenomena are observed:

- the velocity of excitation waves is increased by a negative and decreased by a positive electric field;

- a wave is split into two waves that move in opposite directions if a very high electric field is applied across the evolving medium [253];

- crescent waves are formed, not commonly observed in the evolution of the BZ reaction without a field [93];

- stabilisation and destabilisation of wave fronts [146];

- an alternating electric field generates a spiral wave core that travels within the medium; the trajectory of the core depends on the field frequency and amplitude [249].

Computer simulations with the BZ medium confirm that

- waves do not exist in a field-free medium but emerge when a negative field is applied [206];

- an electric field causes the formation of waves that change their sign with a change in concentration, and an applied constant field induces drift of vortices [194];

- externally applied currents cause the drift of spiral excitation patterns [268].

It is also demonstrated that by applying stationary two-dimensional fields to a reaction-diffusion system one can obtain induced heterogeneity in the reaction-diffusion system and thus increase the morphological diversity of the generated patterns (see e.g. [61]). These findings seem to be universal and valid for all reaction-diffusion systems: applying a negative field accelerates wave fronts; increasing the applied positive field causes wave deceleration, wave-front retardation and eventually wave-front annihilation. Also, a recurrent application of an electric field leads to the formation of complex spatial patterns [254].

A system of methylene blue, sulfide, sulfite and oxygen in a polyacrylamide gel matrix gives us a brilliant example of a medium controlled by the application of an electric field. Typically, hexagonal and striped patterns are observed in the medium. Application of an electric field makes striped patterns dominate in the medium; even orientation of the stripes is determined by the intensity of the electric field [192].

5.2.2 Temperature

Temperature is a key factor in the parameterisation of the space–time dynamics of reaction-diffusion media. It was shown that temperature is a bifurcation parameter in a closed non-stirred BZ reactor [174]. By increasing the temperature of the reactor one can drive the space–time dynamics of the reactor from periodic oscillations (0–3°C) to quasi-periodic oscillations (4–6°C) to chaotic oscillations (7–8°C). Similar findings are reported in simulation experiments on discrete media [7], where a lattice node's sensitivity can be considered as an analogue of temperature.

5.2.3   Substrate's structure

Modifications of the reagent concentrations and structure of the physical substrate contribute to shaping the space–time dynamics of reaction-diffusion media. Thus, for example, by varying the concentration of malonic acid in a BZ medium one can achieve

- the formation of target waves;

- the annihilation of wave fronts;

- the generation of stable propagating reduction fronts [146].

By changing the substrate we can achieve transitions between various types of patterns formed, e.g. transitions between hexagonal and striped patterns. This, however, could not be accomplished easily 'on-line', during the execution of a computational process, or even between two tasks; the whole computing device should be 're-fabricated', so we do not consider this option prospective.

Convection is yet another useful factor governing the space–time dynamics of reaction-diffusion media. Thus, for example, convection second-order waves, generated in collisions of excitation waves in the BZ medium, may travel across the medium and affect, e.g. annihilate, existing sources of wave generation [241].

5.2.4   Illumination

Light was one of the first [153] and still remains one of the most attractive, see the overview in [231], methods of controlling the spatio-temporal dynamics of reaction-diffusion media (this clearly applies to light-sensitive reaction-diffusion systems such as the BZ reaction with appropriate catalyst). Thus, by applying light of varying intensity we can control the medium's excitability [64] and the excitation dynamics in the BZ medium [47, 117], wave velocity [243] and pattern formation [299]. Of particular interest to the implementation of programmable logical circuits is the experimental evidence of light-induced back-propagating waves, wave-front splitting and phase shifting [308].

5.3   Programming with reaction rates

Consider a cellular-automaton model of an abstract reaction-diffusion excitable medium. Let a cell x of a two-dimensional lattice take four states: resting ∘, excited (+), refractory (−) and precipitate ⋆, and update their states in discrete time t depending on the number $\sigma^t(x)$ of excited neighbours in its eight-cell neighbourhood as follows (Fig. 5.2):

- A resting cell x becomes excited if $0 < \sigma^t(x) \leq \theta_2$ and precipitates if $\theta_2 < \sigma^t(x)$.

- An excited cell 'precipitates' if $\theta_1 < \sigma^t(x)$ or otherwise becomes refractory.

- A refractory cell recovers to the resting state unconditionally, and the precipitate cell does not change its state.

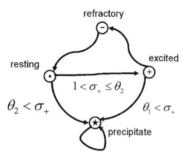

Figure 5.2: Cell state transition graph for cellular-automaton model of precipitating reaction-diffusion medium.

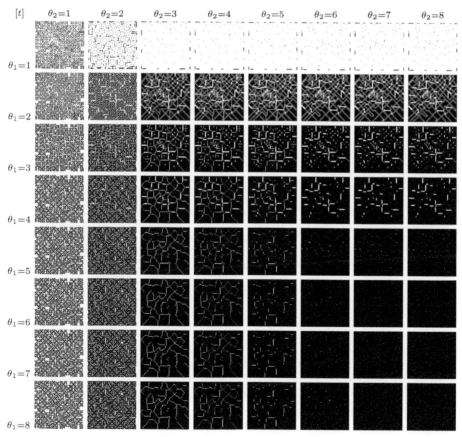

Figure 5.3: Final configurations of reaction-diffusion medium for $1 \leq \theta_1 \leq \theta_2 \leq 2$. Resting sites are black, precipitate is white.

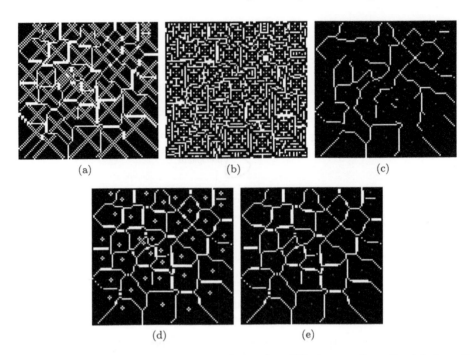

(a)   (b)   (c)

(d)   (e)

Figure 5.4: Exemplary configurations of reaction-diffusion medium for (a) $\theta_1 = 2$ and $\theta_2 = 3$, (b) $\theta_1 = 4$ and $\theta_2 = 2$, (c) $\theta_1 = 7$ and $\theta_2 = 3$, (d) $\theta_1 = 3$ and $\theta_2 = 3$, (e) $\theta_1 = 4$ and $\theta_2 = 3$. Resting sites are black, precipitate is white.

Initially we perturb the medium, excite it in several sites, thus inputting data. Waves of excitation are generated, they grow, collide with each other and annihilate as a result of the collision. They may form a stationary inactive concentration profile of a precipitate, which represents the result of the computation. Thus, we can only be concerned with reactions of precipitation:

$$+ \rightarrow^{k_1} \star$$

and

$$\circ \boxplus + \rightarrow^{k_2} \star,$$

where k_1 and k_2 are inversely proportional to θ_1 and θ_2, respectively. Varying θ_1 and θ_2 from 1 to 8, and thus changing precipitation rates from the maximum possible to the minimum, we obtain various kinds of precipitate patterns, as shown in Fig. 5.3.

Most of the patterns produced, see enlarged examples in Fig. 5.4a–c, are relatively useless (at least there is no sensible interpretation of them).

Precipitate patterns developed for relatively high ranges of reaction rates ($3 \leq \theta_1, \theta_2 \leq 4$) represent discrete Voronoi diagrams (a given 'planar' set, represented by sites of initial excitation, is visible in pattern $\theta_1 = \theta_2 = 3$ as white dots inside the Voronoi cells) derived from the set of initially excited sites, see Fig. 5.4d and

e. This example demonstrates that by externally controlling precipitation rates we can force the reaction-diffusion medium to compute a Voronoi diagram.

5.4   Programming with excitability

In this section we show how to re-program three-valued logical gates — in a geometrically constrained excitable medium — by tuning the excitability of the medium's elements [188]. A logical value is represented by the presence or absence of a wave front at a specified location; a computation is performed when two or more wave fronts interact with each other. For example, a threshold of state switching is employed in the design of chemical wave logical gates in [283]; there the computation is based on the fact that an excitable wave in the BZ reaction is generated only when a critical nucleation size is exceeded. Thus, two excitation waves (input variables) facilitate the formation of the third wave (output variable). Clearly, one can also design a XOR gate in some excitable media, where colliding waves annihilate each another, e.g. in a neural dendritic tree [104].

The Boolean logical gates were ingeniously implemented in laboratory experiments [283] and rigorously tested in numerical simulations [190, 138, 9, 261]. Binary designs inflict fairly artificial restrictions on the suitability of excitable media for tasks of reasoning, machine learning, data mining and dealing with uncertainties. To demonstrate that the limitations are artificial, we implement three-valued logical operations in cellular-automaton and numerical (FitzHugh–Nagumo) models of an excitable medium.

5.4.1   Cellular-automaton and numerical models of excitable-medium gates

We construct a T-shaped interaction gate as shown in Fig. 5.5a and simulate the behaviour of the gate in a discrete automaton network $\mathcal{A}$ and a two-dimensional FitzHugh–Nagumo model $\mathcal{F}$ of an excitable medium [188].

In an automaton network $\mathcal{A}$, every automaton (shown by ○ in Fig. 5.5a), except the end automata of the input and output branches and the input–output 'intersection' automaton (shown by ● in Fig. 5.5a), updates its state, in discrete time, depending on the states of its two neighbours. The junction automaton has three neighbours. Each automaton takes three states: resting (○), excited (+) and refractory (−), and updates its state in discrete time by the following rule. A resting automaton takes the excited state if at least one of its neighbours is excited (model $\mathcal{A}_0$), or exactly one of its neighbours is excited (model $\mathcal{A}_1$). An excited automaton takes the refractory state unconditionally, and a refractory automaton takes the resting state independent of the states of its neighbours. We have also introduced a heterogeneous automaton network $\mathcal{A}_2$, where every automaton apart from the junction automaton changes its state as in model $\mathcal{A}_0$ but the junction automaton executes transition $○ \mapsto +$ only if exactly two of its neighbours are excited.

In numerical experiments, with model $\mathcal{F}$, we used the FitzHugh–Nagumo reaction

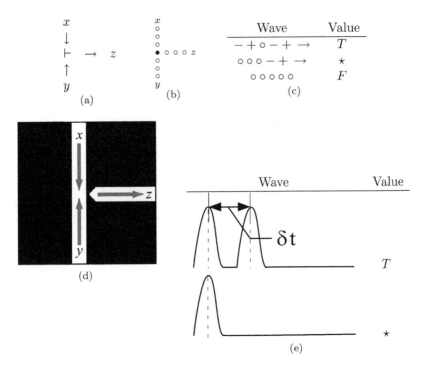

Figure 5.5: Interaction gate. (a) Orientation of the input, x and y, and output, z, channels. (b) Discrete automaton $\mathcal{A}$: construction of the gate. (c) Representation of the logical values by discrete excitation waves (impulses) in $\mathcal{A}$. (d) Geometry of the gate in FitzHugh–Nagumo-based model $\mathcal{F}$; excitable domain is white, passive diffusive domain is black. (e) Representation of logical values by excitation waves (impulses) in $\mathcal{F}$; distance δt is a control parameter for logical gates; each wave is represented by the inhibitor concentration profile.

diffusion equations [99], with diffusing activator u and immobilised inhibitor v:

$$\tau \frac{\partial u}{\partial t} = -\gamma(ku(u-\alpha)(u-1)+v) + D\nabla^2 u,$$

$$\frac{\partial v}{\partial t} = \gamma u,$$

which is an experimentally verified model of an excitable chemical medium [191]. The equations were numerically integrated using an alternating direction implicit method [225]. A logical gate structure is represented by the excitable domain of the medium, surrounded by a non-excitable diffusive domain, as shown in Fig. 5.5d, where an output channel's narrowing end is orientated towards the input channel but separated via a diffusive field gap. The parameters used are $\gamma = 1$ in an excitable field and zero otherwise, $\tau = 0.03$, $k = 3.0$, $\alpha = 0.02$ and $D = 0.00045$; the grid size is 300×300, the width of the input and output channels is 0.6, the integration time step $\delta t = 0.0025$, the space step is 0.02 and the gap width is 0.14.

$\vee_L$	T	F	$\star$
T	T	T	T
F	T	F	$\star$
$\star$	T	$\star$	$\star$

(a)

$\wedge_L$	T	F	$\star$
T	T	F	$\star$
F	F	F	F
$\star$	$\star$	F	$\star$

(b)

$\wedge_S$	T	F	$\star$
T	T	F	T
F	F	F	F
$\star$	T	F	$\star$

(c)

$\boxdot_{A_1}$	T	F	$\star$
T	F	T	$\star$
F	T	F	$\star$
$\star$	$\star$	$\star$	F

(d)

$\boxdot_{\mathcal{F}_1}$	T	F	$\star$
T	T	$\star$	$\star$
F	$\star$	F	F
$\star$	$\star$	F	$\star$

(e)

$\boxdot_{\mathcal{F}_2}$	T	F	$\star$
T	T	$\star$	T
F	$\star$	F	F
$\star$	T	F	T

(f)

Figure 5.6: Compositions implemented in discrete and continuous models of excitable media. (a) Łukasiewicz disjunction, model $\mathcal{A}_0$. (b) Łukasiewicz conjunction, models $\mathcal{A}_2$ and $\mathcal{F}_4$. (c) Sobociński conjunction, model $\mathcal{F}_3$. (d) Composition implemented in model $\mathcal{A}_1$. (e) Composition implemented in model $\mathcal{F}_1$. (f) Composition implemented in model $\mathcal{F}_2$.

In an excitable medium $\mathcal{F}$ an excitable wave is generated only when the local concentration of the activator u exceeds a certain threshold. Also, a wave travelling along the input channel causes the activator to diffuse into an area surrounding the channel. A wave is generated in the output channel only when the concentration of the activator diffusing from the input channels, passing through the diffusive gap between the input and the output channels, exceeds this threshold. We can arrange for such a gap between the input and the output channels to be a certain width so that an activator diffusing from a single travelling wave will not be enough to generate a new wave in the output channel. However, if two waves collide at the junction point then the concentration of the activator exceeds the threshold and an output wave is formed.

Logical values of a three-valued logic were encoded by excitation waves (impulses) as follows: no impulses represent FALSE F, two impulses TRUTH T and one impulse NONSENSE $\star$ (Fig. 5.5c and e). When both input channels receive values T or $\star$ the waves representing the values are generated in phase and wave fronts of two variables are at the same distance from the intersection with the output channel. In discrete models the distance between two impulses in the same wave train representing T was one automaton, while in continuous models the following values $\delta\tau$ were adopted: $2.15 \leq \delta t \leq 3.83$ (model $\mathcal{F}_1$), $3.83 < \delta t \leq 3.98$ (model $\mathcal{F}_2$), $3.98 < \delta t \leq 4.33$ (model $\mathcal{F}_3$) and $4.33 \leq \delta t$ (model $\mathcal{F}_4$). In the condition of $\delta t < 2.15$ the second impulse cannot exist because of refractory conditions induced by the first impulse [188].

5.4.2   Three-valued operators

We demonstrated that Łukasiewicz conjunction, disjunction and negation, and Sobosiński conjunction are implemented in excitable media $\mathcal{A}$ and $\mathcal{F}$.

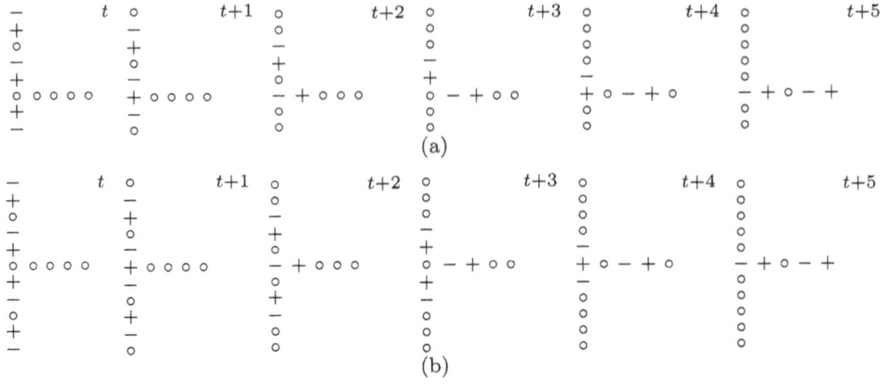

Figure 5.7: Examples of gate $\vee_L$ dynamics in model $\mathcal{A}_0$. (a) $T\vee_L\star = T$, (b) $T\vee_L T = T$.

All possible compositions implemented in T-shaped gates in models $\mathcal{A}$ and $\mathcal{F}$ are shown in Fig. 5.6. Let us explain them in detail.

5.4.3 Composition $\vee_L$

This composition is realised only in model $\mathcal{A}_0$. When just one input channel receives value T or $\star$ the waves travel along the channels undisturbed and reach the output channel; therefore, we have $T \vee_L F = T$ and $\star \vee_L F = \star$. No part of the gate is self-excitable, so $F \vee_L F = F$. When two single waves, $\star$, meet at the intersection • (Fig. 5.5b), the automaton • is excited, and then excites its neighbour in the output channel; thus, $\star \vee_L \star = \star$. Corresponding waves in double impulses interact similarly, so $T \vee_L \star = T$. Two examples of gate dynamics for input values T and $\star$ are shown in Fig. 5.7.

5.4.4 Composition $\Box_{A_1}$

In model $\mathcal{A}_1$ the junction automaton • does not become excited when two of its neighbours are excited; therefore, two colliding wave fronts will annihilate without spreading onto the automata of the output channel. Therefore, $T\Box_{A_1} T = \star\Box_{A_1} \star = F$ (see example in Fig. 5.8b). In the interaction of T and $\star$ signals, the first impulse of signal T is annihilated by the impulse representing $\star$, so we get $T \Box_{A_1} \star = \star$ (Fig. 5.8a).

This gate $\Box_{A_1}$ represents a composition of two gates — Lukasiewicz equivalence and negation as shown in Fig. 5.9, $\text{NOTEQ}(x, y) = \neg \leftrightarrow (x, y)$.

5.4.5 Composition $\wedge_L$

This composition is implemented in both discrete model $\mathcal{A}_2$ and continuous model $\mathcal{F}_4$. In the $\mathcal{A}_2$ model the junction automaton • is excited only when exactly two of its neighbours are excited (all other automata behave in the same way as those in

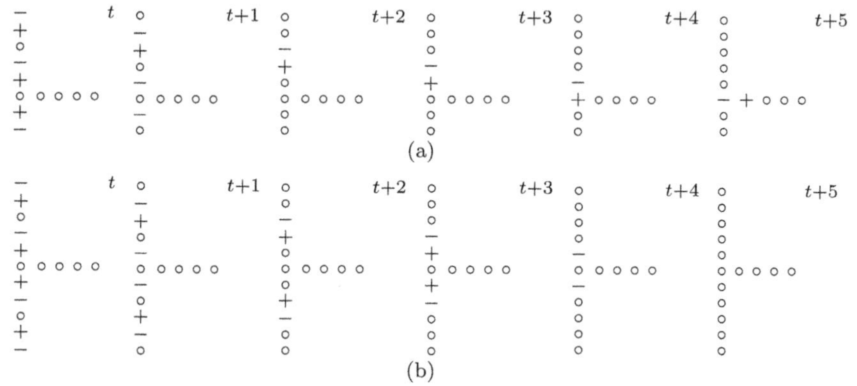

Figure 5.8: Examples of gate $\boxdot_{\mathcal{A}_1}$ dynamics in model $\mathcal{A}_1$. (a) $T \boxdot_{\mathcal{A}_1} \star = \star$, (b) $\star \boxdot_{\mathcal{A}_1} \star = F$.

$\leftrightarrow_{\mathrm{L}}$	T	F	$\star$
T	T	F	$\star$
F	F	T	$\star$
$\star$	$\star$	$\star$	T

(a)

x	$\neg_{\mathrm{L}} x$
T	F
F	T
$\star$	$\star$

(b)

Figure 5.9: Components of NOTEQ gate composition represented by $\boxdot_{\mathcal{A}_1}$: (a) Lukasiewicz equivalence $\leftrightarrow_{\mathrm{L}}$, (b) Negation $\neg_{\mathrm{L}}$.

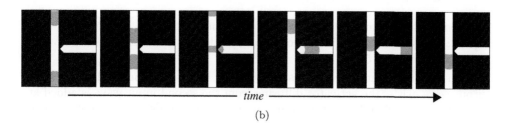

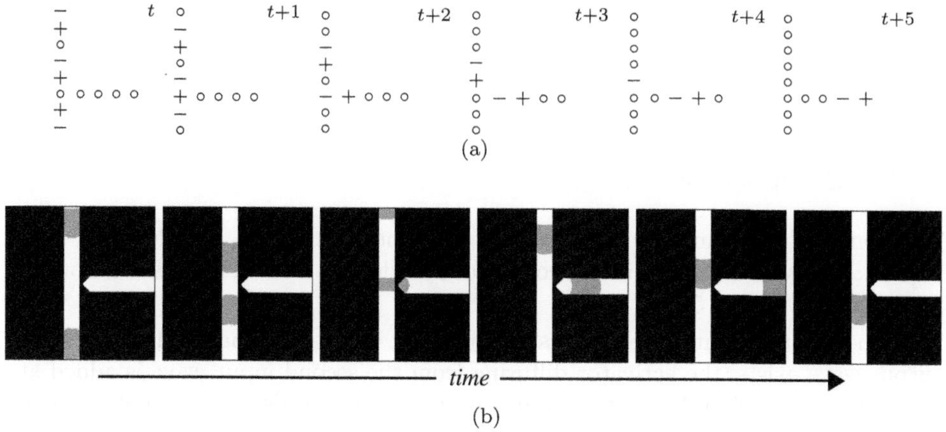

Figure 5.10: Example of gate $\wedge_{\mathrm{L}}$ dynamics: (a) $T \wedge_{\mathrm{L}} \star = \star$ in model $\mathcal{A}_2$. (b) $T \wedge_{\mathrm{L}} \star = \star$ in model $\mathcal{F}_4$. In (b) an excitable domain is white, a passive domain is black and an impulse wave is grey.

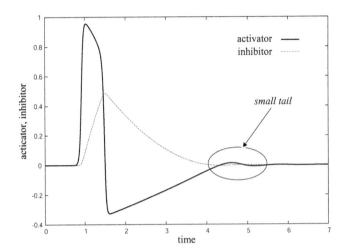

Figure 5.11: Structure of travelling wave in model $\mathcal{F}$. Concentration of the activator is shown by a solid line, the inhibitor by a dashed line. A small secondary wave at the tail of the primary wave is encircled.

model $\mathcal{A}_0$). Therefore, the excitation spreads to the output branch only when two wave fronts meet at the intersection automaton. So, when one of the input variables is F the output is F independent of another input variable: $T \wedge_L F = \star \wedge_L F = F$. When two single waves collide a single wave is generated at the output channel: $\star \wedge_L \star = \star$ and $T \wedge_L T = T$. For inputs T and $\star$ only the first impulse of signal T will be 'supported' by the impulse of the signal $\star$, so $T \wedge_L \star = \star$ (Fig. 5.10a). In the continuous model $\mathcal{F}_4$, the basic mechanism is similar to the discrete model $\mathcal{A}_2$. In the output channel the concentration of the activator exceeds the threshold only when two waves arrive at the junction simultaneously, so $T \wedge_L \star = \star$ (Fig. 5.10b).

5.4.6 Composition $\wedge_S$

This composition is implemented only in continuous medium $\mathcal{F}_3$. As demonstrated in [188], the wave described in the continuous model has a small tail wave following the larger wave (Fig. 5.11).

If the second input wave reaches the junction whilst the small tail wave of the output wave exists, the activator diffusing from the second input wave is added at the output channel and thus the total concentration exceeds the threshold. Then, the second output wave is generated. In the model $\mathcal{F}_3$ δt, the time interval between two input waves satisfies the above condition for timing. Therefore, when one of the input variables is T and the other is $\star$ the output is $T : T \wedge_S \star = T$ (Fig. 5.12a). On the other hand, in the model $\mathcal{F}_3$, the activator diffusing from the single input wave or train of a couple of input waves does not exceed the threshold, so $T \wedge_S F = F$ (Fig. 5.12b).

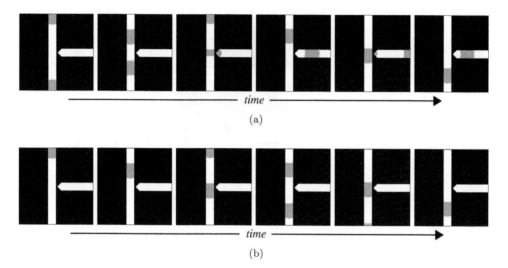

Figure 5.12: Example of gate $\wedge_S$ dynamics in model $\mathcal{F}_3$: (a) $T \wedge_S \star = T$, (b) $T \wedge_S F = F$.

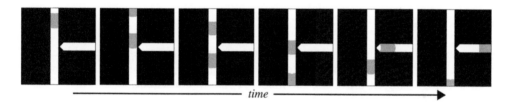

Figure 5.13: Example of gate $\boxdot_{\mathcal{F}_1}$ and $\boxdot_{\mathcal{F}_2}$ dynamics in model $\mathcal{F}_1$: $T \boxdot_{\mathcal{F}_1} F = \star$.

5.4.7 Compositions $\boxdot_{\mathcal{F}_1}$ and $\boxdot_{\mathcal{F}_2}$

They are rather nonsensical compositions, by products of our computational experiments, bearing just phenomenological value. These compositions are implemented in continuous media $\mathcal{F}_1$ and $\mathcal{F}_2$.

When the input wave passes the junction the activator diffuses into the gap, and then diffuses into the output channel. If the second input wave passes while a certain amount of activator still remains at the junction then the total amount of the activator at the output channel exceeds the threshold. In the models $\mathcal{F}_1$ and $\mathcal{F}_2$, the time interval $\delta t \leq 3.98$ between two waves of the two-wave train, representing T, allows for the activator to multiply, and therefore we have $T \boxdot_{\mathcal{F}_1} F = T \boxdot_{\mathcal{F}_2} F = \star$ (Fig. 5.13). On the other hand, a result of operation $T \boxdot \star$ depends on δt: $\wedge_L$ and $\wedge_S$. In model $\boxdot_{\mathcal{F}_1}$, the dynamics of waves implementing $T \wedge_{\mathcal{F}_1} \star$ is similar to that of $\boxdot_{\mathcal{F}_4}$. Wave dynamics of $\boxdot_{\mathcal{F}_2}$ is similar to $\boxdot_{\mathcal{F}_3}$; therefore, $T \boxdot_{\mathcal{F}_1} \star = \star$ and $T \boxdot_{\mathcal{F}_2} \star = T$.

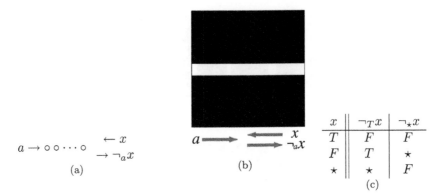

x	$\neg_T x$	$\neg_\star x$
T	F	F
F	T	$\star$
$\star$	$\star$	F

(a) $\quad a \rightarrow \circ \circ \cdots \circ \quad \begin{array}{c} \leftarrow x \\ \rightarrow \neg_a x \end{array}$

(b) $\quad a \xrightarrow{} \quad \begin{array}{c} \xleftarrow{} x \\ \xrightarrow{} \neg_a x \end{array}$

(c)

Figure 5.14: Negation gate in excitable medium. (a) Negation gate structure in the model $\mathcal{A}$: x is an input variable, a is a reference signal, $\neg_a x$ is x negated in interaction with reference signal a. (b) Negation gate structure in the model $\mathcal{F}$: an excitable domain is white, a passive domain is black, and the representation is almost the same as in automaton model. (c) Possible 'negation' tables constructed for reference signals T and $\star$.

5.4.8　Negation

For an excitable medium to be universal it must implement a negation additional to other gates, e.g. conjunction or negation. The gate $\Box_{\mathcal{A}_1}$ represents Lukasiewicz negation (Fig. 5.9b) at once. So, the model $\mathcal{A}_1$ is self-consistent.

To implement negation in other models we can employ a one-channel construction (Fig. 5.14a and b) where constant T or constant $\star$ are generated at one end and value x is fed to another end; at the same end the negated value is read. Compositions implemented by reference signals T and $\star$ are shown in Fig. 5.14c. Some examples of signal interaction implemented in the automaton model $\mathcal{A}$ are shown in Fig. 5.15, and those realised in the continuous model $\mathcal{F}$ in Fig. 5.16.

5.4.9　Logic of excitability

Experimenting with cellular-automaton and numerical models of excitable media, we demonstrated that several types of three-valued logical operations can be implemented in the interaction of excitation wave fronts travelling along quasi-one-dimensional channels arranged in a T-shaped structure.

Three-valued gates constructed in cellular-automaton models can be seen as an expansion of 'conventional' binary logic gates. Thus, representing TRUTH by a single excitation wave $+-$ and FALSE by the absence of a wave, we can demonstrate that model $\mathcal{A}_0$ implements Boolean disjunction, model $\mathcal{A}_1$ exclusive disjunction and model $\mathcal{A}_2$ conjunction. Conditions of cell excitation in models $\mathcal{A}$ can be reformulated as follows: a resting cell is excited if it has $\lfloor \frac{k}{2} \rfloor$ or $\lceil \frac{k}{2} \rceil$ excited neighbours in model $\mathcal{A}_0$; $\lfloor \frac{k}{2} \rfloor$ excited neighbours in model $\mathcal{A}_1$; and $\lceil \frac{k}{2} \rceil$ excited neighbours in

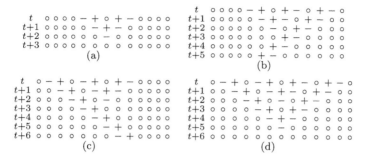

$$
\begin{array}{llllllllllllll}
t   & \circ & \circ & \circ & \circ & - & + & \circ & + & - & \circ & \circ & \circ & \circ \\
t+1 & \circ & \circ & \circ & \circ & \circ & - & + & - & \circ & \circ & \circ & \circ & \circ \\
t+2 & \circ & \circ & \circ & \circ & \circ & \circ & - & \circ & \circ & \circ & \circ & \circ & \circ \\
t+3 & \circ & \circ & \circ & \circ & \circ & \circ & \circ & \circ & \circ & \circ & \circ & \circ & \circ \\
\end{array}
$$

(a)

$$
\begin{array}{llllllllllllll}
t   & \circ & \circ & \circ & \circ & - & + & \circ & + & - & \circ & + & - & \circ \\
t+1 & \circ & \circ & \circ & \circ & \circ & - & + & - & \circ & + & - & \circ & \circ \\
t+2 & \circ & \circ & \circ & \circ & \circ & \circ & - & \circ & + & - & \circ & \circ & \circ \\
t+3 & \circ & \circ & \circ & \circ & \circ & \circ & \circ & + & - & \circ & \circ & \circ & \circ \\
t+4 & \circ & \circ & \circ & \circ & \circ & \circ & + & - & \circ & \circ & \circ & \circ & \circ \\
t+5 & \circ & \circ & \circ & \circ & \circ & + & - & \circ & \circ & \circ & \circ & \circ & \circ \\
\end{array}
$$

(b)

$$
\begin{array}{llllllllllllll}
t   & \circ & - & + & \circ & - & + & \circ & + & - & \circ & \circ & \circ & \circ \\
t+1 & \circ & \circ & - & + & \circ & - & + & - & \circ & \circ & \circ & \circ & \circ \\
t+2 & \circ & \circ & \circ & - & + & \circ & - & \circ & \circ & \circ & \circ & \circ & \circ \\
t+3 & \circ & \circ & \circ & \circ & - & + & \circ & \circ & \circ & \circ & \circ & \circ & \circ \\
t+4 & \circ & \circ & \circ & \circ & \circ & - & + & \circ & \circ & \circ & \circ & \circ & \circ \\
t+5 & \circ & \circ & \circ & \circ & \circ & \circ & - & + & \circ & \circ & \circ & \circ & \circ \\
t+6 & \circ & \circ & \circ & \circ & \circ & \circ & \circ & - & + & \circ & \circ & \circ & \circ \\
\end{array}
$$

(c)

$$
\begin{array}{llllllllllllll}
t   & \circ & - & + & \circ & - & + & \circ & + & - & \circ & + & - & \circ \\
t+1 & \circ & \circ & - & + & \circ & - & + & - & \circ & + & - & \circ & \circ \\
t+2 & \circ & \circ & \circ & - & + & \circ & - & \circ & + & - & \circ & \circ & \circ \\
t+3 & \circ & \circ & \circ & \circ & - & + & \circ & + & - & \circ & \circ & \circ & \circ \\
t+4 & \circ & \circ & \circ & \circ & \circ & \circ & - & + & - & \circ & \circ & \circ & \circ \\
t+5 & \circ & \circ & \circ & \circ & \circ & \circ & \circ & - & \circ & \circ & \circ & \circ & \circ \\
t+6 & \circ & \circ & \circ & \circ & \circ & \circ & \circ & \circ & \circ & \circ & \circ & \circ & \circ \\
\end{array}
$$

(d)

Figure 5.15: Dynamics of negation gate in automaton model $\mathcal{A}$ of an excitable medium. (a) Reference signal $\star$, $x = \star$, $\neg_\star x = F$. (b) Reference signal $\star$, $x = T$, $\neg_\star x = F$. (c) Reference signal T, $x = \star$, $\neg_T x = \star$. (d) Reference signal T, $x = T$, $\neg_T x = F$

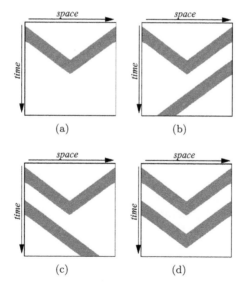

(a) (b)

(c) (d)

Figure 5.16: Dynamics of a negation gate in a continuous model $\mathcal{F}$ of an excitable medium. Space–time plot of waves travelling along horizontal centerline of the excitable domain in Fig. 5.14b. Waves are shown in grey. (a) Reference signal $\star$, $x = \star$, $\neg_\star x = F$. (b) Reference signal $\star$, $x = T$, $\neg_\star x = F$. (c) Reference signal T, $x = \star$, $\neg_T x = \star$. (d) Reference signal T, $x = T$, $\neg_T x = F$

model $\mathcal{A}_2$, where k is the number of neighbours ($k = 2$ for cells along the channels, and $k = 3$ for a junction site). That is,

by decreasing the excitability of the T-shaped excitable medium we achieve a transition from disjunction to exclusive disjunction to conjunction.

Model $\mathcal{A}$ is capable of processing variables of n-valued logic as well as variables of binary and three-valued logic. Assuming that a train of $0 \leq m \leq n$ waves represent arithmetical m, we can demonstrate that model $\mathcal{A}_0$ implements operation $z = \max(x, y)$, $\mathcal{A}_1$ operation $z = \mathrm{abs}(x - y)$ and $\mathcal{A}_2$ operation $z = \min(x, y)$.

Three-valued gates constructed in continuous models also can be seen as an expansion of Boolean logic gates. Representing TRUTH by a single excitation wave and FALSE by the absence of a wave, we can demonstrate Boolean disjunction and conjunction. A Boolean operation with the arrangement of fields and parameters introduced here returns Boolean conjunction. On the other hand, Boolean disjunction can be realised by narrowing the gap width between the input and output channels, or by increasing the excitability; in this condition an activator diffusing from a single wave at the input channel to the output channel exceeds the threshold and then a new output wave is generated [189].

Also, model $\mathcal{F}$ is capable of processing variables of n-valued logic. Assuming a train of $0 \leq m \leq n$ waves, we can demonstrate that model $\mathcal{F}_4$ implements operation $z = \min(x, y)$ and $\mathcal{F}_3$ operation $z = \min(x, y) + 1$. If the field arrangement or parameters are set similarly to a Boolean disjunction gate, then operation $z = \max(x, y)$ is implemented.

5.5   Conclusion

Sluggishness, the narrow range of computational tasks solved, and a perceived lack of effective control are usually seen as the main disadvantages of existing prototypes of reaction-diffusion computers. In this chapter we briefly outlined some ways of externally controlling, tuning and ultimately programming spatially extended chemical devices. We have also indicated how to 'switch' a reaction-diffusion computer, with a fixed set of reactions but variable reaction rates, between several domains of problems, and thus make it more 'omnivorous'. This uncovers possible ways of developing a framework of programmability of excitable-medium processors, and thus may enormously enlarge the spectrum of potential tasks solved by reaction-diffusion excitable processors in general [10]. Future applications of excitable and reaction-diffusion chemical media will not be limited by tasks of image processing or simple robot navigation but will expand into the fields of fuzzy logic and reasoning, artificial intelligence and designs of sophisticated many-valued logical circuits.

Chapter 6

Silicon reaction-diffusion processors

This chapter introduces semiconductor reaction-diffusion computing devices that can be implemented on conventional silicon large scale integrated (LSI) circuits (chips).

Natural systems give us examples of amorphous, unstructured devices, capable of fault-tolerant information processing, particularly with regard to the massive parallel spatial problems that digital processors have difficulty with. For example, reaction-diffusion chemical systems have the unique ability to efficiently solve combinatorial problems with natural parallelism [7]. In liquid-phase parallel reaction-diffusion chemical processors, both the data and the results of the computation are encoded as concentration profiles of the reagents. The computation is performed via the spreading and interaction of the wave fronts. In experimental chemical processors, data are represented by local disturbances in the concentrations, and computation is accomplished via the interaction of waves caused by the local disturbances. The reaction-diffusion chemical computers operate in parallel since the chemical medium's micro-volumes update their states simultaneously, and the molecules diffuse and react in parallel — similar parallelism is observed in cellular automata. Various reaction-diffusion systems can be modelled in terms of cellular automata, including the Belousov–Zhabotinsky (BZ) reaction [7, 109], chemical systems exhibiting Turing patterns [311] and precipitating systems for the computation of Voronoi diagrams [2, 4]. A two-dimensional cellular automaton is particularly well suited for the coming generation of massively parallel machines, in which a very large number of separate processors act in parallel. If an elemental processor in the cellular automaton is constructed from a smart processor and a photosensor, various cellular-automaton algorithms can easily be used to develop intelligent image sensors.

Implementing reaction-diffusion systems in hardware, i.e. LSI circuits, has several advantages, and hardware reaction-diffusion systems are very useful in simulating reaction-diffusion phenomena, even if the phenomena never occur in nature. This implies that a hardware system is a possible candidate for developing an artificial reaction-diffusion system that is superior to a natural system. For instance,

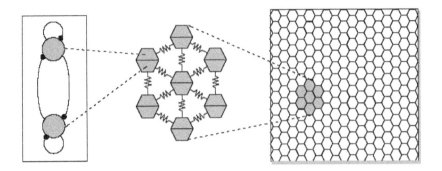

Figure 6.1: Basic construction of reaction-diffusion chip. Left: reaction circuit; middle: hexagonal grid of reaction circuits that are connected by diffusive links; right: reaction-diffusion chip.

hardware reaction-diffusion systems can operate at much faster speeds than actual reaction-diffusion systems. The velocity of chemical waves in a BZ reaction is about 5–6 mm/min [284], while that of a hardware reaction-diffusion system will be over a million times faster than that of the BZ reaction, independent of system size. This property is useful for the developers of reaction-diffusion-based applications; this will make reaction-diffusion processors benefit from high-speed operations, and indeed this may be one key feature that makes silicon-based reaction-diffusion processors viable realistic alternatives when compared to their natural reaction-diffusion counterparts (at least for some specific tasks). It was these properties that encouraged us to develop these reaction-diffusion chips.

A reaction-diffusion chip consists of reaction circuits that emulate elementary reactions occurring in micro-volumes of a chemical medium and diffusion devices that imitate diffusive links between the micro-volumes (diffusion of chemical species). Figure 6.1, left, shows a schematic diagram of self- and reciprocal interactions between two micro-volumes of a chemical medium. The reaction circuits are arranged on a hexagonal grid (Fig. 6.1, middle) and each circuit is connected to its neighbouring circuits through the diffusion devices (links). Reaction-diffusion chips were predominantly designed by digital, analogue or mixed-signal complementary metal–oxide–semiconductor (CMOS) circuits of cellular neural networks (CNNs) or cellular automata.

Electrical cell circuits were designed to implement several cellular-automaton and CNN models of reaction-diffusion systems [23, 36, 60, 175, 235, 258, 34], as well as fundamental reaction-diffusion equations [33, 39, 77, 144, 252]. Each cell is arranged on a two-dimensional orthogonal or hexagonal grid and is connected to adjacent cells through diffusive coupling devices that transmit a cell's state to its neighbouring cells. For example, an analogue–digital hybrid reaction-diffusion chip [36] was designed for emulating a conventional cellular-automaton model for BZ reactions [109]; a full-analogue reaction-diffusion chip that emulates BZ reactions has also been designed and fabricated [39]. A precipitating system for computation of Voronoi diagrams [2, 4] was also implemented on an analogue–digital hybrid reaction-diffusion chip [34]. An all-digital reaction-diffusion processor [175] was designed on the basis of a multiple-valued cellular-automaton model of excitable lattices [7]. Furthermore, a reaction-diffusion cellular-automaton processor for complex image processing has been proposed in [38]; it performs quadrilateral-object extraction based on serial and parallel cellular-automaton algorithms. An analogue cell circuit was implemented to simulate spatially discrete Turing reaction-diffusion systems [77]. Blueprints of non-CMOS reaction-diffusion chips have been developed, namely, for a reaction-diffusion device based on minority-carrier transport in semiconductor devices [33] and a single-electron reaction-diffusion device [209].

In the chapter, we study how to construct an artificial reaction-diffusion system using a solid-state medium, i.e. silicon. Some applications using the solid-state reaction-diffusion system that could rival conventional digital computers are introduced. In Sect. 6.1, we study how to obtain a spatially discrete reaction-diffusion cellular-automaton system suitable for LSI implementation. Sections 6.2 and 6.3 introduce practical reaction-diffusion chips based on digital and analogue cellular automata.

6.1 Modelling reaction-diffusion LSI circuits

Reaction-diffusion phenomena are usually observed in liquid- or gas-state media, while CMOS circuits are conventionally implemented on solid-state media. So, there exists an important problem here — how to implement the principles of liquid reaction-diffusion systems in solid-state media? In this section, we briefly discuss one possible way by introducing a spatially discrete reaction-diffusion cellular-automaton system.

Chemical reactions are formulated in terms of temporal differences in the concentrations of chemical species. For example, if substance x is dissolved in water, the temporal difference in the concentration of x is expressed by an ordinary differential equation (ODE) as

$$\frac{d[x]}{dt} = -k[x], \tag{6.1}$$

where $[x]$ represents the concentration and k the rate constant. Although Eq. (6.1) is a linear ODE, most chemical reactions, including dissipative and autocatalytic reactions in natural systems, will be formulated by non-linear ODEs with the right-hand side of Eq. (6.1) represented by a polynomial of $[x]$. Non-linear chemical reactions with multiple chemical species are thus represented by a set of non-linear ODEs as

$$\frac{d[x_i]}{dt} = f_i([x_1], [x_2], \dots, [x_N]), \quad i = 1, 2, \dots, N, \tag{6.2}$$

where N is the number of species and f_i represents the non-linear reactive functions that depend on several different reactive species x_i.

The BZ reaction can be represented by Eq. (6.2). One well-known model of the BZ reaction is referred to as the two-variable Oregonator [199]. The dynamical

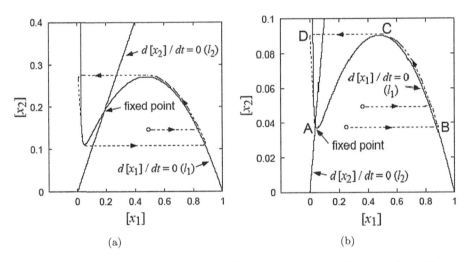

(a)   (b)

Figure 6.2: Nullclines and trajectories of Oregonator operating in (a) oscillatory mode ($a = 1$) and (b) excitatory mode ($a = 3$).

relations between variables, representing concentrations of chemical species, are given by

$$\frac{d[x_1]}{dt} = \frac{1}{\tau}\left([x_1]\,(1-[x_1]) - a\,[x_2]\,\frac{[x_1]-b}{b+[x_1]}\right), \tag{6.3}$$

$$\frac{d[x_2]}{dt} = [x_1] - [x_2], \tag{6.4}$$

where $[x_1]$ and $[x_2]$ represent the abridged concentrations of $HBrO_2$ and Br^- ions, respectively, while τ, a and b represent the reaction parameters. The value of τ is generally set at $\tau \ll 1$ since the reaction rate of $HBrO_2$ ions is much higher than that of Br^- ions. The nullclines of the Oregonator, where $d[x_1]/dt = 0$ and $d[x_2]/dt = 0$, are given by

$$[x_2] = \frac{[x_1]\,([x_1]+b)(1-[x_1])}{a\,([x_1]-b)} \quad (\equiv l_1), \tag{6.5}$$

$$[x_2] = [x_1] \quad (\equiv l_2). \tag{6.6}$$

The point of crossing these two nullclines l_1 and l_2 represents a fixed point of the Oregonator.

Figure 6.2 plots the nullclines and trajectories of the Oregonator with typical parameter values ($\tau = 10^{-2}$ and $b = 0.02$). The value of the parameter a is set at 1 (Fig. 6.2a) and 3 (Fig. 6.2b). Depending on the position of the fixed point, the Oregonator exhibits oscillatory or excitatory behaviour. When $a = 1$, the fixed point is located on nullcline l_1 at which $d[x_2]/d[x_1] > 0$. In this case, the Oregonator exhibits limit-cycle oscillations (Fig. 6.2a). The oscillations represent periodic oxidation–reduction phenomena in the BZ reaction. On the other hand, the fixed point is located on nullcline l_1 at which $d[x_2]/d[x_1] < 0$ when $a = 3$. Under these conditions, the Oregonator exhibits excitatory behaviour (Fig. 6.2b) and is stable at the fixed point as long as an external stimulus is not applied.

Three states are introduced in the Oregonator according to the oscillation phase: inactive or resting (A), active (B → C) and refractory periods (D → A), as shown in Fig. 6.2b. The inactive, active and refractory states represent a depletion in Br^- ions, an autocatalytic increase in $HBrO_2$ ions (oxidation of the catalyst) and a depletion in Br^- ions (reduction of the catalyst), respectively. When the Oregonator is inactive, it is easily activated (A → B) by external stimuli. Then, it turns to the refractory state (C → D). During the refractory state, the Oregonator cannot be activated even if external stimuli are applied.

It should be noted that Eqs. (6.1) to (6.4) represent the time difference in chemical species at a *point* in the reaction space. If the spatial distribution of the chemical species is not uniform, the species will diffuse according to the gradient of the concentration of the species. Such a reaction-diffusion system with multiple chemical species is described by a set of partial differential equations as

$$\frac{\partial [x_i](\mathbf{r}, t)}{\partial t} = D_i \nabla^2 [x_i] + f_i\Big([x_1], [x_2], \ldots, [x_N]\Big), \tag{6.7}$$

where $\mathbf{r}$ represents the space, ∇^2 the spatial Laplacian and D_i the diffusion constant. A two-variable reaction-diffusion system on a two-dimensional plane, which is referred to as a basic reaction-diffusion system, is described in terms of Eq. (6.7) as

$$\frac{\partial [u](x, y, t)}{\partial t} = D_u \Big(\frac{\partial^2 [u]}{\partial x^2} + \frac{\partial^2 [u]}{\partial y^2}\Big) + f_u\big([u], [v]\big), \tag{6.8}$$

$$\frac{\partial [v](x, y, t)}{\partial t} = D_v \Big(\frac{\partial^2 [v]}{\partial x^2} + \frac{\partial^2 [v]}{\partial y^2}\Big) + f_v\big([u], [v]\big), \tag{6.9}$$

where (x, y) represents the space and $([u], [v])$ the concentrations of two-different chemical species [199].

Figure 6.3 is a schematic showing an alternative construction of the basic reaction-diffusion system. It consists of a two-dimensional array of chemical oscillators, e.g. Oregonators, where each oscillator is locally interconnected. An oscillator located at position (i, j) has two system variables $[u_{i,j}]$ and $[v_{i,j}]$. The dynamics are defined as

$$\frac{d[u_{i,j}]}{dt} = f_u\big([u_{i,j}], [v_{i,j}]\big) + g_{i,j}^u, \tag{6.10}$$

$$\frac{d[v_{i,j}]}{dt} = f_v\big([u_{i,j}], [v_{i,j}]\big) + g_{i,j}^v, \tag{6.11}$$

where functions f_u and f_v represent the non-linear chemical interactions between $[u_{i,j}]$ and $[v_{i,j}]$, and where $g_{i,j}^u$ and $g_{i,j}^v$ represent the external inputs to the oscillator. External inputs are applied to an oscillator so that activities $[u_{i,j}]$ and $[v_{i,j}]$ can diffuse throughout the two-dimensional array of oscillators. Such inputs are given

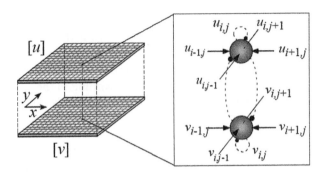

Figure 6.3: Discrete representation of two-dimensional reaction-diffusion system — array of oscillators (left), where each chemical oscillator (right) is locally connected to its closest neighbours. An oscillator located at position (i, j) is represented by a system of two variables $[u_{i,j}]$ and $[v_{i,j}]$.

by a five-point approximation of the Laplacian on the two-dimensional rectangular grid as

$$g_{i,j}^u = D_u \frac{[u]_{i-1,j} + [u]_{i+1,j} + [u]_{i,j-1} + [u]_{i,j+1} - 4[u]_{i,j}}{h^2},$$

$$g_{i,j}^v = D_v \frac{[v]_{i-1,j} + [v]_{i+1,j} + [v]_{i,j-1} + [v]_{i,j+1} - 4[v]_{i,j}}{h^2},$$

where h represents the distance between neighbouring oscillators. The dynamics of a single oscillator, see Eqs. (6.10) and (6.11), thus represents the discrete expression of Eqs. (6.8) and (6.9). This expression allows us to understand the relation between natural reaction-diffusion systems and the physical hardware structure. Namely, two-dimensional reaction-diffusion phenomena can be imitated on solid-state media, e.g. LSIs, where numerous hardware oscillators are regularly arranged on LSIs with diffusive coupling among the local oscillator circuits.

Figure 6.4 shows numerical solutions to Eqs. (6.10) and (6.11) using an Orego-nator with standard parameters. The non-linear reaction functions $f_u(\cdot)$ and $f_v(\cdot)$ in Eqs. (6.10) and (6.11) are replaced with the right-hand sides of Eqs. (6.3) and (6.4), respectively, with the transformation of system variables ($[x_1] \rightarrow [u_{i,j}]$ and $[x_2] \rightarrow [v_{i,j}]$). Each oscillator was set in the excitatory mode ($a = 3$), and the values of the rest of the parameters were $h = 0.01$, $D_u = 5 \times 10^{-4}$, $D_v = 0$, $\tau = 10^{-2}$ and $b = 0.02$. The solution was numerically obtained by solving the ODEs with the fourth-order Runge–Kutta method. At each side of the square reaction space, the von Neumann boundary condition was applied as

$$\nabla[u] = \nabla[v] = (0, 0), \tag{6.12}$$

where $\nabla = (\partial/\partial x, \partial/\partial y)$; e.g. the values of $[u_{0,j}]$ and $[u_{N+1,j}]$ are treated as those of $[u_{1,j}]$ and $[u_{N,j}]$, respectively. In Fig. 6.4, the values of $v_{i,j}$ are represented on a grey scale ($v_{i,j} = 0$: black, $v_{i,j} = 1$: white). Several oscillators adjacent to the inactive oscillators were initially set in a refractory state (left-hand side of the white

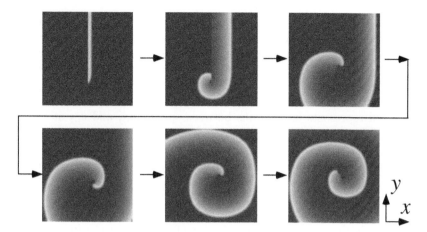

Figure 6.4: Typical numerical results of two-dimensional reaction-diffusion system using Oregonator. Each oscillator was set in the excitatory mode ($a = 3$).

bar in top-left snapshot in Fig. 6.4). The inactive oscillators adjacent to the white bar were suppressed by the adjacent oscillators in the refractory state (oscillators in white bar). The inactive oscillators then took an active, inactive or refractory state, depending on the degree of refractiveness. When the inactive oscillators were in an active or inactive state, the tip of the strip-like wave front rotated inwards, resulting in the generation of spiral patterns that are typically observed in BZ reactions.

We have another choice to implement reaction-diffusion systems on LSIs; i.e. digital implementation of reaction-diffusion systems based on von Neumann architectures. If we convert source codes that solve the above-mentioned reaction-diffusion equations to hardware description language (HDL), the reaction-diffusion processor will easily be designed and fabricated. However, in the case of von Neumann processors, there exist critical problems if considered as reaction-diffusion computing devices. The time required for the computation increases exponentially as the reaction space increases and, due to the complexity and vast power consumption, its compact and low-power implementation was proved to be impossible to date. Analogue implementations of reaction-diffusion systems where reaction ODEs are solved in parallel by the circuit's physical phenomena will certainly solve the above problems. In the next section, we describe examples of a cellular-automaton-based implementation where simplified features of the reaction's dynamics are used instead of analogue ODEs.

6.2 Digital reaction-diffusion chips

A cellular automaton is a discrete dynamical system whose behaviour is completely specified in terms of finite local interactions [281, 2, 226], and is thus suitable for LSI implementation [40, 273]. The discrete expression of basic reaction-diffusion models, introduced in Sect. 6.1, has significant similarity to the cellular-automaton system [279, 73]. First, it consists of a number of identical cells (processors) and local connections between these cells. Second, these cells are regularly arranged on a two-dimensional rectangular grid. Thus, the basic reaction-diffusion model, see Eqs. (6.8) and (6.9), can be replaced with a cellular-automaton model by assuming that

- each cell represents interactions between species $u_{i,j}$ and $v_{i,j}$ at the specific point (i,j),

- the concentrations of chemical species $([u_{i,j}], [v_{i,j}])$ are represented by the values of system variables in each cell and

- local connections between cells are used to input updated values of system variables in the same way as the diffusion of chemical species in a real reaction-diffusion system.

6.2.1 Implementing a multiple-valued reaction-diffusion cellular-automaton model

Gerhardt *et al.* proposed a cellular-automaton model that used a two-dimensional array of digital oscillators (cells) to imitate the Oregonator [109]. In the model, a cell corresponds to a point in the BZ reaction space, e.g. an architecture of a chemical medium is imitated by arranging cells on a grid, and assigning to cell (i,j) two types of variables that represent concentrations of $HBrO_2$ and Br^- ions at site (i,j): $[u_{i,j}]$ and $[v_{i,j}]$, respectively. The variable $[u_{i,j}]$ takes binary values, and $[v_{i,j}]$ takes multiple values as $n\ dv$ $(dv = 1/N,\ n = 0, 1, \ldots, N-1)$.

Figure 6.5 plots a circulative state diagram of a single cell. The reaction states are discriminated by the values of $[u_{i,j}]$ and $[v_{i,j}]$ and are represented by the grey boxes in Fig. 6.5a. A cell can be activated $([u_{i,j}] = 0 \rightarrow 1)$ only when $[u_{i,j}] = 0$ and $0 \leq [v_{i,j}] < v_{\text{th}}$ by collective activations in its neighbouring cells, as transition $F^{(0)}$ or $F^{(1)}$ in Fig. 6.5b. During a refractory state $([u_{i,j}] = 0$ and $v_{\text{th}} \leq [v_{i,j}] \leq (N-1)\ dv)$, the cell cannot be activated by its neighbouring cells. Once the cell is activated, the value of $[v_{i,j}]$ is increased as the discrete time step increases, see label 'B' in Fig. 6.5b. When the value of $[v_{i,j}]$ reaches its maximum value $(N-1)\ dv$, label 'C' in Fig. 6.5b, the cell is de-activated and the value of $[v_{i,j}]$ is decreased as the time step increases, label 'D' to label 'A' in Fig. 6.5b.

A CMOS cell circuit (reaction cell) is designed to implement the cellular-automaton model of the BZ reaction. Each reaction cell is placed on a two-dimensional hexagonal grid and is connected with adjacent cells through diffusive-coupling devices. The diffusive-coupling device is used to transmit a cell's state to its neighbouring cells. To ensure computational accuracy, digital circuits are used for storing a cell's states (values of $[u_{i,j}]$ and $[v_{i,j}]$). Analogue circuits are used for designing a compact cell circuit and for determining the subsequent cell states.

The cell circuit consists of a digital memory circuit and a transition-decision (TD) circuit. The value of variable U_i is stored in the TD circuit as a binary value, while the value of variable V_i is stored in a conventional N-bit up–down (UD) shift

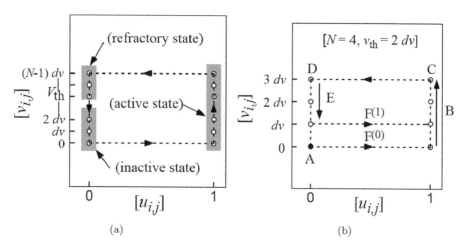

Figure 6.5: State diagram of digital cellular automaton for Oregonator: (a) discrete BZ model, reaction states are discriminated by values of $[u_{i,j}]$ and $[v_{i,j}]$ and represented by grey boxes, (b) example of state transitions for $N = 4$.

register (Fig. 6.6a) as

$$s_i^{(n)} = \begin{cases} \text{logical '1'} & (V_i \geq n \, dv), \\ \text{logical '0'} & (V_i < n \, dv), \end{cases}$$

where $n = 1, 2, \ldots, N$. The up or down operation of the shift register is determined using the value of U_i. When U_i is set at logical '1' (or '0'), the circuit operates as an up (or down) counter, which results in the value of V_i increasing (or decreasing) with the clocks (CLK).

Figure 6.6b shows the part of the TD circuit that compares the cell's states $s_i^{(1)}$ and $s_i^{(2)}$ with those of the cell's adjacent neighbours (U_i^1 to U_i^6) using analogue operations. The circuit consists of a νMOS differential amplifier acting as a multi-input, and thus a variable-threshold, comparator. Each circuit has six coupling devices consisting of the floating-gate capacitors of the νMOS transistors and accepts six adjacent inputs (U_i^1 to U_i^6) via the capacitors. Using νMOS transistors we make the cell circuit very compact compared with ordinary logic circuits that provide the same functions [259]. The output of the comparator of the ith cell (VOUT$_i$) is found using

$$\text{VOUT}_i = \Theta\left(M_i + \text{VM} - (s_i^{(1)} + s_i^{(2)} + 0.5)\right),$$

where $\Theta(\cdot)$ represents the step function, M_i the number of active cells around the ith cell, VM the logical input that determines the excitatory or oscillatory mode and $s_i^{(1),(2)}$ the output of the ith shift register. The output becomes logical '1' when $M_i \geq NV_i + 1 - \text{VM}$.

The output of the comparator is connected with the other part of the TD circuit, i.e. a cell-state memory (Fig. 6.6b). It consists of a latch and additional control

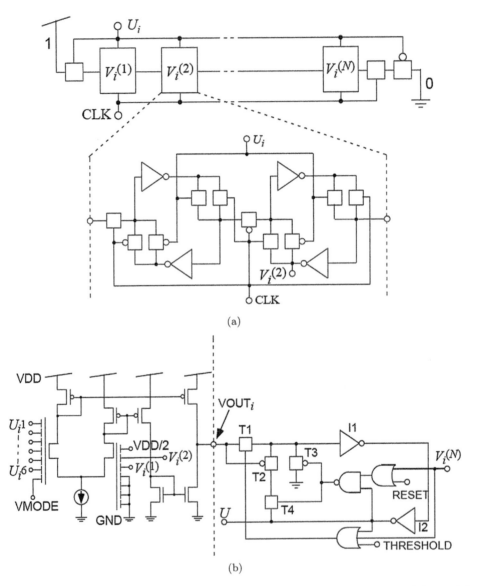

Figure 6.6: Cell circuit structure: (a) N-bit up–down shift register, (b) νMOS comparator (left) and cell-state memory (right).

circuits. The latch stores a cell state U_i according to the state diagram of the BZ model. The operation of the state diagram shown in Fig. 6.5b is implemented into the cellular-automaton circuit. The parameter of the cell transition and the refractory threshold V_{th} are controlled by the voltage input VM to the comparator and the voltage input VT that is connected to terminal $s_i^{(1)}$ or $s_i^{(2)}$ in the UD shift register within the same (ith) cell.

The latch circuit in Fig. 6.6b consists of two inverters, I1 and I2, and two transfer gates, T2 and T4. When the RESET or $s_i^{(N)}$ terminal is set to logical '1', the cell state U_i becomes logical '0' because transfer gate T3 is turned on. The latch circuit receives the output of the comparator (VOUT$_i$) through transfer gate T1. The gate T1 is turned on — the latch is ready to store the value of VOUT$_i$ — when the cell is active, U_i = logical '1', $s_i^{(N)}$ = logical '1' or VT = logical '1'.

Suppose that the cell state is $(U_i, V_i) = ($'0', '0'$)$ at $t = 0$ (label 'A' in Fig. 6.5b). When VM is set at logical '0' (excitatory mode), the cell state does not change with the clocks as long as $M_i = 0$ because VOUT$_i$ is always logical '0'. When $M_i > 0$, U_i becomes logical '1' (transition $F^{(0)}$ in Fig. 6.5b), and the value of V_i is increased by the up counter (label 'B' in Fig. 6.5b) because U_i is set to '1'. When V_i reaches its maximum value, $s_i^{(N)}$ becomes logical '1' (label 'C' in Fig. 6.5b) and the value of U_i is set to logical '0' (label 'D' in Fig. 6.5b) by T3. Then, V_i is decreased by the down counter (label 'E' in Fig. 6.5b) because U_i is set to logical '0'. If $M_i = 0$ during the down-count operation, the cell state (U_i and V_i) is settled at ('0', '0'). During the down-count operation, U_i becomes '1' again when $M_i \geq NV_i + 1$ and VT is set at logical '1'. This transition is regulated by the values of V_i because VT is connected to the terminal $s_i^{(1)}$ or $s_i^{(2)}$. If VT is connected to the terminal $s_i^{(2)}$ (or $s_i^{(1)}$), the transition $F^{(1)}$ (or $F^{(0)}$) in Fig. 6.5b can occur when $M_i \geq 2$ (or $M_i \geq 1$).

On the other hand, if VM is set at logical '1' (oscillatory mode), the cell state is not settled at $(U_i, V_i) = ($'0', '0'$)$ because the output VOUT$_i$ becomes logical '1' when $V_i = 0$. The subsequent operation is the same as the excitatory mode. Table 6.1 summarises when the cell transition occurs with respect to the parameters VM and VT. Entries with values 0, 1 and 2 indicate the number of active adjacent cells (M_i) required to implement the cell transition. Entries marked with '–' indicate that a transition does not occur for any values of M_i.

Table 6.1: Transition rules for excitatory and oscillatory operation modes with $N = 4$.

V_i	VM = '0' (excitatory mode)		VM = '1' (oscillatory mode)	
	VT = $s_i^{(1)}$	VT = $s_i^{(2)}$	VT = $s_i^{(1)}$	VT = $s_i^{(2)}$
$3dv$	–	–	–	–
$2dv$	–	–	–	–
dv	–	2	–	1
0	1	1	0	0

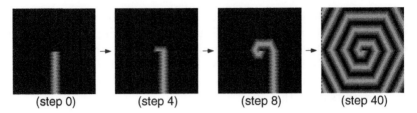

(step 0)   (step 4)   (step 8)   (step 40)

Figure 6.7: Excitatory mode-lock operations of reaction-diffusion chip.

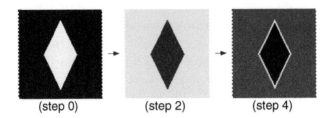

(step 0)   (step 2)   (step 4)

Figure 6.8: Oscillatory operations of reaction-diffusion chip.

Figure 6.7 shows an example operation of the cellular-automaton circuit (50 by 50 cells) in excitatory mode ($N = 3$, VM = logical '0', VT was connected to $s_i^{(1)}$). A 0.6-μm double-poly CMOS process was adopted in the simulations. Initial values were set by transmitting binary sequences serially to the digital memory circuits and TD circuits through row–column selectors. Each snapshot was constructed from the serial data representing the cell states by stopping system clocks at steps 0, 4, 8 and 40. Each cell state was represented in grey scale ($V_i = 0$: black, $V_i = (N - 1)dv$: white). In the circuit's initial state, cells adjacent to inactive cells were in a refractory period (step 0 in Fig. 6.7). The inactive cells adjacent to the light-coloured strip in Fig. 6.7, step 0, were suppressed by adjacent cells in the refractory period (cells in the light-coloured strip). The inactive cells then entered an active, inactive or refractory period, depending on the degree of the refractory condition. When the inactive cells were in an active or inactive period, the tip of the bar rotated inwards (steps 4 to 8 in Fig. 6.7), resulting in the generation of the mode lock (spiral patterns) typically observed in the BZ reaction (step 40 in Fig. 6.7). A hexagonal distortion of the propagating waves was generated by interactions between adjacent cells.

Figure 6.8 presents yet another example of a cellular-automaton circuit operating in oscillatory mode ($N = 4$, VM = logical '1', VT was connected to $s_i^{(2)}$). The diamond pattern was given to the circuit as an initial state (step 0). The initial pattern became inverted after two iterations (step 2); the edges of the initial pattern were then extracted after two more iterations (step 4), which represented the typical results of the famous demonstration (edge detection) of the BZ reaction.

Excitable-lattice cellular-automaton model and its CMOS circuit

The excitable lattice [7] is a cellular-automaton model in which each cell is regularly arranged on a two-dimensional grid. Each cell has eight neighbours and each cell state is updated in a discrete time step, as in a conventional cellular automaton. The cellular-automaton model has a multiple-valued state variable, and takes excited (EXC), refractory (REF) and resting (RES) states. Figure 6.9b shows the state diagram. The dynamics is given by

$$x^{t+1} = \begin{cases} \text{EXC}, & (x^t = \text{RES}) \wedge (s^t(x) \geq 1), \\ \text{REF}, & (x^t = \text{EXC}), \\ \text{RES}, & (x^t = \text{RES}) \wedge (s^t(x) = 0) \vee (x^t = \text{REF}), \end{cases} \tag{6.13}$$

where $s^t(x)$ is the number of excited cells among the neighbours (Fig. 6.9a), x^t and x^{t+1} representing a cell's current and subsequent states, respectively. A cell becomes excited when it is in the resting state and the number of excited cells among its neighbours is larger than 1. Then, the state changes from the excited to the refractory state in the subsequent step. In the next iteration, the state is changed from the refractory to the resting state. A cell is stable in its resting state as long as the number of excited cells among the cell's neighbours is zero.

The excitable lattice mimics the fundamental spatio-temporal properties of the BZ system, i.e. excitable wave propagation and annihilation, on the lattice. Figure 6.9c illustrates examples of the lattice's operation. Initially (step 0), two corner cells (○) are excited and the remaining cells are resting (●). At step 1, neighbouring cells of the two corner cells are also excited (○) and the states of the corner cells are changed from the excited to the refractory state (⋆). Other cells are still in their resting states (●) at this step. At step 2, resting cells adjacent to excited cells are excited (○), and the states of the corner cells are changed from the refractory to the resting state (●). Consequently, waves whose fronts are represented by groups of excited cells (○) propagate on the lattice. At this step, two waves of excitation generated by two corner cells collide at the centre of the lattice. At step 3, since the neighbouring cells of the centre cell along the direction of the excited waves are resting (●) or in the refractory state (⋆), no waves propagate in this direction and the state of the centre cell is changed from excited to refractory (⋆). This cell's state is further changed from the refractory to the resting state (●) at step 4. Consequently, the wave front disappears at this centre cell. Therefore, excited waves travelling on this lattice are annihilated at sites of their collision.

Now let us implement this excitable lattice using standard CMOS circuits. Since a cell has three states, the cell state can be represented by two-bit binary values, as shown in Fig. 6.10a. The cell dynamics is represented by a binary transition rule, Fig. 6.10b, where (q_1, q_2) and (d_1, d_2) represent a cell's current and subsequent states, respectively, and s_1 is a logical '1' when the number of excited cells in the cell's neighbourhood is larger than 1. Otherwise, s_1 is logical '0'. From Fig. 6.10b, we obtain

$$d_1 = s_1 \bar{q}_1 \bar{q}_2, \quad d_2 = q_1 \bar{q}_2. \tag{6.14}$$

Figure 6.10c illustrates the resulting cell circuit. Logic functions given by Eq. (6.14) are implemented by using NAND gates. To store the cell state (d_1, d_2), two D-type

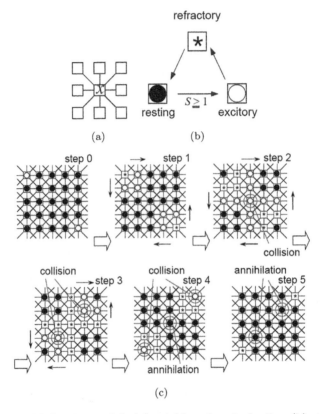

Figure 6.9: Excitable lattice model: (a) neighbourhood of cell x, (b) state-transition diagram of cell x, where S is the number of excited cells in the neighbourhood of x, (c) series of subsequent configurations of excitable lattice.

flip-flops (D-FFs) were used per cell circuit. Since Fig. 6.10b indicates that q_1 of each cell is logical '1' only when the cell is excited, the output EXC of each cell can be represented by q_1. Now, s_1 is obtained by logical OR operation of output EXC (q_1) of each cell's eight neighbours.

A prototype reaction-diffusion chip that implements 16×16 cells was fabricated using a 1.5-μm double-poly double-metal n-well CMOS process (MOSIS, vendor: AMIS). Figure 6.11 shows the layout of a cell circuit, including a photodetector (PD in the figure) and additional switching circuits for re-set and read-out operations. All circuit areas except for photodetectors were masked by top metal. The resulting cell size was $261\lambda \times 299\lambda$ ($\lambda = 0.8\ \mu$m).

Spatio-temporal patterns of the fabricated reaction-diffusion chip can be recorded by the following read-out circuitry. Each cell in the chip was located beneath each wire crossing row and column buses, and was connected to a common output wire through a transfer gate. The gate connects the cell's output to the common wire when both the row and the column buses are active. Thus, a cell's output (EXC)

MSB	LSB	State
1	0	EXC
0	1	REF
0	0	RES

(a)

s_1	$(q_1,$	$q_2)$	$(d_1,$	$d_2)$	Transition
1	0	0	1	0	RES → EXC
x	1	0	0	1	EXC → REF
x	0	1	0	0	REF → RES
0	0	0	0	0	RES → RES

(b)

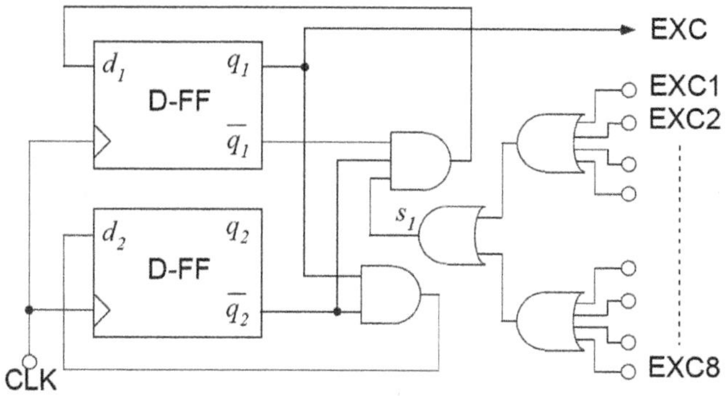

(c)

Figure 6.10: Binary cell circuit: (a) binary representation of cell states; (b) binary transition rule, where s_1 is '1' if $S \geq 1$, (q_1, q_2) is cell's current state, (d_1, d_2) is cell's subsequent state; (c) binary cell circuit.

appeared on the common output wire when the cell was selected by activating the corresponding row and column buses simultaneously. A binary stream from the common output wire can be obtained by selecting each cell sequentially. Using a conventional displaying technique, the binary stream was re-constructed on a two-dimensional display. Figure 6.12 shows the snapshots of the recorded movie. In the experiment, the supply voltage was set at 5 V, and the system clock was set at low frequency (2.5 Hz) so that 'very slow' spatio-temporal activities could be observed visually (the low frequency was used only for the visualisation, and was not the upper limit of the circuit operation). Pin-spot lights were applied to several cells at top-left and bottom-right corners of the chip. The circuit exhibited that two excitable waves of excited cells triggered by the corner cells propagated towards the centre and disappeared when they collided. This result suggests that if a more microscopic process was used and thus a large number of cells were implemented, one can observe the same complex (BZ-like) patterns.

Constructing Voronoi diagram in reaction-diffusion circuits

Here we discuss a reaction-diffusion chip modelling a chemical processor for a well-known NP-complete problem of computational geometry — computation of a Voronoi diagram. To design a silicon analogue of reaction-diffusion 'wet-ware' processors we must firstly uncover an abstract mechanic of computation in exper-

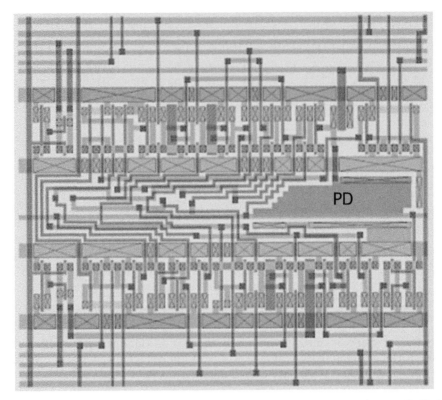

Figure 6.11: Cell layout with 1.5-μm double-poly double-metal n-well CMOS process (MOSIS, vendor: AMIS). The circuit size is $261\lambda \times 299\lambda$, $\lambda = 0.8$ μm. See also Fig. 9.26 in colour insert.

imental chemical devices, and only then transfer key rules of micro-volume state transitions to electronic schemes of elementary units of reaction-diffusion chips. A cellular automaton can be used as a device transient between real-life chemistry and silicon chips.

A reaction-diffusion cellular-automaton model $\mathcal{A}_1$ has been introduced in Chap. 2. Despite being pleasantly naturalistic, the model $\mathcal{A}_1$ is presently too complicated for LSI implementation. So, let us use a simplified model $\mathcal{A}_2$ obtained from $\mathcal{A}_1$ by assuming that every cell takes just four states, resting $\circ$, excited $+$, refractory $-$ and precipitate $\#$. A cell in $\mathcal{A}_2$ updates its state as follows:

$$
x^{t+1} = \begin{cases}
+, \text{if } x^t = \circ \text{ and } 0 < \eta_x^t \leq 4 \\
\#, \text{ if } ((x^t = \circ \text{ or } x^t = +) \text{ and } \eta_x^t > 4) \text{ or } x^t = \# \\
-, \text{ if } x^t = + \text{ and } \eta_x^t \leq 4 \\
\circ, \text{ otherwise}
\end{cases}
\tag{6.15}
$$

Model $\mathcal{A}_2$ was discussed in detail in [81], so we do not study it in detail here, but instead move straight to the silicon implementation.

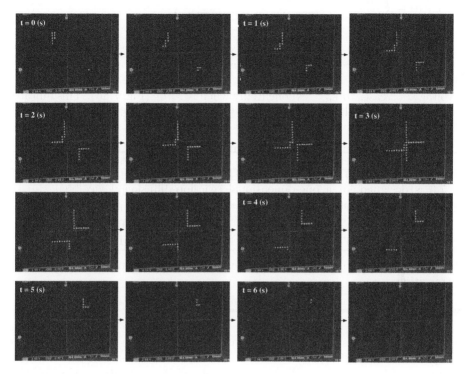

Figure 6.12: Snapshots of recorded movie obtained from fabricated reaction-diffusion chip: white dots represent excited cells, where EXC is logical '1'. See also Fig. 9.27 in colour insert.

Now, let us design electrical circuits implementing model $\mathcal{A}_2$ for Voronoi-diagram computation and skeletonisation. Each cell of the cellular automaton should be implemented on a unit (cell) circuit, which results in a two-dimensional array of the cell circuits. To increase the integration density of the cell circuits, each cell circuit has to be designed as compactly as possible. Therefore, an analogue–digital hybrid circuit for the cell is employed here.

To store the four types of cell states in model $\mathcal{A}_2$, i.e. resting ∘, excited +, refractory − and precipitate #, a two-bit static memory (two D-type flip-flop circuits: D-FFs) was used for each cell. The cell state is encoded in two-bit binary values, as shown in Table 6.2. A cell's state encoded in the memory is updated according to the current state of the cell and the number of excited cells amongst its neighbours.

The cell transition (6.15) is represented by a transition table shown in Table 6.3. In the table, q_1 and q_2 represent the present state of a cell in two-bit binary values (see Table 6.2 for translation), s_n is a binary value that becomes logical '1' when the number of surrounding excited cell exceeds n (otherwise s_n is logical '0') and d_1 and d_2 represent the subsequent state of the cell after the transition. If a cell is in its excited state $[(q_1, q_2) = ('1', '1')]$ and the number of excited cells among

Cell state	q_1	q_2
Resting (∘)	'0'	'0'
Refractory (−)	'0'	'1'
Precipitate (#)	'1'	'0'
Excited (+)	'1'	'1'

Table 6.2: Translation table of the cell states.

Number of excited neighbours		Current state		Subsequent state		Transition
> 0 (s_0)	> 4 (s_4)	q_1	q_2	d_1	d_2	$t \rightarrow t + \Delta t$
×	'1'	'1'	'1'	'1'	'0'	$+ \rightarrow \#$
×	'0'	'1'	'1'	'0'	'1'	$+ \rightarrow -$
×	'1'	'0'	'0'	'1'	'0'	$\circ \rightarrow \#$
'1'	'0'	'0'	'0'	'1'	'1'	$\circ \rightarrow +$

Table 6.3: Translation table of the cell states: '×' represents conditions not affecting the transition.

the neighbours exceeds 4 (s_4 = '1'), the subsequent cell state is set at precipitate $[(d_1, d_2) = ('1', '0')]$. If $(q_1, q_2) = ('1', '1')$ and s_4 = '0', the subsequent cell state is refractory $[(d_1, d_2) = ('0', '1')]$. If the current cell is resting $[(q_1, q_2) = ('0', '0')]$ and s_4 = '1', the subsequent cell state is precipitate $[(d_1, d_2) = ('1', '0')]$. If $(q_1, q_2) = ('0', '0')]$ and s_0 = '1', the subsequent cell state is excited $[(d_1, d_2) = ('1', '1')]$. Otherwise, no transition occurs. This indicates that the transition occurs only when the current cell state matches the four cases above, i.e. when $d_1 + d_2 =$ logical '1'.

Now, we can design a logic circuit that determines the subsequent state of a cell. We call the circuit a 'transition-decision circuit' (TD circuit). From Table 6.3, we easily obtain

$$
\begin{aligned}
d_1 &= s_4 q_1 q_2 + s_4 \bar{q}_1 \bar{q}_2 + s_0 \bar{s}_4 \bar{q}_1 \bar{q}_2 \\
&= s_4 q_1 q_2 + \bar{q}_1 \bar{q}_2 (s_0 + s_4), \\
d_2 &= \bar{s}_4 (q_1 q_2 + s_0 \bar{q}_1 \bar{q}_2).
\end{aligned}
$$

Figure 6.13 shows a basic construction of the TD circuit that is implemented by conventional logic gates. A cell state is changed by the number of its surrounding 'excited' cells. In order to give the information about excitation to its neighbours, the TD circuit produces a signal EXC_i, where EXC_i becomes logical '1' only when $(q_1, q_2) = ('1', '1')$ (otherwise, EXC_i is '0'). Here, s_n is a binary value that represents whether the number of surrounding excited cells exceeds n or not. To obtain s_n signals, the number of surrounding excited cells must be counted. Model $\mathcal{A}_2$ requires both s_0 and s_4 signals because of the conditions $\eta_x^t > 0$, $\eta_x^t \leq 4$ and $\eta_x^t > 4$.

To detect them, a current-mode analogue circuit can be used (Fig. 6.14), instead of digital counters. The circuit consists of a conventional digital to analogue (DA) converter and current-mode comparator. In the circuit, dimensions of all MOS transistors are identical. The DA converter receives EXC_i signals from its eight

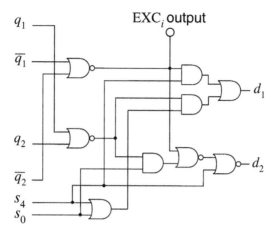

Figure 6.13: Transition-decision circuit.

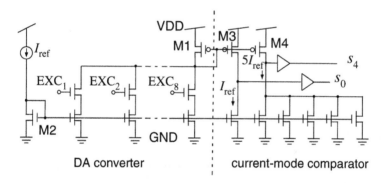

Figure 6.14: DA converter and current-mode comparator.

neighbours (EXC_1, EXC_2, ..., EXC_8). The current of M1 is obtained by multiplying I_{ref} by the number of EXC_is = '1' because the current of M2 ($= I_{ref}$) is copied to all nMOS transistors whose source terminal is connected to the ground, and nMOS transistors receiving EXC_is act as current switches. The current of M1 is copied to M3 and M4. If the current of M3 is larger than that of M2 ($= I_{ref}$), s_0 becomes '1', and if the current of M4 is five times larger than that of M2 ($= 5I_{ref}$), s_4 becomes '1'. This circuit is quite compact when compared with conventional digital counters. Therefore, it is useful for the large-scale integration of reaction-diffusion cell circuits.

The outputs of the TD circuit (d_1, d_2) are directly connected to the memory circuits (D-FFs). Now, assume that the current cell state stored in the D-FFs (q_1, q_2) satisfies one of the conditions listed in Table 6.3. A clock signal ϕ is given to the D-FFs through a clock control circuit that provides ϕ to the memory circuits only when $d_1 + d_2 =$ '1'. Thus, the D-FFs capture input data (d_1, d_2) and update

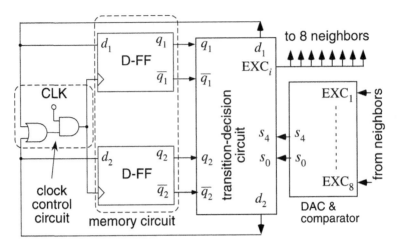

Figure 6.15: Cell circuit for model $\mathcal{A}_2$.

the current state (q_1, q_2) to (d_1, d_2). Figure 6.15 shows the whole construction of a cell circuit that consists of a TD circuit, DA converter, current-mode comparator, memory circuits and clock control circuit.

A prototype cellular-automaton reaction-diffusion chip that implements 16×16 cells was fabricated using a 0.5-μm double-poly triple-metal n-well CMOS process (MOSIS, vendor: AMIS). The chip can accept parallel optical inputs, which is very useful for parallel image-processing applications. Figure 6.16 shows the layout of a cell circuit, including a photodetector (simple pn junction between p-substrate and n-diffusion) and additional switching circuits for re-set and read-out operations. All circuit areas except for photodetectors were masked by top metal. The resulting cell size was $297\lambda \times 338\lambda$ ($\lambda = 0.3\ \mu$m).

Spatio-temporal patterns of the fabricated reaction-diffusion chip were recorded by the following read-out circuitry. Figure 6.17 shows the experimental setup. Initial patterns were given by optical images through a microscope, where bright areas were set as 'excited' states and dark areas were set as 'resting' states. In the fabricated reaction-diffusion chip, each cell circuit was located beneath a wire crossing row- and column-address buses, and was connected to two common output wires through a transfer gate. This gate connects the cell's output to the common wires when both the row and column buses are active. Thus, a cell's outputs (d_1 and d_2) appeared on the common output wires when the cell was selected by activating the corresponding row and column buses simultaneously. One could obtain a binary stream from the common output wires by selecting each cell sequentially. Using a conventional displaying technique with an external address encoder that produces both analogue and digital addressing signals, the binary streams of d_1 and d_2 were re-constructed on analogue oscilloscopes operating in x–y Lissajous modes where z [brightness at (x, y)] accepts d_1 or d_2. Therefore, spatio-temporal patterns of d_1 and d_2 can be observed as brightness on the display.

Figure 6.18 shows snapshots of the recorded display. Each bright spot represents

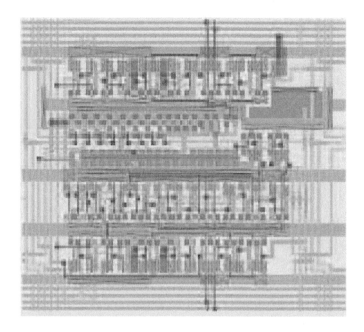

Figure 6.16: Cell (fabricated in MOSIS AMIS 0.5-μm 2-poly-3-metal SCMOS) layout including a photodetector, TD circuit, DA converter, current-mode comparator, two-bit memory circuits and clock control circuit. The circuit's size is $297\lambda \times 338\lambda$.

a cell where d_1 or d_2 is logical '1'. In the experiment, the supply voltage was set at 5 V, and the system clock was set at low frequency (2.5 Hz) so that 'very slow' spatio-temporal activities could be observed visually (the low frequency was used only for the visualisation, and was not the upper limit of the circuit operation). First, bar-shape lights were applied to cells at the left- and right-hand sides of the chip. When tested, the circuit exhibited the expected results, i.e. two excitation waves triggered by the excited cells at the edges propagated towards the centre and precipitated when they collided.

Figure 6.19 shows more snapshots with different initial conditions. In this experiment, pin-spot lights were applied to two cells at the top-left and bottom-right corners of the chip. Two excitable waves were generated by these trigger inputs. When they collided, diagonal cells were precipitated, as expected. These results indicate that the fabricated chip could produce a Voronoi diagram, although the spatial resolution is low in the prototype chip.

Figures 6.20 and 6.21 show examples of a skeletonisation operation — computation of an internal Voronoi diagram — of 'T'- and '+'-shaped images, respectively. To input the initial images to the reaction-diffusion chip, a glass mask was made where 'T' and '+' areas are exactly masked off. Therefore, cells under the 'T' and '+' masks were initially resting and the remaining cells were initially excited. At its equilibrium, skeletons of the 'T' and '+' shapes were successfully obtained.

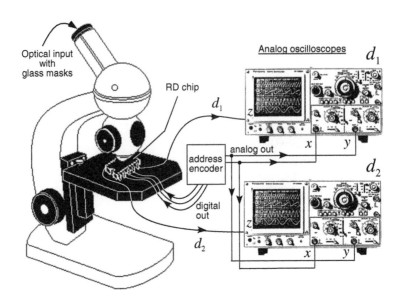

Figure 6.17: Experimental setup.

6.2.2 A quadrilateral-object composer for binary images using reaction-diffusion-based cellular automata

Here, we discuss the design of the image processing using reaction-diffusion processors with complementary object reduction and enlargement operations. We can expand this idea into a cellular automaton that detects quadrilateral objects, such as box areas filled with the same pixel values, from binary images. Then, we study serial and parallel architectures to implement the system on LSIs.

Chemical reactions are formulated in terms of temporal differences in the concentrations of the chemical reagents, at specific points in the reaction space. If the spatial distribution of the chemicals is not uniform, the reagents will diffuse according to the reagents' concentration gradients. This diffusive-reaction system, with multiple chemical reagents, is the primary component of chemical processors today. According to a basic reaction-diffusion system model, it is clear that chemical processors can be regarded as two-dimensional arrays of excitable units. In the BZ chemical processor, generally, three circulative states are introduced according to each unit's dynamical phase, i.e. resting state (depletion in the Br^- ion), excited state (autocatalytic increase in the $HBrO_2$ ion, oxidation of the catalyst) and refractory state (depletion in the Br^- ion, reduction of the catalyst). When a unit is resting, it is easily excited by the neighbouring units or external stimuli. Then, it turns to the refractory state. Being in the refractory state, the unit cannot be excited even with external stimuli.

Figure 6.22 illustrates the basic ideas of object scaling in an excitable reaction-

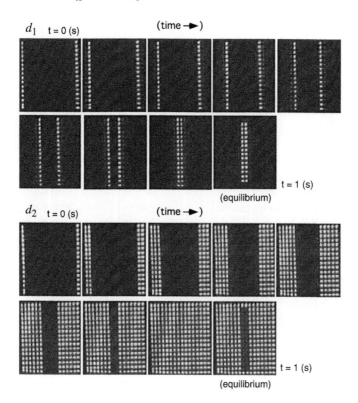

Figure 6.18: Spatio-temporal patterns on the fabricated reaction-diffusion chip (Voronoi-diagram operation with two bars).

diffusion processor. Here the refractory period for each unit is assumed to be infinity. When an object (background) is defined by the black (white) area, on which each unit is in the resting (refractory) state, the black area will be reduced as the time increases, because units in resting states can easily be excited by the surrounding excited units that exist along the boundary between the object and background, as shown in Fig. 6.22a. This process is called 'erosion' in the chemical processor. Conversely, a 'dilation' process occurs when an object (background) is defined by a white (black) area, on which each unit is in the resting (refractory) state, as shown in Fig. 6.22b.

Let us consider the task of detection of some particular object among multiple objects in an incident image, in order of the object size. Figure 6.23 shows a possible sequential algorithm for a reaction-diffusion processor. First, the initial input (i) in Fig. 6.23 is incident upon the processor. The input includes two circular objects, where each object is defined by a black (resting) area. As the time increases, the size of each object is reduced on the processor due to erosion. Now assume that the smaller circular object is about to disappear at $t = \Delta t_1$, step (ii) in Fig. 6.23, while the larger one is about to disappear at $t = \Delta t_1 + \Delta t_2$, step (iii) in Fig. 6.23. At

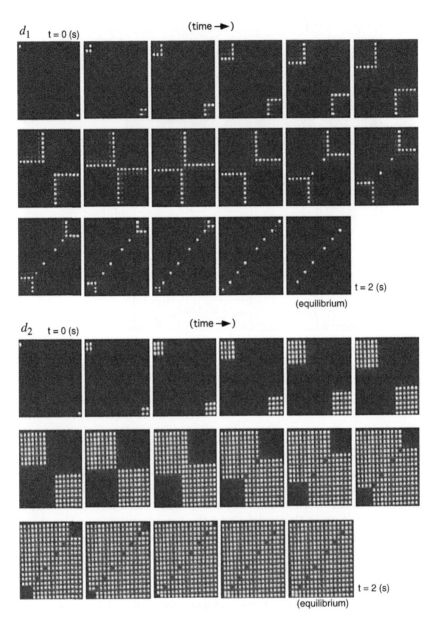

Figure 6.19: Spatio-temporal patterns on the fabricated reaction-diffusion chip (Voronoi-diagram operation with two points).

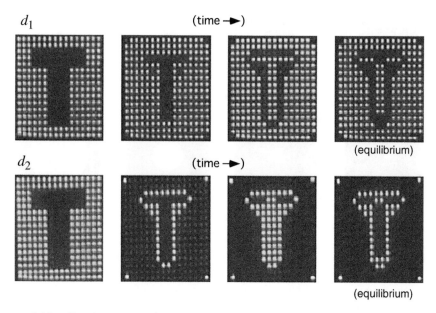

Figure 6.20: Spatio-temporal patterns on the fabricated reaction-diffusion chip (skeletonisation operation with 'T' shape).

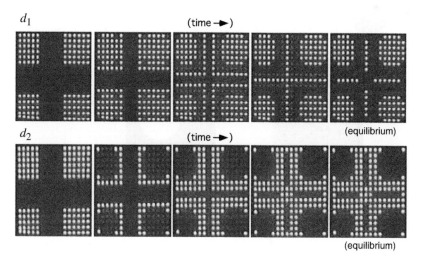

Figure 6.21: Spatio-temporal patterns on the fabricated reaction-diffusion chip (skeletonisation operation with '+' shape).

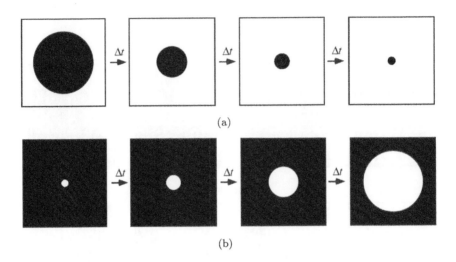

(a)

(b)

Figure 6.22: Image processing (scaling) in excitable chemical processors: (a) reduction of a circle (erosion of black area), where a circular domain of micro-volumes in the resting state is reduced with time; (b) enlargement of a point (dilation of white area), where a point-wise domain of excited and refractory micro-volumes increases with time.

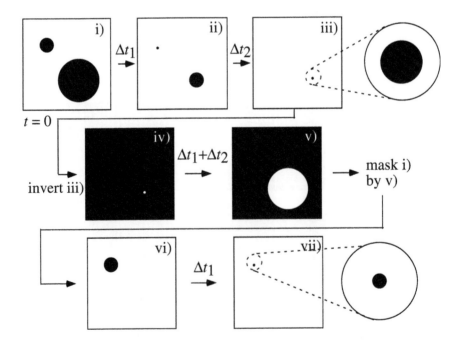

Figure 6.23: Object scaling in a chemical processor. An excitable chemical processor performs erosion and dilation with very high (but molecular-level) resolution.

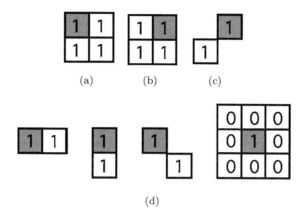

Figure 6.24: Basic templates for reaction-diffusion cellular automaton implementing image-processing operations: (a) erosion I, (b) erosion II and (c) $xy2$ filter; (d) shape-estimation templates for erosion I, templates x, y, $xy1$ and box, respectively.

step (iii), the shape of the circle is left unchanged, as long as the resolution of the unit array is higher than the display resolution. This is thus the smallest and most efficient representation of the circular object in the reaction-diffusion processor. So, we can consider the reaction-diffusion processor to be a video memory that has very high – of molecular level — resolution. Consequently, at step (iii), only one circular object (the larger one) exists on the reaction-diffusion processor, which indicates that one can easily pick up the object by just memorising the coordination of the small circle (point) on the processor as well as $\Delta t_1 + \Delta t_2$. To restore the original object, we simply invert the image — step (iii) in Fig. 6.23, which results in the image shown at step (iv). At this point, the reaction-diffusion processor automatically starts performing dilation. When $\Delta t_1 + \Delta t_2$ have passed since the start of dilation, the circular object is restored in the processor to its original size, see step (v) in Fig. 6.23. To detect the small circular object in (i), we perform further erosions with the initial inputs (vi) obtained from the mask of (i) by (v). When Δt_1 time steps have passed since the start of erosion, we see the same results, i.e. the most efficient representation of the small circular object, as seen for the larger circular object (iii). So, by memorising the coordinates of the point on the processor and Δt_1, we can pick up the small object. For a large number of objects in an incident image, by just repeating the process above, one can obtain a list of both the coordinates and the time, which corresponds to a compressed representation of all of the circular objects.

Based on the memory model of chemical computers described in the preceding section, we construct a sequential cellular-automaton system that is suitable for LSI implementation. For cellular-automaton implementation, we have to consider that a cellular automaton's spatial resolution is not 'infinite', unlike the chemical processors. For instance, during erosion, the original shapes of the objects will be changed significantly due to the finite resolution of cellular-automaton operations. Thus, here we limit the shape of the objects to 'quadrilateral objects' by using seven asymmetric templates and one symmetric template. Figure 6.24 shows all the templates, where the centre cell is represented by a grey box.

Figure 6.25 shows example operations of quadrilateral-object extraction on a sequential cellular automaton. First, a binary input image is given to the cellular automaton, step 0 in Fig. 6.25a. Then, the cellular automaton implements erosion repeatedly with the 'erosion I' template shown in Fig. 6.24a. At step 3 in Fig. 6.25a, two objects, both represented by clusters of '1', are about to disappear. Subsequently, one can estimate the smallest object type in the reduced image (step 3) by using four filters, i.e. 'x', 'y', '$xy1$' and a 'box', as shown in Fig. 6.24c. In this example, the '$xy1$' and 'box' templates can be matched, Fig. 6.25b. If no templates can be matched, the smallest object must match template '$xy2$', as shown in Fig. 6.24b. In this case, we have to re-start the erosion process using the 'erosion II' template shown in Fig. 6.24b.

Once the smallest object type is determined, we can estimate the maximum 'box' area that can be stored within the original object from the number of steps in the applied template, as shown in Fig. 6.25c. At this point, we must memorise the object's box information. It should be noticed that this box is the largest box in the initial image, step 0 in Fig. 6.25a. To find the second largest box, the initial image is masked with this box, either as a result of dilation, or by simply recalling the box information, as shown in Fig. 6.25d.

Now, let us design a cellular-automaton processor that implements the operations of quadrilateral-object extraction. Figure 6.26 shows a block diagram of the whole cellular-automaton system, which consists of four blocks: (i) a two-dimensional cell-circuit array, (ii) flag collection circuits, (iii) template-matching circuits and (iv) a controller.

The cell-circuit array acts as an erosion processor for extracting quadrilateral objects, and a dilation processor for re-constructing objects that have been removed from the initial image. In the above operations, all cells work synchronously as a cellular automaton. The controller manages these blocks and creates quadrilateral-object information from the erosion time steps and the results of the template matching.

Flag collection circuits, which consist of OR circuits, determine the condition when all cell states are at a logical '0'. The template-matching circuits, with three line buffers and template circuits, perform pattern matching on each row of the cell-circuit array at each time step. Because a matching process with 'x', 'y', '$xy1$', '$xy2$' and 'box' templates is performed only once during the output period, we have separated the erosion (dilation) cellular-automaton processor from the template-matching cellular-automaton circuits with the line buffer.

Figure 6.27 shows a cell circuit with four subcircuits: (i) cross-bar switches for changing the input signals, (ii) AND and OR circuits for erosion and dilation templates, (iii) multiplexers for changing the direction of the data transfer and (iv) cell-state-memory (flip-flop) circuits. By using signals from the controller (ϕ_1, ϕ_2, sel-sig0 → sel-sig2), these subcircuits can operate in three modes: (i) retrieval of the initial image mode, (ii) an operating erosion and dilation cellular-automaton mode and (iii) a data-output mode.

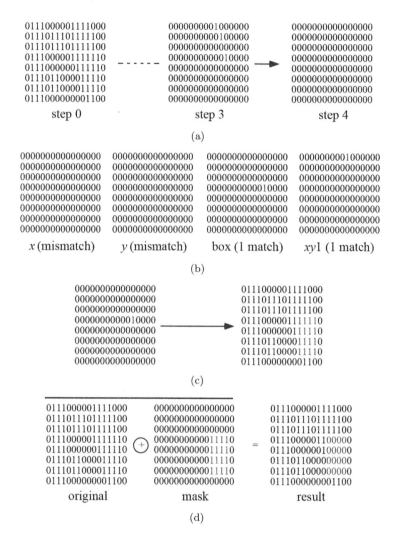

Figure 6.25: Example operations of proposed cellular automaton. (a) Initial erosion: convolution of initial values with template 'erosion I' until all cell states take value 0; (b) shape detection: convolution of the image obtained at step 3 of initial erosion with templates 'x', 'y', 'box' and '$xy1$'; (c) maximum quadrilateral-object composition: three steps with 'erosion I' are shown, estimation of original rectangular unit from the matched box in (b); (d) masking original image with the estimated box in (c).

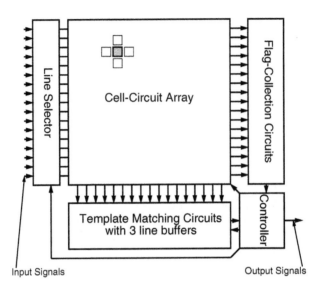

Figure 6.26: Block diagram of a quadrilateral-object composer for binary images with reaction-diffusion cellular automata.

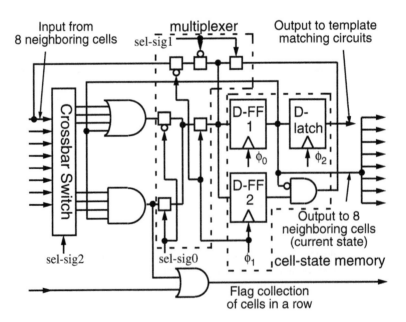

Figure 6.27: Cell circuit with four subcircuits.

When sel-sig1 is set to logical '0', ϕ_0 supplies a CLK signal that starts from the logical '0', and ϕ_1 is set to a logical '0'. The cell circuits for each row also act as shift registers. If the initial data is put into the left-hand side of the cell arrays in Fig. 6.26, the memory circuits of all the cells can obtain the initial data sequentially. After this operation, when ϕ_0 and ϕ_1 are set to logical '1', the initial data can be stored. When sel-sig0 is set to '0' (or '1'), and ϕ_0 receives a CLK signal, the cell circuit performs erosion (or dilation) on the cellular-automaton cell. To obtain an instant when the cell states are all '0', the controller recognises each subsequent cell state as a flag at each time step. By setting ϕ_0 to '0', the current cell state is stored in the D-latch for output to the template circuits. We use sel-sig1 and ϕ_1 for logical operations between D-FF1 and D-FF2. This operation removes objects from the initial image, while sel-sig2 controls the cross-bar switches that can change the template window.

The system has five operation periods. In the first period, the initial image is loaded in the cell array by shift registers. In the second period, the system erodes the cell array and decides whether the subsequent state of all cells is '0'. When all cell states are '0', the third period starts. The system reads the current states of all cells at each line and performs template matching. From the results of this operation and the erosion times, information about the objects can be obtained. During the fourth period, the cell array dilates the resultant states to restore the object size. Then, the objects are removed from the initial image by logical operations between the memory circuits of the cell. If the system cannot extract any objects by repeating the first to fourth periods, the fifth period is processed once and the above four-period sequence is repeated. If the system still cannot extract any objects, the system operations are completed.

We designed a LSI layout of a single-cell circuit using a 0.25-μm 1-poly 5-metal n-well scalable CMOS process (MOSIS, vendor: TSMC). Figure 6.28 shows the layout, including the logic circuits and pass gates in Fig. 6.27, as well as a photodetector (simple pn junction between p-substrate and n-diffusion) and additional switching circuits for re-set and read-out operations. All circuit areas except for photodetectors were masked by top metal. The resulting cell size was approximately 15 by 15 μm^2 ($125\lambda \times 125\lambda$, $\lambda = 0.12$ μm).

Now, let us explore the operations of the chip for binary images, by using Verilog HDL simulations [70]. In the simulations, all pass gates in Fig. 6.27 were replaced with conventional logic circuits. The operation flow is as follows:

1. Transfer the input binary image to the cell-circuit array and repeatedly perform 'erosion I'. The number of erosion processes is counted, and the number at which all the cell values become zero ($\equiv s_c$) is memorised by the peripheral controller. Note that D-FF1 stores the cell values just before the moment when all the cell values become zero.

2. Transfer the cell values in rows of three to the template-matching circuits and sequentially perform x-, y-, $xy1$- and box-template matching. The number of matching processes is counted, and the number at which all the cell values become zero ($\equiv x_c$, y_c, $xy1_c$ and box$_c$) is stored in the matching circuits.

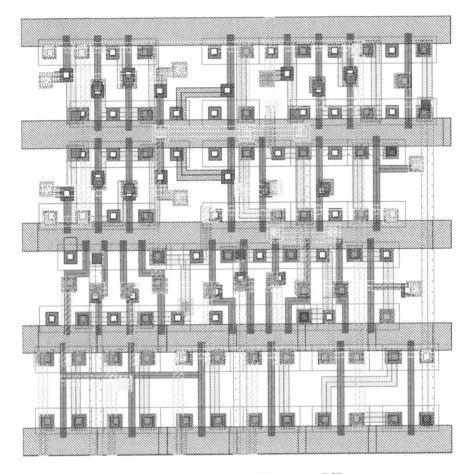

Figure 6.28: Single-cell layout on LSI.

3. When the row values do not match the above templates, the cell-circuit array starts the erosion process again by repeating template 'erosion II'. The number of the processes is counted, and the number at which all the cell values become zero ($\equiv$ is$_c$) is memorised by the peripheral controller. Then, perform $xy2$-template matching, as described in the preceding process (3).

4. Generate a quadrilateral mask by (i) the position of the resting active cell ('1') and (ii) the values of x_c, y_c, $xy1_t$ and box$_c$, and create subsequent binary input images, masked by this quadrilateral box.

5. Repeat tasks (1)–(4) until s$_c$ or is$_c$ becomes zero.

6. Finally, produce the quadrilateral objects (vertical and horizontal rectangles and squares) from the resting active-cell position and the values of x_c, y_c and box$_c$.

(a)

(b)

Figure 6.29: Simulation results for 194 × 48 image: (a) original quantised input image given to cellular-automaton LSI, (b) detected quadrilateral objects. The maximum boxes are detected in order.

Figure 6.29 shows the result for an input bit-map image. The input image that consisted of 194×48 pixels (Fig. 6.29a) was composed of 256 quadrilateral objects, as shown in Fig. 6.29b. The bit-map image occupied 9312 bits (1164 bytes) of memory, while the objects used 1024 bytes (8-bit address of four corners × 256 boxes). The output contains only rectangles and squares, since slanted objects were removed during the last process (6) described above. Also, we examined object extraction for a natural image shown in Fig. 6.30a. The image was quantised, and given to the cellular-automaton LSI. Figure 6.30b shows the results. The maximum boxes were correctly detected in order, as predicted. The input bit-map image that consisted of 181 × 238 pixels was composed of 1020 quadrilateral objects. The bit-map image occupied 43 078 bits (5385 bytes) of memory, while the objects used 4080 bytes (8-bit address of four corners × 1020 boxes).

One of the most important application targets for the chip is a CAD system for VLSIs. Conventional VLSI CAD tools use polygons to represent device structures. However, recent VLSIs include not only polygon patterns but also graphical patterns, consisting of a large number of dots, usually imported from image files such as JPEGs, to implement complex analogue structures. In the mask-manufacturing process, exposing a large number of dot patterns is quite a time-consuming task. Recently, electron beam (EB) lithography systems that can expose wide areas through a quadrilateral window have been produced on a commercial basis. The proposed LSI can produce efficient stream files from binary image files that can easily be handled by the new EB systems, by developing simple software that converts the box format, produced by the proposed LSI, to a conventional stream format.

(a)   (b)

Figure 6.30: Simulation results for 181×238 image; (a) original grey-scale image, (b) resultant image with detected quadrilateral objects; the maximum boxes are detected in order.

6.3 Analogue reaction-diffusion chips

Analogue implementation of non-linear systems is usually troublesome because the non-linearity required for the system must be realised by the device's intrinsic non-linear characteristics. In the case of a silicon CMOS LSI, these are exponential functions. Combining the functions, we can obtain various non-linear functions. However, we are always confronted with many practical problems in analogue LSIs, e.g. their limited dynamic range, thermal fluctuations, device mismatch, electromagnetic interference and so on. Consequently, the number of non-linear systems that can naturally be implemented in analogue circuits is limited. In this section, we study how to utilise a device's intrinsic non-linear characteristics to construct analogue reaction-diffusion-based LSI systems.

6.3.1 Analogue reaction-diffusion chip with Oregonator

Here we study an analogue cell that is qualitatively equivalent to the Oregonator introduced in Sect. 6.1. First, let us consider the following dynamics of a cell:

$$\frac{\mathrm{d}[x_1]}{\mathrm{d}t} = \frac{1}{\tau}\Big(-[x_1] + f([x_1] - [x_2], \; \beta_1)\Big), \tag{6.16}$$

$$\frac{\mathrm{d}[x_2]}{\mathrm{d}t} = -[x_2] + f([x_1] - \theta, \; \beta_2), \tag{6.17}$$

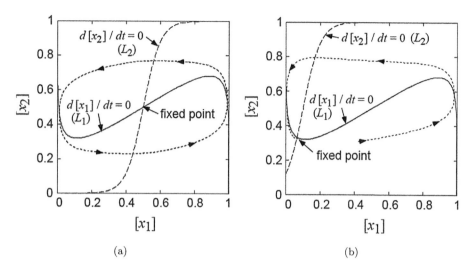

Figure 6.31: Nullclines and trajectories of analogue cell operating in (a) oscillatory mode ($\theta = 0.5$), limit-cycle oscillation, and (b) excitatory mode ($\theta = 0.14$), excitation.

where $f(\cdot)$ represents a sigmoid function defined by

$$f(x, \beta) = \frac{1 + \tanh \beta x}{2}. \tag{6.18}$$

The cell dynamics is designed so that the shapes of nullcline and velocity flows ($[\dot{x}_1], [\dot{x}_2]$) are qualitatively equivalent to that of the Oregonator, where the cubic nullcline (l_1 in Fig. 6.2) is approximated by a nullcline of Eq. (6.16) as

$$[x_2] = [x_1] - \beta_1^{-1} \tanh^{-1}(2[x_1] - 1) \ (\equiv L_1), \tag{6.19}$$

while the linear nullcline (l_2 in Fig. 6.2) is approximated by a nullcline of Eq. (6.17) as

$$[x_2] = f([x_1] - \theta, \beta_2) \quad (\equiv L_2). \tag{6.20}$$

An analogue cell, whose dynamics is described by Eqs. (6.16) and (6.17), is very suitable for implementation in analogue LSIs because the sigmoid function can easily be implemented on LSIs by using differential-pair circuits.

The cell exhibits qualitatively equivalent behaviour to the Oregonator, as we can see from Fig. 6.31. The values of the parameters are $\tau^{-1} = 10$, $\beta_1 = 5$ and $\beta_2 = 10$. When $\theta = 0.5$, the fixed point exists on nullcline L_1 where $d[x_2]/d[x_1] > 0$, and the system exhibits limit-cycle oscillations (Fig. 6.31a). The system exhibits excitatory behaviour (Fig. 6.31b), on the other hand, when the fixed point exists on nullcline L_2 where $d[x_2]/d[x_1] < 0$ (Fig. 6.31b).

Let us introduce cell dynamics into the basic reaction-diffusion model for the purpose of constructing a two-dimensional analogue cellular-automaton system.

The dynamics of the cellular automaton is obtained by substituting the right-hand terms of Eqs. (6.16) and (6.17) for the non-linear reactive functions $f_u(\cdot)$ and $f_v(\cdot)$ in Eqs. (6.10) and (6.11), and by transforming the system variables ($[x_1] \rightarrow [u_{i,j}]$ and $[x_2] \rightarrow [v_{i,j}]$). The resulting dynamics of an analogue cell is

$$\frac{d[u_{i,j}]}{dt} = \frac{1}{\tau}\left(-[u_{i,j}] + f([u_{i,j}] - [v_{i,j}], \beta_1)\right) + g_{i,j}^u, \qquad (6.21)$$

$$\frac{d[v_{i,j}]}{dt} = -[v_{i,j}] + f([u_{i,j}] - \theta, \beta_2) + g_{i,j}^v, \qquad (6.22)$$

where $g_{i,j}^u$ and $g_{i,j}^v$ represent external inputs to the cell (interactions between cell and neighbouring cells) defined in Sect. 6.1.

Figure 6.32 shows the spatio-temporal activities of the analogue cellular automaton given by Eqs. (6.21) and (6.22) with 50×50 cells ($\beta_1 = 5$, $\beta_2 = 10$, $h = 0.01$ and $D_v = 0$) where the values of $v_{i,j}$ are represented on a grey scale ($v_{i,j} = 0$: black, $v_{i,j} = 1$: white). The von Neumann boundary condition was applied at the side of the square reaction space. When $\tau^{-1} = 10^2$ and $\theta = 0.14$, at which the cell exhibits excitatory behaviour, the two-dimensional cellular-automaton system produced spiral patterns (Fig. 6.32a), as can be observed in the basic reaction-diffusion system with the Oregonators (Fig. 6.4). In the simulation, diffusion coefficient D_u was set at 5×10^{-4}, and the initial states of cells were set at the same states as in Fig. 6.4. The results indicate that the analogue cellular automaton is qualitatively equivalent to the basic reaction-diffusion system with the Oregonators, since the excitatory properties of the analogue cells are inherently the same as those of the Oregonators.

Figure 6.32b shows the dynamic behaviour of the analogue cellular automaton with $D_u = 10^{-3}$ and $\theta = 0.5$, at which the cell exhibits oscillatory behaviour. The initial values of the cells were randomly chosen as $[u_{i,j}] = \text{RAND}[0, 1]$ and $[v_{i,j}] = \text{RAND}[0, 1]$, where $\text{RAND}[0, 1]$ represents random real numbers between 0 and 1. The cellular automaton produced two-dimensional phase-lagged stable synchronous patterns called 'mode lock', due to weak coupling between the neighbouring cells. When $D_u > 10^{-3}$, all cells exhibited synchronous oscillation (no spatial patterns were produced).

Now let us simplify the analogue cell for LSI implementation and test the analogue cell's suitability for constructing analogue LSIs. The resulting cell circuit consists of just a capacitor and two operational-transconductance amplifiers (OTAs). Consequently, the circuit can easily be implemented on silicon LSIs using conventional CMOS technology.

When the rate constant of Eq. (6.16) is much larger than that of Eq. (6.17), the differential term of Eq. (6.16) can be neglected ($\tau \ll 1$), as explained in Sect. 6.1. On the other hand, Eq. (6.17) with $\beta_2 \rightarrow \infty$ forces the value of variable $[x_2]$ to be 0 when $[x_1] \leq \theta$ and $[x_2] \rightarrow 1$ when $[x_1] > \theta$. Thus, if variable $[x_2]$ is forced to have a value within the range $[0, 1]$, the temporal difference in Eq. (6.17) can approximately be represented by binary values. Consequently, we can obtain a new

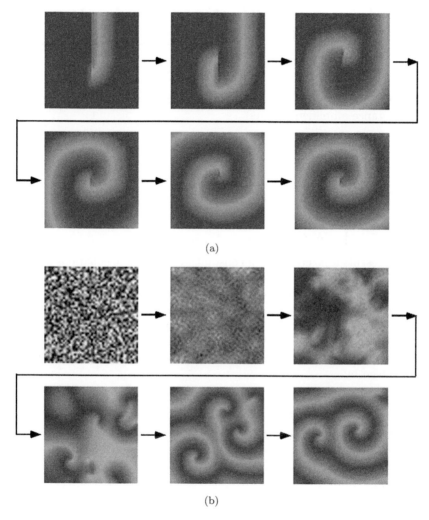

Figure 6.32: Numerical results for reaction-diffusion system with (a) excited cells, $D_u = 5 \times 10^{-4}$, $\theta = 0.14$, and (b) oscillatory cells, $D_u = 10^{-3}$, $\theta = 0.5$.

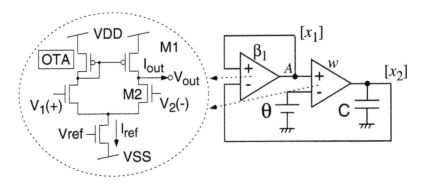

Figure 6.33: Analogue cell circuit consisting of single capacitor and two operational-transconductance amplifiers (OTAs).

dynamical equation from Eqs. (6.16) and (6.17) as

$$[x_1] = f([x_1] - [x_2], \beta_1) \tag{6.23}$$

$$\frac{d[x_2]}{dt} = \begin{cases} w & (if \ [x_1] > \theta), \\ -w & (else), \end{cases} \tag{6.24}$$

where w represents a positive, small constant. We use these equations to design an analogue cell circuit that could easily be implemented on LSIs.

Figure 6.33 illustrates a circuit diagram for the analogue cell that consists of a single capacitor and two operational transconductance amplifiers (OTAs) labelled β_1 and w. The construction of a single OTA is represented by the dashed ellipse in Fig. 6.33. It transduces differential input voltages $(V_1 - V_2)$ to current $[-I_{ref}:I_{ref}]$. When all the transistors are operated in their subthreshold region, the output current is given by

$$I_{out} = I_{ref} \tanh \frac{\kappa(V_1 - V_2)}{2V_T}, \tag{6.25}$$

where κ represents the effectiveness of the gate potential and $V_T \equiv kT/q = 26$ mV at room temperature (k is Boltzmann's constant, T the temperature and q the charge of an electron) [165]. This OTA can also be used in the open-circuit mode as a differential-voltage amplifier. Its open-circuit voltage gain at $V_1 \approx V_2$ is given by

$$A = \frac{\kappa V_0}{2V_T}, \tag{6.26}$$

where V_0 represents the early voltage of output transistors M1 and M2. Since the output voltage of the OTA is saturated to its supply voltage VDD (or VSS) when $V_1 > V_2$ (or $V_1 > V_2$), its input–output characteristics are approximately represented by a piece-wise linear function as

$$V_{out}(x) \equiv F(x) = \begin{cases} VDD & (x \gg 0), \\ A\,x & (x \approx 0), \\ VSS & (x \ll 0), \end{cases} \tag{6.27}$$

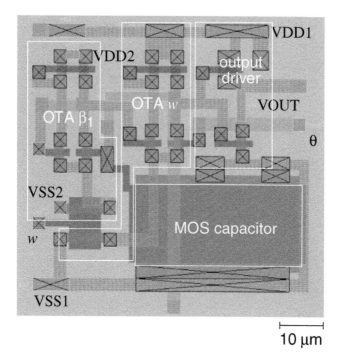

Figure 6.34: Layout for analogue cell designed with 1.5-μm CMOS rule (cell size is 70×70 μm^2).

where $x \equiv V_1 - V_2$.

From Eqs. (6.25) and (6.27), we obtain the dynamics of the cell circuit as

$$[x_1] = F([x_1] - [x_2]), \tag{6.28}$$

$$C\frac{\mathrm{d}[x_2]}{\mathrm{d}t} = I_{\mathrm{ref}} \tanh \frac{\kappa([x_1] - \theta)}{2V_T}, \tag{6.29}$$

where F is the piece-wise linear function defined in Eq. (6.27). In Fig. 6.33, OTA β_1 serves as the function of Eq. (6.28), while capacitor C and OTA w receiving voltage θ produce the dynamics for Eq. (6.29). Equation (6.29) can further be simplified as

$$C\frac{\mathrm{d}[x_2]}{\mathrm{d}t} = \begin{cases} I_{\mathrm{ref}} & ([x_1] > \theta), \\ -I_{\mathrm{ref}} & (else), \end{cases} \tag{6.30}$$

when $\kappa/V_T \gg 1$. The output current of OTA w becomes 0 when the voltage of output node $[x_2]$ equals the supply voltage (VDD or VSS). The value of $[x_2]$ is thus restricted within the range [VDD, VSS]. Note that Eqs. (6.28) and (6.30) are qualitatively equivalent to (6.23) and (6.24), respectively.

Figure 6.34 shows the layout of a cell circuit obeying the MOSIS 1.5-μm standard n-well CMOS rule. This device consists of four blocks: OTA β_1, OTA w, MOS

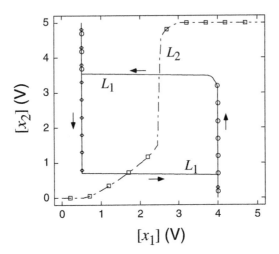

Figure 6.35: Numerically obtained nullclines of analogue cell circuit.

capacitor C and the output driver (inverter). The size of the analogue cell is 70×70 μm^2. The supply voltages of OTA β_1 are labelled as VDD2 and VSS2, while those of OTA w and the driver are labelled as VDD1 and VSS1. The well and substrate are common to all blocks, and are connected to VDD1 and VSS1, respectively. The following results were obtained from extracted circuits from the device layout (Fig. 6.34) with actual parasitic capacitances. A MOSIS 1.5-μm CMOS technology file with a transistor model of BSIM3 (level 8) and ngspice (rework-14) were used to evaluate the circuit with an actual device layout. The size of the MOS transistors was set at $W/L = 5\lambda/10\lambda$ for transistors providing source currents of the OTAs and $W/L = 5\lambda/2\lambda$ for the others ($\lambda = 0.8$ μm). Capacitor C was replaced with an nMOS capacitor (lower-right rectangle in Fig. 6.34).

Figure 6.35 plots the nullclines of the cell circuit. The supply voltages VDD1, VDD2, VSS1 and VSS2 were set at 5 V, 4 V, GND and 0.5 V, respectively. Bias voltage V_{ref} was set at 2 V, while the threshold θ was set at 2.5 V so that the circuit would exhibit oscillatory behaviour. In the figure, hysteresis curve L_1 representing the inverse 'N' characteristic corresponds to the nullcline of Eq. (6.19), while L_2 represents the nullcline of Eq. (6.20). Time courses for $[x_1]$ and $[x_2]$ are plotted in Fig. 6.36. Stiff oscillations were obtained as observed in a typical well-stirred BZ reaction. The stiffness and frequency of the oscillations can be controlled with the reference voltages of the OTAs (V_{ref} in Fig. 6.33). In the simulation, the reference voltage for OTA β_1 was the same as that for OTA w.

According to Eqs. (6.28) to (6.30), $[x_2]$ should only increase if $[x_1]$ is greater than the threshold θ. In addition, Eq. (6.30) indicates that the decreasing rate of $[x_2]$ should be equal to the increasing rate. This implies a triangular-wave oscillation in $[x_2]$ and a square-wave oscillation in $[x_1]$. However, the results clearly show that $[x_2]$ was increasing while $[x_1]$ was equal to 0.5 V ($< \theta$). In the circuit, OTA w cannot produce an ideal sigmoid-type nullcline L_2, as seen in Fig. 6.35, due

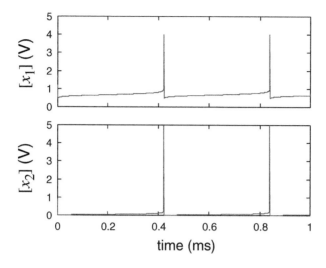

Figure 6.36: Time course for cell circuit's system variables $([x_1], [x_2])$.

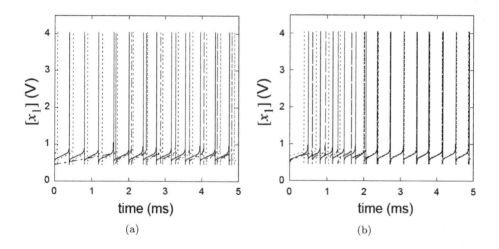

(a)   (b)

Figure 6.37: Time course for three analogue cells coupled by pass transistors with (a) very weak and (b) weak connections.

to the unsaturated operation of the source nMOS transistor. Therefore, $[x_1]$ and $[x_2]$ change very slowly when $[x_2]$ is lower than the threshold voltage of an nMOS transistor (V_{th}) and $[x_1] < \theta$, compared with $[x_2] > V_{th}$ and $[x_1] > \theta$. This operation generates the stiff oscillation rather than triangular- and square-wave oscillations. Note that if full-range OTAs had been used instead of the five-transistor OTA employed here, such triangular and square oscillations would have been obtained.

Figure 6.37 shows the synchronising phenomena for three cell circuits. These

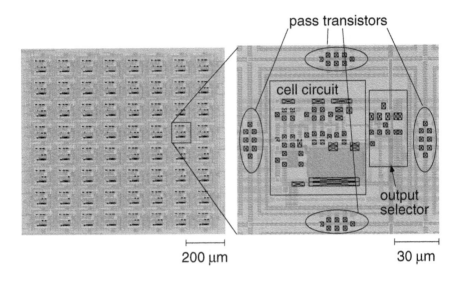

Figure 6.38: Layout of 8×8 analogue cells designed with 1.5-μm CMOS rule (array size: $970 \times 1050 \ \mu$m^2).

cell circuits were locally connected by pass transistors, instead of linear resistors, because if the qualitative behaviours of the system had not been changed by non-linearity, they would have reduced the circuit size significantly. The connection strength was thus controlled by the common gate voltage (V_g) of these transistors. In the simulation, the bias condition was the same as in the previous experiments, and the output node [x_1] of each cell was connected to that of two neighbours with minimum-size nMOS (pass) transistors. The output node [x_2] was not connected anywhere. When $V_g = 0$, each circuit oscillated independently (Fig. 6.37a). As $V_g \to$ VDD, these circuits tended to exhibit synchronising oscillatory behaviours. Figure 6.37b plots the gradual synchronising behaviour ($V_g = 0.9$ V). Initially, the oscillators were not synchronised. After $t = 2$ msec, they were.

A two-dimensional array of cell circuits (8×8 cells) was designed, as we can see in Fig. 6.38. The size of the array was $970 \times 1050 \ \mu$m^2 using the 1.5-μm CMOS design rule. The cell circuits were regularly arranged on a two-dimensional rectangular grid. Each cell was locally connected to its four neighbours with four pass transistors (non-linear resistors). Figure 6.39 has snapshots at $t = n\Delta t$ ($\Delta t = 20 \ \mu$sec) of spatio-temporal outputs of the two-dimensional array. In the simulation, V_g was set at 1.5 V with an open boundary condition. The remaining parameters were the same as in the previous experiments. In the photographs, voltages [x_1]$_{i,j}$ are represented on a grey scale ($v_{i,j} = 0$: black, $v_{i,j} = 4$ V: white). The initial values of the cell circuits were randomly chosen as [$u_{i,j}$] = RAND[0.5 V, 4 V] and [$v_{i,j}$] = RAND[0, 5 V]. The distribution became almost uniform at $t = 5\Delta t$. During $t = 5\Delta t \to 50\Delta t$, the cell circuit located at the bottom-right corner was oscillating by chance. Depending on the initial conditions, the position of a cell such as this that survives will be changed. The surviving cell's neighbours were

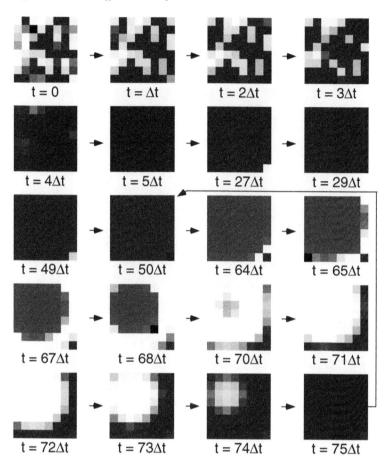

Figure 6.39: Generation of active travelling waves in two-dimensional reaction-diffusion chip.

stimulated through pass transistors. Consequently, the neighbours were excited at $t = 64\Delta t$, and chain excitation was observed during $t = 65\Delta t \rightarrow 75\Delta t$. After $t = 75\Delta t$, the 'seed' cell at the bottom-right corner was still oscillating, which induced a subsequent chain reaction. Consequently, this array produced two-dimensional phase-lagged stable synchronous patterns. When $V_g > 1$ V, all cells exhibited synchronous oscillations (no spatial patterns were produced). The results indicate that although a simplified cell circuit and non-linear resistors (pass transistors) were used, the analogue system is qualitatively equivalent to the basic reaction-diffusion system with the Oregonators, since the excitatory properties of the analogue cells are inherently the same as those of the Oregonators.

A prototype reaction-diffusion chip was fabricated with the 1.5-μm CMOS process (MOSIS, vendor: AMIS). Figure 6.40 is a micrograph of the reaction-diffusion chip that includes an array of cell circuits (example in Fig. 6.38). An

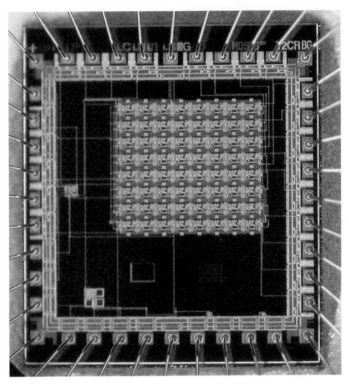

Figure 6.40: Micrograph of fabricated reaction-diffusion chip. See also Fig. 9.28 in colour insert.

array of 8×8 cell circuits was implemented on a 2.3×2.3 mm^2 die.

Figure 6.41 plots experimental results for a reaction circuit. The upper and lower graphs plot time courses of voltages $[x_1]$ and $[x_2]$ in the circuit. The supply voltages of an OTA of β_1 were set at VDD = 4 V and VSS = 0.5 V, while those of the remaining OTA (w) were set at VDD = 5 V and VSS = GND. Threshold θ was set at 2.5 V so that the circuit exhibited oscillatory behaviours. As expected, the circuit exhibited the same qualitative behaviours as in the simulation results (Fig. 6.36), i.e. stiff non-linear oscillations. In the experiment, V_{ref} was set just above the threshold voltage of an nMOS transistor, i.e. the OTA was driven by a current of about 10^{-6} A.

The operations of the 8×8 reaction cell array were recorded with the following read-out circuitry. Each cell in the chip was located beneath each wire crossing row and column buses, and was connected to a common output wire through a transfer gate. The gate connects the cell's output to the common wire when both the row and the column buses are active. The cell's output $[x_2]$ is amplified by an inverter (output driver) in the prototype chip. Thus, a cell's quantised output appeared on the common output wire when the cell was selected by activating the corresponding row and column buses simultaneously. One could obtain a binary

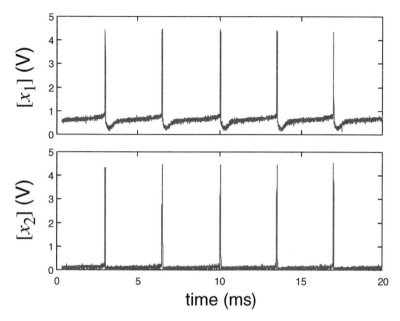

Figure 6.41: Experimental results for reaction circuit operating in its oscillatory mode.

stream from the common output wire by selecting each cell sequentially. Using a conventional displaying technique, the binary stream was re-constructed on a two-dimensional display. Figures 6.42 to 6.44 present the recorded results. Each white dot represents an inactive cell where $[x_2] <$ VDD/2. In the experiment, V_{ref} was set at the threshold voltage of an nMOS transistor, i.e. the OTAs were driven by the subthreshold current, so that 'very slow' spatio-temporal activities could be observed visually. The remaining parameters were the same as in the experiments in Fig. 6.41. When V_g (common gate voltage of pass transistors) was set at 0, each cell oscillated independently (Fig. 6.42). Snapshots were taken at intervals of 200 msec. The collective activities of cell clusters were observed by increasing V_g. Figure 6.43 presents an example where some spiral (mode-lock) patterns of cell clusters were observed ($V_g = 0.8$ V). Snapshots were taken at intervals of 100 msec. The active-cell clusters and cores of the spirals are superimposed on the figure by white curves and white circles, respectively. Although observing 'beautiful' spirals as in Fig. 6.32 is difficult because of the small number of cells, the appearance and disappearance of small sections of spiral waves were successfully observed. Under strong lateral connections ($V_g = 1.2$ V), large clusters of cells were synchronised, as we can see in Fig. 6.44. Due to mismatches of fabricated devices, the cells were not synchronised even when $V_g > 1.2$ V.

These results indicate the expected operations of the reaction-diffusion chip. The next challenge is to develop reaction-diffusion chips with a more microscopic process because a large number of reaction cells must be implemented on a chip to

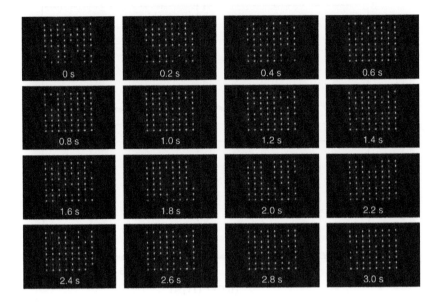

Figure 6.42: Spatio-temporal patterns on the fabricated reaction-diffusion chip (no lateral connections between cells).

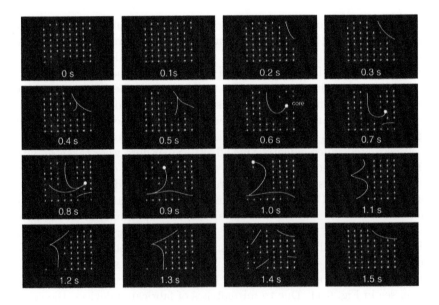

Figure 6.43: Spiral patterns on reaction-diffusion chip (weak connections between cells).

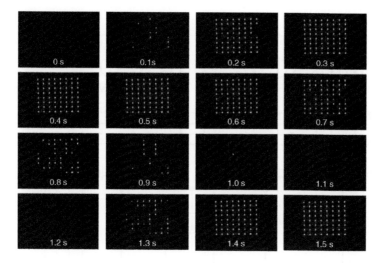

Figure 6.44: Synchronisation of cells on reaction-diffusion chip (strong connections).

observe complex (BZ-like) patterns. Furthermore, at this stage, each reaction cell should have optical sensors for parallel inputs. In fact, the optical input of information was previously used at the pioneering experiments with image processing in light-sensitive BZ systems [154].

6.3.2 Implementing Turing-type system in silicon

A Turing system is a reaction-diffusion-based system that generates stationary, spatially periodic patterns commonly termed Turing patterns [196]. In simple terms, it consists of two chemicals that control the synthesis rate of each other. If the chemicals react and diffuse in an appropriate way — the inhibitor must diffuse more quickly in the system than the activator — spatial patterns of their concentrations can arise from an initial distribution caused by a small disturbance. This phenomenon can be applied to image pre-processing such as enhancement and restoration of textured patterns.

Here, we study a CMOS LSI that imitates the dynamics of Turing systems. The LSI, a Turing chip, consists of regularly arrayed cell circuits, i.e. reaction cells and diffusion cells: a reaction cell imitates a local reaction, and a diffusion cell imitates diffusion between adjacent reaction cells. We design cell circuits that make use of the transfer characteristic of MOS-transistor differential pairs. In these circuits, the concentrations of two chemicals are represented by two voltage signals. As a first step towards the LSI, we combine the reaction cells and diffusion cells into a ring to construct a circuit that imitates a one-dimensional Turing system.

A Turing system is a two-variable system with a linear reaction function (a linear function suffices for the pattern formation in Turing systems). The equations for

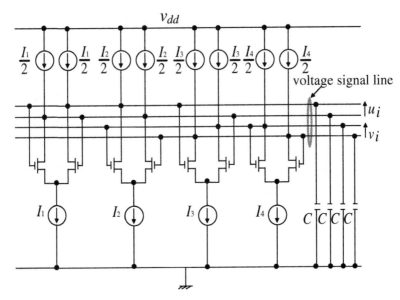

Figure 6.45: Reaction cell circuit consisting of four differential pairs and four integral capacitors.

this system are given by

$$\frac{\partial u}{\partial t} = D_u \nabla^2 u + au - bv, \qquad (6.31)$$

$$\frac{\partial v}{\partial t} = D_v \nabla^2 v + cu - dv, \qquad (6.32)$$

where u and v are the concentrations of the two substances, and D_u and D_v are diffusion coefficients of the substances. Parameters a, b, c and d in the reaction terms determine the dynamics of the system. If all reaction parameters are set positive $(a, b, c, d > 0)$, variable u acts as an activator, while variable v acts as an inhibitor. Under appropriate conditions, i.e. $a < d$, $bc > ad$ and $D_v \gg D_u$, the system produces Turing patterns [196].

As described in Sect. 6.1, we regard variables u and v as voltage signals, to construct the Turing system on an LSI chip. The LSI consists of reaction cell circuits that imitate reaction and diffusion cell circuits that imitate diffusion. Both cell circuits are regularly arrayed on a chip and interconnected by wires that carry the voltage signals; each reaction cell is connected with its neighbouring reaction cells through the diffusion cells.

Figure 6.45 shows a reaction cell circuit that implements the point dynamics of Eqs. (6.31) and (6.32), except for diffusive terms. It consists of four differential pairs and four integral capacitors C. The two variables are represented by differential voltages u and v on the signal line. The four differential pairs correspond to the four reaction terms in Eqs. (6.31) and (6.32), i.e. the differential pair to the far left

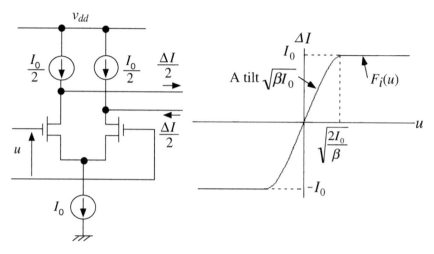

Figure 6.46: Transfer characteristic of a differential pair.

corresponds to the reaction term au, the second to $-bv$, the third to cu and the furthest right to $-dv$. The transfer characteristic of a differential pair (Fig. 6.46) occurs as

$$\Delta I \equiv F(u) = \begin{cases} k_1 u \sqrt{1 - u^2/(2k_2^2)} & \text{for } -k_2 < u < k_2, \\ k_1 k_2/\sqrt{2} & \text{for } u > k_2, \\ -k_1 k_2/\sqrt{2} & \text{for } u < -k_2, \end{cases} \tag{6.33}$$

where ΔI is the output current and the positive coefficients k_1 and k_2 are a function of MOS-transistor gain factor β and bias current I_0 and are given by $k_1 = \sqrt{\beta I_0}$ and $k_2 = \sqrt{2I_0/\beta}$. The dynamics of the reaction cell circuit can therefore be expressed as

$$\begin{aligned} C\frac{du}{dt} &= F_1(u) - F_2(v), \\ C\frac{dv}{dt} &= F_3(u) - F_4(v), \end{aligned}$$

where $F_1(u)$ to $F_4(v)$ are the transfer functions [corresponding to $F(u)$ in Eq. (6.33)] of the four differential pairs. The transfer functions are linear when the values of the voltages u and v are small (see Fig. 6.46), so the reaction terms in Eqs. (6.31) and (6.32) can be implemented. The reaction parameters a, b, c and d can be controlled by adjusting the MOS-transistor gain factor and the bias current in each differential pair.

Three examples of the cell operation are shown in Fig. 6.47 on the u–v phase plane, with the values of the reaction parameters. A circuit simulator HSPICE was used with 1.2-μm CMOS device parameters. Depending on the reaction parameters, the circuit can be oscillatory or stationary. The bias current was set to 50 μA for all four differential pairs. The capacitance of the integral capacitor was set to 10 pF.

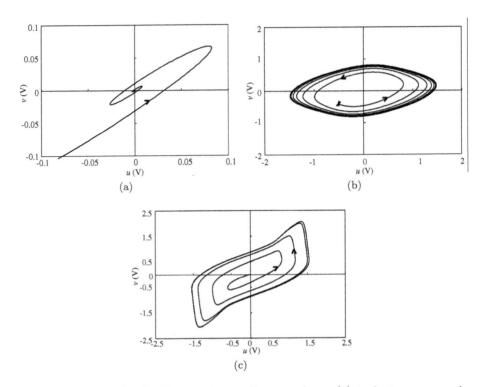

Figure 6.47: Example of cell operation on the u–v plane: (a) trajectory converging to the origin on the u–v phase plane (the ratio between reaction parameters a:b:c:d = 1:1.2:1:1.1); (b) limit cycle and a trajectory converging to the cycle (a:b:c:d = 1:2.3:1:0.7); (c) limit cycle and a trajectory converging to the cycle (a:b:c:d = 1:1:1:0.7).

The reaction parameters were set by adjusting the MOS-transistor gate width in each differential pair. For generating Turing patterns, we use the circuit under stationary conditions.

Figure 6.48 shows the diffusion cell circuit consisting of four differential pairs. The circuit has two pairs of differential signal lines to couple with two adjacent reaction cells; the left-hand pair (u_i and v_i) are connected to the signal lines of one reaction cell, and the right-hand pair (u_{i+1} and v_{i+1}) are connected to the other reaction cell. The circuit operates as a pair of floating resistors and conducts differential currents (ΔI_u and ΔI_v) from one side to the other as shown in Fig. 6.48. The currents are proportional to the difference between the signal voltages ($u_i - u_{i+1}$ and $v_i - v_{i+1}$) when the voltage signals are small, so we can imitate diffusion with this cell circuit. We can control the diffusion coefficient by adjusting the MOS-transistor gain factor and the bias current in each differential pair.

Figure 6.49 shows simulation results representing diffusion of a voltage signal through the chain of 50 diffusion cells. A 10-pF capacitor was connected to each connection point of adjacent cells. At time = 0, a step voltage of 0.6 V was applied

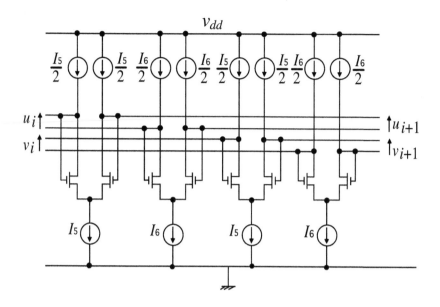

Figure 6.48: Diffusion circuit consisting of four differential pairs.

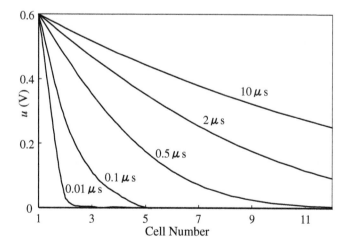

Figure 6.49: Diffusion of a voltage signal through a chain of the diffusion cells. A step voltage of 0.6 V was applied to the first cell, or cell 1.

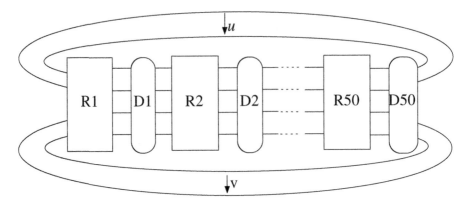

Figure 6.50: One-dimensional Turing system consisting of the reaction cells (R1–R50) and the diffusion cells (D1–D50) linked to a ring.

to the first cell of the chain, i.e. $u_1 = v_1 = 0.6$ V at $t = 0$. The figure shows the distribution of the signal voltage u_i through the chain, with time as a parameter. The voltage signal is transported by diffusion through the chain of cells.

Now, let us combine the reaction cells and diffusion cells into a ring to construct a circuit that imitates a one-dimensional Turing system. The configuration of the ring circuit is illustrated in Fig. 6.50. Figure 6.51 illustrates the results for the circuit with 50 reaction cells and 50 diffusion cells. Initial voltage signals u_i and v_i were given to each reaction cell (Fig. 6.51a), where u_i in each reaction cell is plotted. After a transition time, the circuit stabilised to a final state as shown in Fig. 6.51b. The pattern that is formed by cell voltages u_i is a one-dimensional Turing pattern. The spatial frequency of the pattern can be controlled by the parameters (reaction parameters and diffusion coefficients). Figure 6.51c shows a Turing pattern with another set of parameters.

The circuit acts as a spatial-frequency filtering device and can be used in signal processing such as the restoration and enhancement of one-dimensional texture images. The simulated result is shown in Figs. 6.52 and 6.53. The upper stripe in Fig. 6.52 is a blurred picture to process. To restore this picture, the shaded pattern of the picture was converted into a voltage pattern u_i as depicted in Fig. 6.53 by a dashed curve. Then, the voltage u_i was given to each reaction cell (voltages v_i were set to 0). After some transition time, voltages u_i in the cells stabilised to their final values to form a pattern as shown by a solid curve in Fig. 6.53. By re-converting the voltage pattern to a shaded pattern, we obtain a restored enhanced picture as shown by the bottom stripe in Fig. 6.52. This operation is useful for restoring noisy striped, or spotted, images, e.g. fingerprint patterns and polka-dot patterns.

6.3.3   Implementing Lotka–Volterra systems

The design of non-linear, chaotic oscillators has been the subject of increasing interest during the past few years [69, 228]. Indeed, analogue integrated circuits that

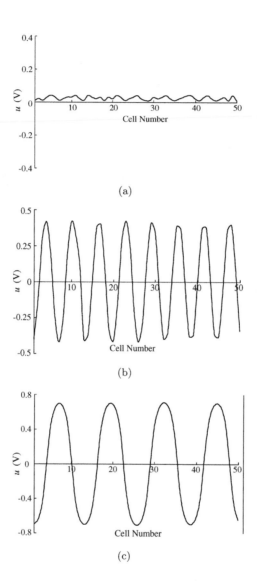

Figure 6.51: Dynamics of ring circuit of 50 diffusive and 50 reaction cells: (a) initial distribution of voltage signals u_i in the ring circuit, cell numbers 0 and 50 denote the identical cell; (b) Turing pattern in the ring circuit, the ratio between reaction parameters is $a{:}b{:}c{:}d = 1{:}2{:}1{:}1.5$, and the ratio between diffusion coefficients is $D_u{:}D_v = 1{:}10$; (c) Turing pattern in the ring circuit with another set of parameters — $a{:}b{:}c{:}d = 1{:}1.2{:}1{:}1.1$ and $D_u{:}D_v = 1{:}10$.

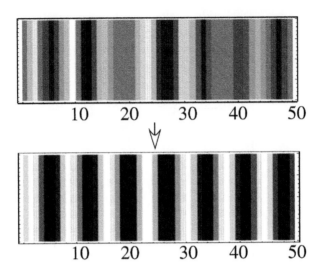

Figure 6.52: Restoration and enhancement of a given one-dimensional pattern. The upper stripe is an initial blurred picture, and the bottom stripe is the restored, enhanced picture. The parameters of the circuit are $a{:}b{:}c{:}d = 1{:}1.2{:}1{:}1.1$ and $D_u{:}D_v = 1{:}3$.

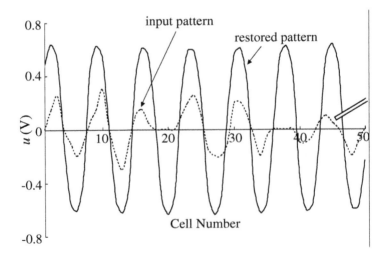

Figure 6.53: The pattern of cell voltages u_i over 50 cells from cell 1 to cell 50. The dashed line corresponds to the blurred input picture in Fig. 6.52, and the solid line corresponds to the restored, enhanced picture in Fig. 6.52.

implement chaotic oscillatory systems provide us with important cues for exploring and discovering novel forms of information processing. Many designs for such oscillators were introduced starting with the use of a coil in Chua's circuit [176] to the use of large blocks such as operational amplifiers [91, 92]. In both cases, the fabrication area was very large. These designs were also dependent on the use of floating capacitors, high supply voltages and high power dissipation, which are not desirable due to the current demand for portability. Here, we study micro-power analogue MOS circuits that exhibit chaotic behaviours with very simple circuit construction [34], to construct novel analogue reaction-diffusion chips.

Although there are numerous simple chaotic equations, e.g. [263, 264, 265, 69], we study the implementation of the Lotka–Volterra (LV) equation. The advantages of the LV system are its simplicity, absence of multiplication terms with non-linear transformation of system variables [37, 35], ease of scaling over a wide range of frequencies and ease of construction. The LV circuit can be designed based on the use of subthreshold MOS FETs and three grounded capacitors in order to realise the LV equation, which is the minimum requirement for the implementation of a chaotic oscillator. This circuit operates on a low supply voltage and all MOS FETs operate in their subthreshold region. In this sense, the circuit overcomes the previously mentioned drawbacks and can be used in manufacturing portable devices.

The LV model is one of the earliest predator–prey models to be based on sound mathematical principles. It forms the basis of many models used today in the analysis of population dynamics [110]. We use an LV model that describes interactions between three species in an ecosystem, i.e. one predator and two prey [182]. In addition to the predation of the prey, the two prey compete with each other for their feeding ground. The dynamics is given by

$$\tau \dot{x}_1 = (1 - x_1 - c\, x_2 - k\, y)x_1, \tag{6.34}$$
$$\tau \dot{x}_2 = (a - b\, x_1 - x_2 - y)x_2, \tag{6.35}$$
$$\tau \dot{y} = (-r + \alpha\, kx_1 + \beta\, x_2)y, \tag{6.36}$$

where x_1 and x_2 represent the prey population, y the predator population and τ the time constant, and the additional terms k, a, b, c, r, α and β are control parameters. The system exhibits stable, periodic and chaotic behaviours that can be controlled by a single parameter r under some parameter constraints [182].

Analogue MOS circuits for LV-type neural networks have already been proposed in [37, 35, 34], where logarithmic transformation of system variables was used to remove the multiplication terms of system variables in the LV system.

By introducing the following variables with scaling constant s,

$$X_1 = s + \ln x_1, \quad X_2 = s + \ln x_2, \quad X_3 = s + \ln y, \tag{6.37}$$

Eqs. (6.34), (6.35) and (6.36) can be transformed into

$$\tau' \dot{X}_1 = s' - \exp(X_1) - c \exp(X_2) - k \exp(Y), \tag{6.38}$$
$$\tau' \dot{X}_2 = as' - b \exp(X_1) - \exp(X_2) - \exp(Y), \tag{6.39}$$
$$\tau' \dot{X}_3 = -rs' + \alpha k \exp(X_1) + \beta \exp(X_2), \tag{6.40}$$

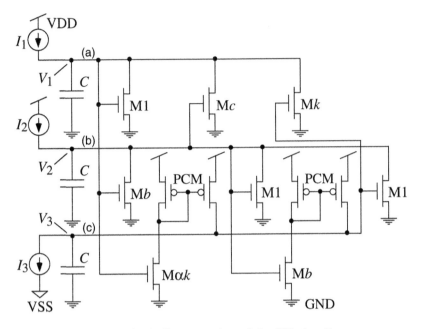

Figure 6.54: Construction of the LV circuit.

where $s' \equiv \exp(s)$ and $\tau' \equiv \tau s'$. This logarithmic transformation has two advantages in analogue MOS implementation:

- the resulting equations (6.38), (6.39) and (6.40) do not have multiplication terms of system variables and can be described by a linear combination of exponential functions, which enables us to design the circuit without an analogue multiplier;

- exponential non-linearity is an essential characteristic of semiconductor devices, which enables us to design a circuit based on the intrinsic characteristics of semiconductors.

Here, we use the exponential current–voltage characteristics of subthreshold MOS FETs [296, 32].

Figure 6.54 is a diagram of the construction of an LV circuit. Applying Kirchhoff's current law (KCL) at nodes (a) and (b) in Fig. 6.54, we obtain

$$C\dot{V}_1 = I_1 - I_0^{(M1)} \exp\left(\frac{\kappa}{V_T}V_1\right) - I_0^{(Mc)} \exp\left(\frac{\kappa}{V_T}V_2\right) - I_0^{(Mk)} \exp\left(\frac{\kappa}{V_T}V_3\right), \quad (6.41)$$

$$C\dot{V}_2 = I_2 - I_0^{(Mb)} \exp\left(\frac{\kappa}{V_T}V_1\right) - I_0^{(M1)} \exp\left(\frac{\kappa}{V_T}V_2\right) - I_0^{(M1)} \exp\left(\frac{\kappa}{V_T}V_3\right), \quad (6.42)$$

where V_i represents the node voltage, $I_{1,2}$ the injecting current, C the capacitance, κ the effectiveness of the gate potential, $V_T \equiv kT/q \approx 26$ mV at room temperature

(k is Boltzmann's constant, T the temperature and q the electron charge) and $I_0^{(\text{M}i)}$ the fabrication parameter of nMOS FET Mi given by

$$I_0 \equiv \mu C_{\text{ox}} \frac{W}{L} \frac{1-\kappa}{\kappa} V_T^2, \tag{6.43}$$

where μ is the electron mobility, C_{ox} the gate capacitance and W and L the channel width and length of MOS FETs [296]. Typical parameter values for minimum-size devices fabricated in a standard 1.5-μm n-well process are $I_0 = 0.5 \times 10^{-15}$ A and $\kappa = 0.6$. Parameter $I_0^{(\text{M}i)}$ is proportional to the channel width (or inversely proportional to the channel length) of MOS FETs. Since the dimensions (width/length) are responsible for the parameters of the LV model, these must be pre-determined before the IC is fabricated. It should be noted that Eqs. (6.41) and (6.42) are valid only when the MOS FETs are saturated. Node voltages V_1 and V_2 are also applied to the gates of MOS FETs Mαk and Mβ, respectively. Because the currents of Mαk and Mβ are copied to node (c) by two pMOS current mirrors (PCMs in Fig. 6.54), the node equation is represented by

$$C\dot{V_3} = -I_3 + I_0^{(\text{M}\alpha k)} \exp\left(\frac{\kappa}{V_T} V_1\right) + I_0^{(\text{M}\beta)} \exp\left(\frac{\kappa}{V_T} V_2\right). \tag{6.44}$$

Equations (6.41), (6.42) and (6.44) become equivalent to Eqs. (6.38), (6.39) and (6.40), respectively, when

$$V_i = \frac{V_T}{\kappa} X_i \ (i = 1, 2, 3), \tag{6.45}$$

$$\frac{I_1}{I_0^{(\text{M}1)}} = s', \quad \frac{I_2}{I_0^{(\text{M}1)}} = as', \quad \frac{I_3}{I_0^{(\text{M}1)}} = rs', \tag{6.46}$$

$$\frac{I_0^{(\text{M}k)}}{I_0^{(\text{M}1)}} = k, \quad \frac{I_0^{(\text{M}b)}}{I_0^{(\text{M}1)}} = b, \quad \frac{I_0^{(\text{M}c)}}{I_0^{(\text{M}1)}} = c, \tag{6.47}$$

$$\frac{I_0^{(\text{M}\alpha k)}}{I_0^{(\text{M}1)}} = \alpha k, \quad \frac{I_0^{(\text{M}\beta)}}{I_0^{(\text{M}1)}} = \beta, \quad \tau' = \frac{CV_T}{I_0^{(\text{M}1)} \kappa}. \tag{6.48}$$

With the original parameter set of the LV model [182], the values of system variables (x_1, x_2 and y) were restricted within interval $[0, 1]$, which resulted in the extent of $(-\infty : s\, V_T/\kappa]$ V for the circuit system variables V_1, V_2 and V_3 obtained by Eq. (6.45). Note that both V_1 and V_2 cannot take negative values due to the limit of the supply voltage ($V_i \geq$ GND). Furthermore, Eqs. (6.41) and (6.42) are valid only when the nMOS FETs are saturated, i.e. $V_1, V_2 \geq 4V_T \approx 0.1$ V at room temperature. Therefore, $s\ (= \ln s')$ must satisfy the condition

$$s > 4\kappa - \ln \min[x_i(t)] \ (i = 1, 2). \tag{6.49}$$

With a large negative V_i ($x_i \approx 0$), this limit is negligible because $\dot{x}_i \approx 0$. Assuming that typical values for I_0 and the maximum subthreshold current are of orders 10^{-15} A and 10^{-7} A, respectively, we can estimate $s \approx 18.4$ ($= \ln 10^8$) from Eq. (6.46). This means that the circuit can emulate Eqs. (6.34) and (6.35) perfectly as long as $\min[x_i(t)] > 1.1 \times 10^{-7}$ ($\kappa = 0.6$ was assumed here).

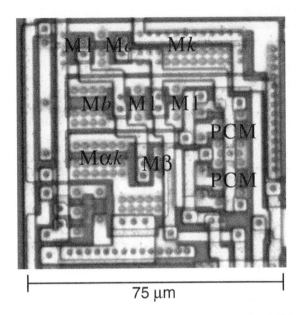

Figure 6.55: Chip micrograph of a fabricated LV circuit (MOSIS, vendor: AMIS, n-well single-poly double-metal CMOS process, feature size: 1.5 μm, total area: 75 μm $\times$ 40 μm).

If we employ an nMOS transistor instead of current source I_3, on the other hand, negative V_3 breaks the isolation of pn junctions between the p-substrate and the drain of the nMOS transistor. Therefore, we have to employ an off-chip nMOS transistor as current source I_3.

A prototype circuit was fabricated with a scalable CMOS rule (MOSIS, vendor: AMIS, n-well single-poly double-metal CMOS process, $\lambda = 0.8$ μm, feature size:

MOS FET	W (μm)	L (μm)
M1	4	1.6
Mb	12	3.2
Mc	4	1.6
Mk	40	1.6
Mαk	20	1.6
Mβ	4	3.2

Table 6.4: Design size of nMOS FETs on LV circuit. Corresponding parameters of LV models in Eqs. (13) to (15) are $k = 10$, $b = 1.5$, $c = 1$, $\alpha k = 5$ and $\beta = 0.5$. Since $\lambda = 0.8$ μm and the feature size is 1.5 μm, all the design sizes are scaled down by 1.5 μm/1.6 μm ($\approx$ 94%) on actual chips.

1.5 μm). Figure 6.55 is a micrograph of the LV circuit. The same parameter set for the LV system ($k = 10$, $b = 1.5$, $c = 1$, $\alpha k = 5$ and $\beta = 0.5$) as used in [182], where stable focus bifurcates into chaotic oscillation via stable period-n cycles. The resulting sizes of nMOS FETs are listed in Table 6.4. The pMOS current mirrors (PCM) were designed with a dimension of $W/L = 4$ μm/1.6 μm. Minimum-size transistors were employed for M1 and PCM to design the circuit so that it was as compact as possible, rather than compensating for device mismatches between transistors. The circuit took up a total area of 75 μm × 40 μm.

Before fabrication, the operation range of variable V_3 was simulated by using ideal current source I_3. Unfortunately, V_3 took both positive and negative values when the given parameter set by Mimura and Kan-on [182] was used. Since finding a good parameter set that ensures $V_3 > 0$ is another subject altogether, an off-chip current source was used as I_3 with the original parameter setup.

In the following experiments, off-chip capacitors ($C = 0.1$ μF) were added. The values of capacitances did not change the qualitative behaviours of the circuit, as long as the value was much larger than that of the gate capacitance of MOS transistors. Time courses for V_1, V_2 and V_3 were sampled simultaneously with an Agilent 4156B. The supply voltage (VDD) was set at 2.5 V. The input currents (I_1, I_2) were fixed at (250, 287) nA. The dynamic behaviours of the fabricated LV circuit were evaluated by changing the rest input current I_3 that corresponded to control parameter r in Eq. (6.36).

Figure 6.56 plots the measurement results. Figure 6.56a and b plot the time courses of system variables V_1, V_2 and V_3 and trajectories on a V_1–V_3 plane, respectively. In this experiment, I_3 was set at 320 nA (compliance was set at –2.5 V). The LV circuit exhibited stable oscillation with period-1 cycles. As predicted by simulations, V_3 took positive and negative values. An off-chip current source (or nMOS transistor) is thus necessary for this circuit with the original parameter set.

In Fig. 6.56c and d, which represent the time courses for system variables and trajectories on the V_1–V_3 plane, respectively, I_3 was set at 360 nA. The LV circuit exhibited stable oscillation with period-2 cycles. Figure 6.56e and f plot the time course of the system variables and trajectories on the V_1–V_3 plane, respectively. In this experiment, I_3 was set at 420 nA. The value for the maximum Lyapunov exponent was 10.1, which indicated that the LV circuit exhibited chaotic oscillation.

According to Mimura and Kan-on [182], as the value of the control parameter r increases, Hopf bifurcation occurs where a stable focus bifurcates to an unstable focus with an enclosing limit cycle. Then, an unstable focus bifurcates to a stable focus. This transition — stable focus → unstable focus with enclosing limit cycle → stable focus — can be observed in the LV circuit when I_3 ($\sim r$) increased. Figure 6.57 is the bifurcation diagram obtained from the LV circuit. The diagram was created as follows:

- when the circuit had stable focus with a given I_3, we plotted a stable value for V_3,

- when the circuit oscillated with a given I_3, we plotted a value for V_3 at which $\dot{V}_3 = 0$.

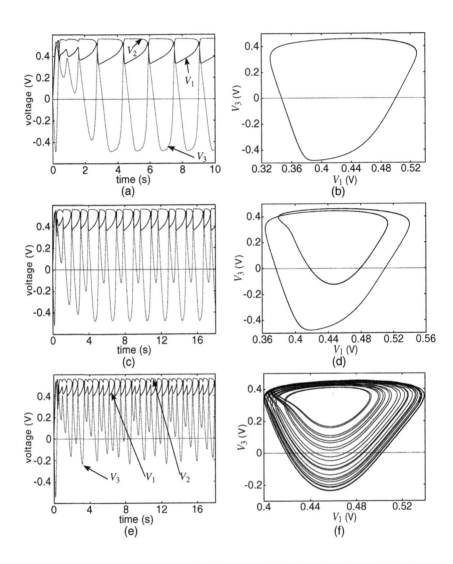

Figure 6.56: Experimental results for fabricated LV circuit: (a) and (c) show time course for system variables V_1, V_2 and V_3, (b) and (d) show trajectories on V_1–V_3 plane, (a) and (b) represent results for $I_3 = 320$ nA, and (c) and (d) results for $I_3 = 360$ nA; (e) and (f) show time courses for system variables V_1, V_2 and V_3 trajectories on V_1–V_3 plane, respectively, when $I_3 = 420$ nA.

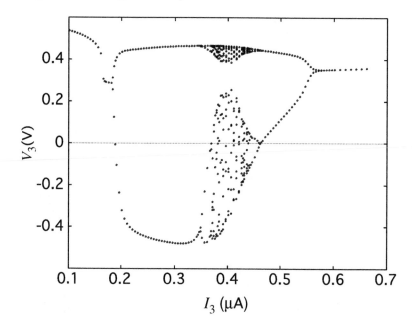

Figure 6.57: Bifurcation diagram obtained from LV circuit.

When $I_3 < 182$ nA, the LV circuit did not oscillate (stable focus). Stable focus bifurcated at $I_3 \approx 182$ nA to stable period-1 cycles. Increasing the value of I_3 led to further bifurcations to period-2 cycles, period-4 cycles and chaotic cycles occurred around 370 A$< I_3 <$450 nA. Finally, unstable focus bifurcated to stable focus again at $I_3 \approx 580$ nA.

The results in Fig. 6.57 indicate that the LV circuit has two important properties:

- although practical subthreshold MOS FETs were used, the bifurcation property was qualitatively consistent with the theoretical prediction,

- the LV circuit exhibited stable oscillation with period n and chaotic cycles over a wide range of I_3, i.e. 182 nA$< I_3 <$580 nA, which allowed it to maintain stable oscillation within a noisy environment, even though subthreshold MOS FETs were used in the circuit.

Now, a large-scale two-dimensional array of chaotic (non-linear) oscillators can easily be incorporated with conventional CMOS technology. Diffusive LV systems, where each LV oscillator is locally connected through diffusive coupling, are known to produce various spatio-temporal patterns [182, 140, 95]. For example, Fig. 6.58 shows spatio-temporal patterns produced by a two-dimensional array of the LV

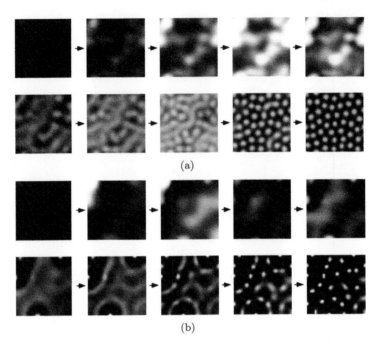

(a)

(b)

Figure 6.58: (a) Spatio-temporal patterns of diffusive LV system: (a) $D \equiv D_1 = D_2$, $D_y/D = 100$, (b) $D_y/D_1 = 100$, $D_2 = 0$; the remaining parameters are the same as in [182]).

system that can be implemented on an analogue LSI. The dynamics is given by

$$\tau \frac{\partial x_1}{\partial t} = D_1 \nabla^2 x_1 + (1 - x_1 - c\, x_2 - k\, y)x_1, \tag{6.50}$$

$$\tau \frac{\partial x_2}{\partial t} = D_2 \nabla^2 x_2 + (a - b\, x_1 - x_2 - y)x_2, \tag{6.51}$$

$$\tau \frac{\partial y}{\partial t} = D_y \nabla^2 y + (-r + \alpha\, kx_1 + \beta\, x_2)y, \tag{6.52}$$

where D_1, D_2 and D_y represent the diffusion coefficients. In the figure, the values of x_2 are represented on a grey scale ($x_2 = 0$: black, $x_2 = 1$: white). When prey y 'diffuses' much faster than predators x_1 and x_2, we will observe stationary non-uniform patterns. Such properties are very useful if we consider the diffusive LV system to be a reaction-diffusion computing medium [7]. Analogue ICs implementing the two-dimensional array of LV oscillators should assist us in exploring and discovering novel reaction-diffusion-based computing in addition to the applications of non-linear-coupled oscillators.

6.3.4   Implementing a Wilson–Cowan oscillator

Wilson–Cowan (WC) oscillators imitate the group activities of cortical neurons [304], which could also be applied to micro-volumes of an excitable medium. The WC

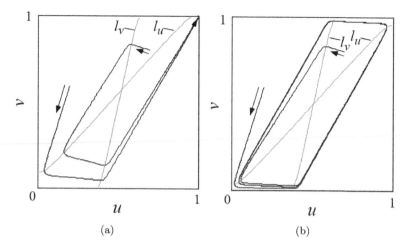

Figure 6.59: Stability of the WC oscillator: (a) fixed point, (b) limit cycle.

equation has a large number of parameters and, therefore, can represent a wide range of dynamics. Here we study analogue reaction-diffusion circuits implementing locally coupled Wilson–Cowan oscillators.

A WC oscillator is represented by a two-variable system described by a set of ordinary differential equations

$$\tau \dot{u} = -u + f_\beta(u - bv - \theta_u), \tag{6.53}$$
$$\tau \dot{v} = -v + f_\beta(au - cv - \theta_v), \tag{6.54}$$

where $f_\beta(x)$ represents a sigmoid function $(1 + \tanh(\beta x))/2$, and u and v are interacting terms of excitatory and inhibitory variables, respectively.

The time evolution of the WC oscillator is determined by the locations of two nullclines on the (u, v) plane, $\dot{u} = 0$ (denoted by l_u) and $\dot{v} = 0$ (l_v), as shown in Fig. 6.59. In Fig. 6.59a, a fixed-point attractor appears if l_u lies on the region of $v > u$, while a limit-cycle attractor is obtained if l_u intersects with segment $v = u$ ($v \geq 0, u < 1$) and, in addition, the slope of l_u is positive and smaller than unity (Fig. 6.59b).

Figure 6.60 shows a basic reaction circuit, hereafter called the WC circuit, that consists of excitatory and inhibitory element circuits. Both circuits have the same structure except for the polarity of connections between elementary units (microvolumes of a chemical medium). The sigmoidal response (f) of the neuron circuit is obtained by a CMOS differential pair consisting of floating-gate transistors. The differential pair also acts as excitatory and inhibitory synapses. It accepts an excitatory input (u) and an inhibitory input (v) through the '+' and '−' terminals, respectively.

Figure 6.61 shows an actual schematic of the WC circuit. It consists of two differential pairs (M1–M4) with floating-gate MOS transistors (M2 and M4), two pMOS current mirrors (pCM) and integrating circuits (C_1, R_1, C_2 and R_2). The

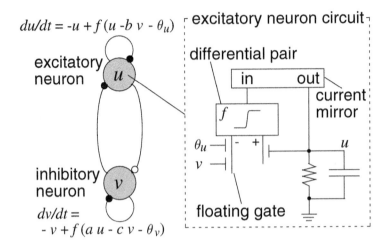

Figure 6.60: Schematic image of basic reaction circuit.

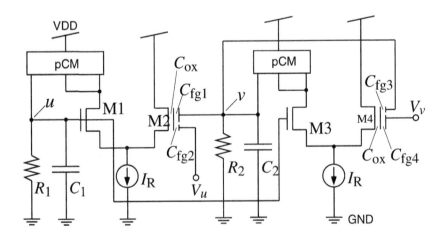

Figure 6.61: Actual schematic of the WC circuit.

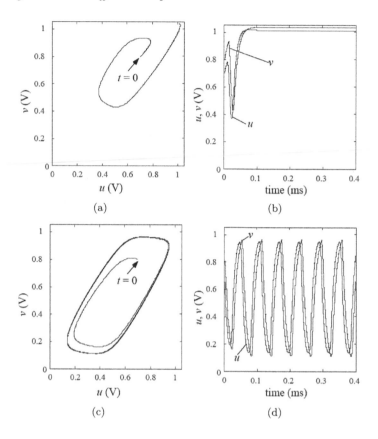

Figure 6.62: Stability of WC oscillator obtained by SPICE simulations.

dynamics of the circuit is expressed by

$$C_1 \dot{u} = -u/R_1 + I_R \ f_\beta(u - bv - \theta_u), \tag{6.55}$$
$$C_2 \dot{v} = -v/R_2 + I_R \ f_\beta(u - (c/a)v - \theta_v/a), \tag{6.56}$$

where u and v are the excitatory and inhibitory voltages, $\beta = \kappa/V_T$, $c/a \equiv C_{\mathrm{fg3}}/(C_{\mathrm{fg3}}+C_{\mathrm{fg4}}+C_{\mathrm{ox}})$, $b \equiv C_{\mathrm{fg1}}/(C_{\mathrm{fg1}}+C_{\mathrm{fg2}}+C_{\mathrm{ox}})$, $\theta_u \equiv C_{\mathrm{fg2}} V_v/(C_{\mathrm{fg1}}+C_{\mathrm{fg2}}+C_{\mathrm{ox}})$ and $\theta_v/a \equiv C_{\mathrm{fg4}} V_v/(C_{\mathrm{fg3}}+C_{\mathrm{fg4}}+C_{\mathrm{ox}})$. Equations (6.55) and (6.56) are equivalent to Eqs. (6.53) and (6.54) if $C_{\mathrm{ox}} \ll C_{\mathrm{fg}i}$ ($i = 1, 2, 3, 4$) and $\beta \to \infty$.

Figure 6.62 shows the SPICE simulation results for the WC circuit. All the parameters of the transistors were assumed to be for a Motorola 1.2-μm CMOS process with $W/L = 3.2 \ \mu$m/2.4 μm ($C_{\mathrm{ox}} \approx 12$ fF). Other parameters are $C_1 = C_2 = 1$ pF, $C_{\mathrm{fg1}} = 0.9$ pF, $C_{\mathrm{fg2}} = 0.1$ pF, $C_{\mathrm{fg3}} = 0.1$ pF, $C_{\mathrm{fg4}} = 0.4$ pF, $R_1 = R_2 = 10$ MΩ, $I_R = 100$ nA, VDD = 3 V and $V_v = 0.5$ V. When $V_u = -1.2$ V ($\theta_u = -0.12$), the WC circuit had a fixed-point attractor, Fig. 6.62a and b, while the circuit had a limit-cycle attractor when $V_u = 0$ V ($\theta_u = 0$), Fig. 6.62c and d. This shows good agreement with the theoretical prediction (Fig. 6.59).

(a)

(b)

Figure 6.63: Numerical simulation results for the two-dimensional reaction-diffusion systems performing edge detection (a) and feature extraction (b).

To construct the reaction-diffusion chip, a number of WC circuits are arranged on a silicon chip and the status nodes (u and v) of each WC circuit are connected with diffusive-reactive coupling ('synaptic') terminals of the neighbouring circuit. For example, the reaction-diffusion circuit can be constructed by connecting the excitatory and inhibitory voltage nodes (u and v) by resistors (D_u and D_v) to make a diffusion network. The resultant reaction-diffusion equations are

$$\frac{\partial u(x,y,t)}{\partial t} = D_u \nabla^2 u - u + f_\beta(u - bv - \theta_u), \tag{6.57}$$

$$\frac{\partial v(x,y,t)}{\partial t} = D_v \nabla^2 v - v + f_\beta(au - cv - \theta_v), \tag{6.58}$$

which represent a two-dimensional reaction-diffusion system with the WC oscillators. Figure 6.63 represents simulation results of the reaction-diffusion system consisting of a 100×100 matrix of WC oscillators. In the simulations, all oscillators were configured to have a fixed-point attractor. As an example, a binary input image was given to the system; that is, the pixel data of the image was encoded to the initial phase of each oscillator ($\phi = \arctan[v(t=0)/u(t=0)]$). When $D_u \leq D_v$, oscillators possessing identical initial phases were attracted to a fixed point ($u \rightarrow 1$ V), grey areas in Fig. 6.63a, while oscillators existing on the boundary of different phases were attracted to the other fixed point ($u \rightarrow 0$ V), black areas in Fig. 6.63a, because of inhibitory interactions between the oscillators. As a result, an outline image was obtained in a stationary state. In contrast, when $D_u \geq D_v$, the system showed oscillatory (non-stationary) behaviours due to the excitatory diffusion effects. By observing at an opportune moment, salient areas in the image could be extracted (feature extraction), as shown in Fig. 6.63b.

Figure 6.64 shows yet more results for the WC reaction-diffusion system with random initial patterns. As described above, the stability of each WC circuit is determined by its neighbours, i.e. the magnitudes of the external inputs (θ_u and θ_v). Therefore, we employ the following equations:

$$\frac{\partial u(x,y,t)}{\partial t} = -u + f_\beta(u - bv - \theta_u + D_u \nabla^2 u), \tag{6.59}$$

$$\frac{\partial v(x,y,t)}{\partial t} = -v + f_\beta(au - cv - \theta_v + D_v \nabla^2 v), \tag{6.60}$$

where the diffusion terms are included in sigmoid functions $f(\cdot)$ so that the terms can fluctuate in θ_u and θ_v. In the simulations, the following parameter set was used: $a = 2.5$, $b = 0.9$, $c = 0.5$, $\theta_u = 0.05$, $\theta_v = 1$, $\beta = 10$ and $D_v/D_u = 50/3$. As shown in Fig. 6.64, this system produced stationary Turing-like patterns. The important feature of the system is that the output striped images (Turing patterns) reflect the initial input images well. Namely, when a striped image is given to the system as an initial image, the system will produce the striped image if D_v/D_u, that determines the output spatial frequency, is set at an appropriate value.

So what will happen when a noisy striped image is given as the original input to the system? Figure 6.65 shows the results. When a striped image with some scratches is presented to a reaction-diffusion processor, the system eliminated all the scratches during the reaction process, and produced striped patterns Fig. 6.65a, as expected. So, we can conclude that the system can restore striped images from

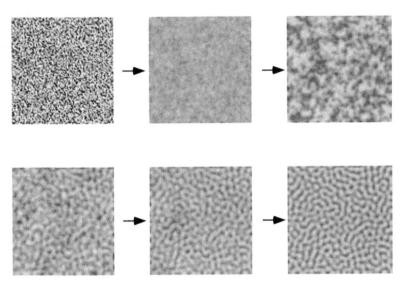

Figure 6.64: Numerical simulation results for the two-dimensional reaction-diffusion systems.

noisy stripes if appropriate values of D_v/D_u are used for a given image. Figure 6.65b shows an example of the application for fingerprint images. The system produced a stable fingerprint pattern without noise and unnatural discontinuities or wrinkles, which implies a future potential of reaction-diffusion chips in engineering applications, e.g. fingerprint re-construction identification systems.

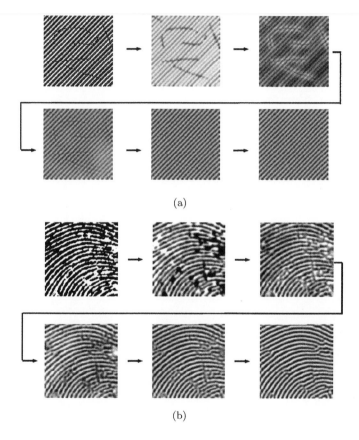

(a)

(b)

Figure 6.65: Numerical simulation results for the two-dimensional reaction-diffusion systems performing image restoration for (a) striped image with scratch and (b) noisy fingerprint image.

Chapter 7

Minority-carrier reaction-diffusion device

In this chapter, reaction-diffusion semiconductor devices that utilise autocatalytic multiplication and diffusion of minority carriers are introduced. We study in computer simulation propagation phenomena of the density waves of minority carriers, and discuss what computational problems can be solved in these types of reaction-diffusion semiconductor devices, as well as the space–time complexity of computation in devices. The practical value of reaction-diffusion chemical systems is significantly reduced by the low speed of travelling waves, which makes real-time computation senseless. The increase in speed will be indispensable for developers of reaction-diffusion computers. Moreover, if a reaction-diffusion system is implemented in integrated circuits, then one would be able to artificially design various types reaction-diffusion spatio-temporal dynamics and thus develop parallel computing processors for novel applications. Based on experimental evidence of reaction-diffusion-like behaviour, namely travelling current density filaments [200], in p-n-p-n devices a novel type of semiconductor reaction-diffusion computing device, where minority carriers diffuse as chemical species and reaction elements are represented by p-n-p-n diodes, was proposed [33].

7.1 Reaction-diffusion computing device with p-n-p-n diode

In forwardly biased p-n junctions, minority carriers are generated in both the areas of the p- and n-type semiconductors, as shown in Fig. 7.1. For p-type semiconductors, the minority carriers are electrons, while they are holes in n-type semiconductors. Once minority carriers are generated, they diffuse among the semiconductor and finally disappear by the re-combination of electrons and holes as

$$
\frac{\partial n(\mathbf{r}, t)}{\partial t} = D_n \nabla^2 n(\mathbf{r}, t) - \frac{n(\mathbf{r}, t)}{\tau},
\tag{7.1}
$$

where n represents the density of minority carriers, $\mathbf{r}$ the space, t the time, $D_n \equiv \mu \, kT/q$ (μ is the mobility of the carriers, k Boltzmann's constant, T the temperature and q the electron charge) and τ the lifetime of the minority carriers [274]. Now

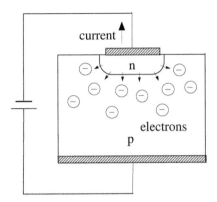

Figure 7.1: Minority-carrier diffusion in forwardly biased p-n junction.

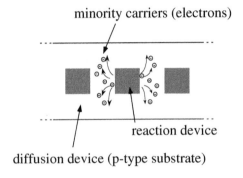

Figure 7.2: Construction of reaction-diffusion device that consists of a number of 'sources' of minority carriers regularly arranged on a common semiconductor die.

we can regard the minority carriers as one chemical substance, i.e. electrons in the p-type semiconductor, because Eq. (7.1) has the same form as a reaction-diffusion equation.

To construct an active reaction-diffusion device, a number of 'sources' of minority carriers must be regularly implemented on a common semiconductor die (substrate), as shown in Fig. 7.2. The source is referred to as a reaction device. Minority carriers produced at a reaction device will travel through the substrate by diffusion and reach the adjacent reaction devices. To generate minority carriers on a diffusive medium, a p-n-p-n device is employed. A p-n-p-n device is a traditional semiconductor device used as triacs, thyristors and so on [274], in each reaction device. The following subsections review the operations of a p-n-p-n device.

7.1.1 A brief review of p-n-p-n device operation

Let us consider the bias condition of a p-n-p-n device shown in Fig. 7.3. In this condition, p^+-n and p-n^+ junctions are forward biased, while n-p junctions are

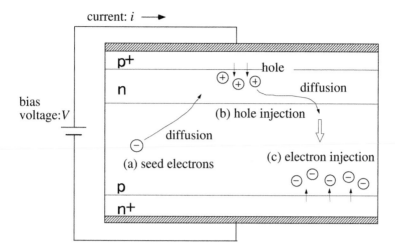

Figure 7.3: Autocatalytic multiplication of minority carriers (electrons) in the p-n-p-n device.

reverse biased. So, the flow of majority carriers, which generates the current flow i, is blocked at this n-p junction. The device is thus in the off state and electrons as minority carriers are generated in the p region. They diffuse into this region and disappear within a given time constant via re-combination with holes.

Assume that a seed electron is given to the p region and it diffuses amongst the region shown (a) in Fig. 7.3. If the electron is not lost via re-combination accidentally and it induces several holes from the p$^+$ region, (b) in Fig. 7.3, each hole also induces multiple electrons from the n$^+$ region, (c) in Fig. 7.3. This results in the autocatalytic multiplication of electrons in the p region. If the p region is fully filled with electrons, the p region is inverted to n-type semiconductor. This causes the transition of the p-n-p-n device from off to on state because the reverse-biased n-p junction becomes just an n-n junction at this state. The initial density of electrons in the region determines a threshold voltage at which the device is turned on. The voltage is called the breakover voltage. Figure 7.4 shows the typical current–voltage characteristics of a p-n-p-n device. If the voltage V exceeds the breakover voltage V_{B}, the device is instantly turned on.

Once the p-n-p-n device is turned on, it cannot be turned off again until the supply voltage V is cut off. When V is set to zero, electrons in the p region (now it is inverted to n type) disappear due to re-combination. As a result, the device returns to the resting (off) state.

7.1.2   A heuristic model of p-n-p-n device

A simple model of a p-n-p-n device for large-scale numerical simulations is introduced here for modelling large-scale p-n-p-n arrays. Let us assume that the p-n-p-n

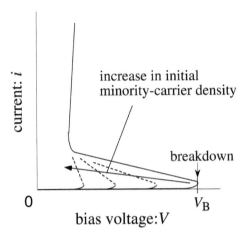

Figure 7.4: Voltage–current characteristic of typical p-n-p-n device.

device is a non-linear resistive device. Its resistance is defined by

$$R(V, n) = \begin{cases} R_{\text{off}}, & \text{if } V < V_{\text{B}}(n) \\ R_{\text{on}} + (R_{\text{off}} - R_{\text{on}}) \exp\left(-\frac{V - V_{\text{B}}(n)}{V_0}\right), & \text{else} \end{cases} \tag{7.2}$$

where V represents the bias voltage, n the density of minority carriers (electrons) in the p region of the p-n-p-n device, R_{off} (or R_{on}) the resistance of the device in off (or on) state and $V_{\text{B}}(n)$ the breakover voltage as a function of n. If $V < V_{\text{B}}(n)$, the device is turned off with a high resistance (R_{off}), while the device is turned on [$R(V, n) \to R_{\text{on}}$] when $V > V_{\text{B}}(n)$, as shown in Fig. 7.5a. Its on–off sensitivity is determined by a constant voltage V_0. This imitates the characteristics of the sudden change of resistance (V/i) in Fig. 7.4.

As described earlier, the breakover voltage is a function of n. The function is defined as

$$V_{\text{B}}(n) = \begin{cases} V_{\text{B}}^{\text{max}}, & \text{if } n < n_{\text{th}} \\ V_{\text{B}}^{\text{min}} + (V_{\text{B}}^{\text{max}} - V_{\text{B}}^{\text{min}}) \exp\left(-\frac{n - n_{\text{th}}}{n_0}\right), & \text{else} \end{cases} \tag{7.3}$$

where $V_{\text{B}}^{\text{max}}$ (or $V_{\text{B}}^{\text{min}}$) represents the maximum (or minimum) breakover voltage, n_{th} the threshold density and n_0 the sensitivity of the change of the breakover voltage. When $n < n_{\text{th}}$, the breakover voltage does not change ($V_{\text{B}} = V_{\text{B}}^{\text{max}}$), while the voltage decreases to $V_{\text{B}}^{\text{min}}$ exponentially when $n > n_{\text{th}}$ (Fig. 7.5b). This imitates the change of the breakover voltage with the value of n in Fig. 7.4.

7.1.3 The reaction-diffusion device using p-n-p-n devices

The reaction device is illustrated in Figs. 7.6 and 7.7. The structure shown in Fig. 7.6 is designed for a one-dimensional reaction-diffusion device that can be fabricated with conventional CMOS technology. Figure 7.6a and b show the top

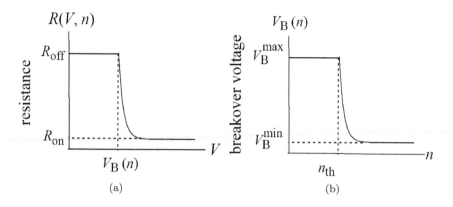

Figure 7.5: Threshold functions of p-n-p-n device model: (a) bias voltage vs. resistance, (b) minority-carrier density vs. breakover voltage.

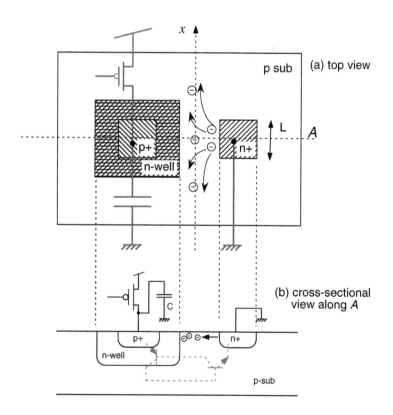

Figure 7.6: Construction of one-dimensional reaction-diffusion device utilising conventional CMOS technology.

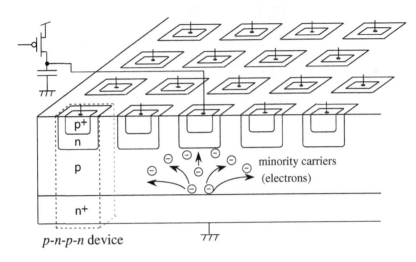

p-n-p-n device

Figure 7.7: Construction of two-dimensional reaction-diffusion device with vertical p-n-p-n devices.

view and a cross-sectional view of the device along with a dashed line A in Fig. 7.6a. A p-n-p-n device is laterally constructed beneath the substrate surface. On the other hand, the device structure shown in Fig. 7.7 is designed for a two-dimensional reaction-diffusion device. A p-n-p-n device is constructed vertically. It will require some special processes for actual fabrication. In both constructions above, minority carriers (electrons) produced at a p region in a p-n-p-n device will travel through the p-type common area by diffusion.

The p-n-p-n device is connected with a capacitor and a pMOS FET acting as a current source. Figure 7.8 shows its equivalent circuit. This combination of a p-n-p-n device, a capacitor and a pMOS FET is an elementary unit of the proposed reaction-diffusion device. This unit is referred to as a 'reaction cell'.

The dynamics of a reaction cell is given by

$$C\frac{dv}{dt} = I_b(v) - \frac{v}{R(v,n)}, \tag{7.4}$$

$$q\frac{dn}{dt} = -q\frac{n}{\tau} + \frac{v}{R(v,n)}, \tag{7.5}$$

where C represents the capacitance, v the capacitor voltage, n the minority-carrier density, q the charge of an electron, R the resistance of the p-n-p-n device defined by Eq. (7.2), $I_b(v)$ the current of the pMOS FET as a function of v and τ the minority-carrier lifetime. The current of the pMOS FET is given by

$$I_b(v) = I_0 \exp\left(\kappa \frac{\text{VDD} - V_{\text{bias}}}{V_T}\right)\left(1 - \exp\left(-\frac{\text{VDD} - v}{V_T}\right) + \frac{\text{VDD} - v}{V_0}\right), \tag{7.6}$$

where VDD represents the supply voltage, V_{bias} the gate voltage of the pMOS FET, I_0 the MOS fabrication parameter, κ the effectiveness of the gate potential

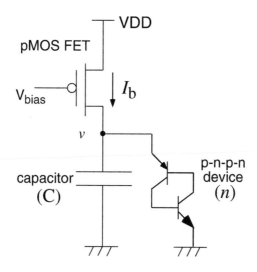

Figure 7.8: Equivalent circuit of a reaction cell.

and $V_T \equiv kT/q \approx 26$ mV at room temperature [296]. Equations (7.4) and (7.5) assume that the capacitor charge decreases by an amount equal to the increased minority carriers. It should be noted that this device imitates a substrate-depleted reaction because the charge of the capacitor is depleted by the p-n-p-n device.

The reaction cell can be oscillatory (astable) or excitatory (monostable) depending on the supply voltage VDD. It is oscillatory if VDD is higher than the breakover voltage V_B of the p-n-p-n device, and excitatory if VDD is lower than V_B. In the oscillatory condition (VDD $> V_B$), the capacitor is charged by bias current $I_b(v)$ and, consequently, the capacitor charge increases until the capacitor voltage v reaches the breakover voltage V_B. When v reaches V_B, the breakover of the p-n-p-n device starts and minority carriers are injected from the n$^+$ region to the p region. Then, the autocatalytic multiplication of minority carriers occurs to turn the device on. The stored charge on the capacitor flows into the device, so the capacitor charge (and therefore the capacitor voltage v) decreases and, consequently, the device is turned off. The reaction cell repeats this cycle and produces oscillatory dynamics. On the other hand, in the excitatory condition (VDD $< V_B$), the capacitor voltage v cannot reach the breakover voltage V_B because v does not exceed the supply voltage VDD — bias current $I_b(v)$ becomes 0 when v increases to VDD. In this condition, the p-n-p-n device turns on only when minority carriers are injected from outside, i.e. from neighbouring cells.

Figure 7.9 shows a one-dimensional reaction-diffusion device with multiple reaction devices on a common substrate. Each reaction device is labeled as Rn (n is the index number), while the diffusion areas are labeled as D. Minority carriers (electrons) produced by each reaction device will travel through the diffusion area (p-subregion) by diffusion and reach adjacent reaction devices. When the reaction devices are closely arranged on the substrate, these minority carriers will induce a chain reaction amongst the reaction devices. One can estimate the minimum

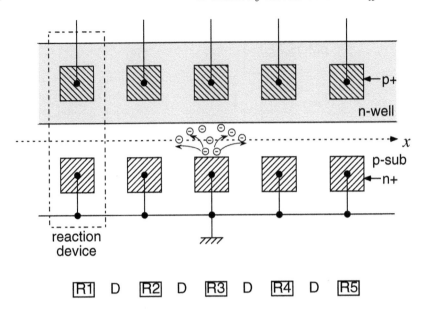

Figure 7.9: Experimental settings for one-dimensional reaction-diffusion device.

distance between adjacent devices to cause the chain reaction as follows.

The dynamics of minority carriers along the x axis in Fig. 7.9 is governed by Eq. (7.1). The impulse response to Eq. (7.1) is

$$g(x) = \frac{1}{\sqrt{4\pi D_n t}} \exp\left(-\frac{x^2}{4D_n t} - \frac{t}{\tau}\right). \tag{7.7}$$

When reaction device R1 turns on at $t = 0$, the minority-carrier concentration around the reaction device increases; see $n(x, 0)$ in Fig. 7.10. The total amount of minority carriers in the p-n-p-n device is assumed to be equal to the charge of the capacitor at $t = 0$. Thus, the minority-carrier distribution $n(x, t)$ $[= n(x, 0)g(x)]$ is obtained as

$$n(x, t) = \frac{N_0}{2} \exp\left(-\frac{t}{\tau}\right) \left[\text{erf}\left(\frac{x + L/2}{\sqrt{4D_n t}}\right) - \text{erf}\left(\frac{x - L/2}{\sqrt{4D_n t}}\right) \right], \tag{7.8}$$

where $N_0 \equiv CVDD/q$ and L represents the length of the reaction device (see Fig. 7.6a).

Let us assume that the reaction device turns on when the total amount of minority carriers in the p-n-p-n device exceeds γN_0. As shown in Fig. 7.10, minority carriers produced by reaction device R1 will impact on the adjacent device R2 when

$$\int_{D+L/2}^{\infty} n(x, t) \mathrm{d}x \geq \gamma N_0, \tag{7.9}$$

where D is the distance between adjacent reaction devices. With a given γ, one can briefly estimate the reaction-diffusion device geometries (L and D). Figure 7.11 shows example plots of Eq. (7.8) with practical device parameters $D_n = 39 \text{ cm}^2/\text{sec}$, $\mu_n = 1500 \text{ cm}^2/\text{V sec}$, $\tau = 1 \text{ }\mu\text{sec}$ and $L = 20 \text{ }\mu\text{m}$.

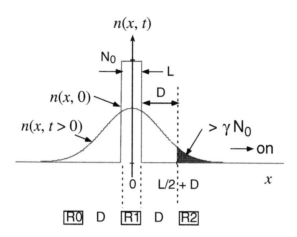

Figure 7.10: Diffusion of minority carriers around an active cell R1.

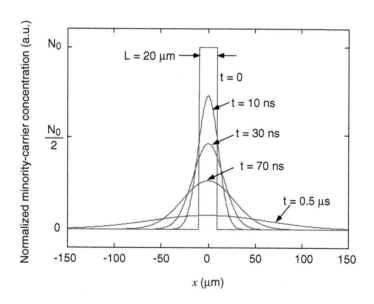

Figure 7.11: Diffusion of minority carriers within a typical semiconductor (silicon) device.

7.2   Numerical simulation results

For simplification of numerical simulation, we can normalise the capacitor voltage
and minority-carrier density in Eqs. (7.4) and (7.5). The resultant equations are

$$\frac{du}{dt} = k_0\, i(u) - \frac{u}{r(u,v)}, \tag{7.10}$$

$$\frac{dv}{dt} = -v + \frac{u}{r(u,v)}, \tag{7.11}$$

where

$$i(u) = 1 - \exp\!\Big(k_1(u-1)\Big) + k_2(1-u), \tag{7.12}$$

$$r(u,v) = \begin{cases} r_{\text{off}}, & \text{if } u < v(n) \\ r_{\text{on}} + (r_{\text{off}} - r_{\text{on}})\exp\!\Big(-\frac{u-v(n)}{\alpha}\Big), & \text{else} \end{cases} \tag{7.13}$$

$$v(n) = \begin{cases} x_b, & \text{if } v < x_s \\ x_f + (x_b - x_f)\exp\!\Big(-\frac{v-x_s}{\beta}\Big), & \text{else} \end{cases}. \tag{7.14}$$

In the following simulations, parameter values obtained from a typical p-n-p-n
device were used. The normalised values are $r_{\text{on}} = 10^{-2}$, $r_{\text{off}} = 10^4$, $\alpha = \beta = 40$,
$x_b = 0.8$ (for oscillatory mode), $x_b \sim 1.2$ (for excitatory mode), $x_f = 0.2$, $x_s = 0.02$,
$k_0 = 0.05$, $k_1 = 25$ and $k_2 = 0.2$. Figure 7.12 illustrates numerical solutions to
Eqs. (7.10) and (7.11). Figure 7.12a shows the relaxation oscillations of a reaction
cell's variables u and v. Figure 7.12b shows the excitatory behaviour of the reaction
cell. The cell settles down in the stable state of $u = 1$ and $v = 0$, and no further
change occurs as long as minority carriers are not injected. In the simulation,
minority carriers were injected from the outside at $t = 50$. Triggered by this
injection, the p-n-p-n device turned on once, and then returned to the stable state.

Now we see the operation of the one-dimensional reaction-diffusion device. The
reaction-diffusion equation was obtained along the x axis in Fig. 7.7. At the position
of each reaction device, the following reaction-diffusion equations were solved:

$$\frac{\partial u}{\partial t} = k_0\, i(u) - \frac{u}{r(u,v)}, \tag{7.15}$$

$$\frac{\partial v}{\partial t} = D_n \frac{\partial^2 v}{\partial x^2} - v + \frac{u}{r(u,v)}. \tag{7.16}$$

In other positions where no reaction device exists, the following equations were
solved:

$$\frac{\partial u}{\partial t} = 0, \qquad \frac{\partial v}{\partial t} = D_n \frac{\partial^2 v}{\partial x^2} - v. \tag{7.17}$$

These equations above were solved by using three-point approximation of the
Laplacian on the one-dimensional grid and the fourth-order Runge–Kutta method.
At each side of the reaction space, the von Neumann boundary condition

$$\nabla[u] = \nabla[v] = (0,0) \tag{7.18}$$

was applied.

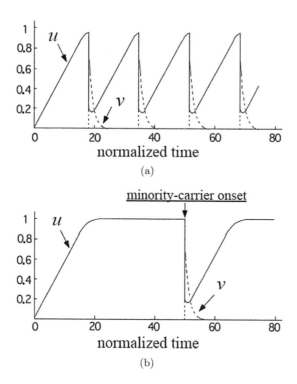

Figure 7.12: Simulation results of reaction cell for oscillatory (a) and excitatory (b) modes.

Figure 7.13 shows a result for the one-dimensional reaction-diffusion device with nine reaction devices. The parameters were $x_b = 1.2$ (excitatory mode), $\gamma = 0.1$ and $D_n = 20 \times 10^{-6}$. In Fig. 7.13, the horizontal and vertical axes represent the normalised space and time, respectively. The position of the reaction device is indicated by Rn $(n = 1, 2, \ldots, 9)$ at the bottom of Fig. 7.13.

At an initial state $(t = 0)$, a centre reaction device produced minority carriers. The carriers diffused around the reaction device and, at $t \approx 25$, its adjacent devices turned on (were activated). They also produce minority carriers; then, at $t \approx 50$ (80), their adjacent reaction devices are activated. The propagating waves are produced in the form of propagation activations of the reaction devices.

Very slow decay of the minority-carrier concentration was observed in a reaction device being activated by its adjacent device. Figure 7.14 shows the time course of the concentration of the reaction device. The normalised refractory period was of the order of 1. It was approximately a hundred times as long as the propagating time between the adjacent reaction devices. Note that the decay time can be controlled by ejecting minority carriers from the p-subregion.

A two-dimensional reaction-diffusion device was also designed by arranging the reaction cells on a plane. The same set of reaction-diffusion equations as Eqs. (7.15), (7.16) and (7.17), where the diffusion terms were expanded to two dimensions x and

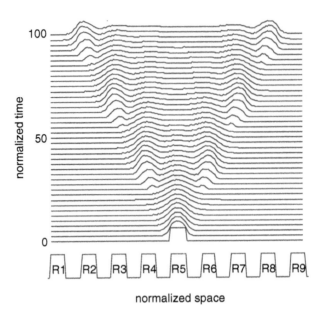

Figure 7.13: Active wave propagation on a one-dimensional reaction-diffusion device.

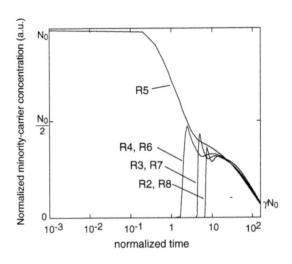

Figure 7.14: Temporal changes of concentration of minority carriers during wave propagation.

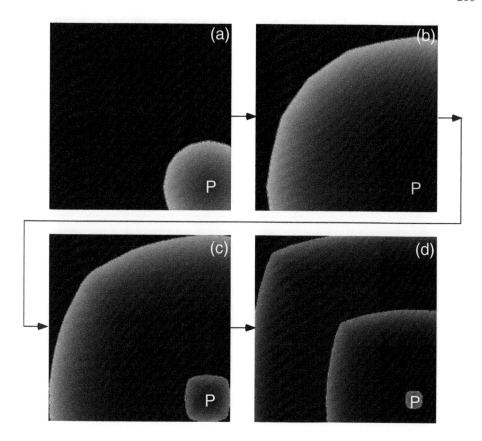

Figure 7.15: Simulation results of the two-dimensional reaction-diffusion device producing target patterns.

y, were solved by five-point approximation of the Laplacian on the two-dimensional grid and the fourth-order Runge–Kutta method. At each side of the reaction space, we applied the von Neumann boundary condition. Figure 7.15 shows a result for a device with 200×200 excitatory reaction cells. The spatial density of minority carriers is represented in grey scale ($v = 0$: black, $v = 1$: white). With periodic injection of minority carriers at a point (P in Fig. 7.15), the reaction-diffusion device produced spreading concentric waves of minority carriers. These results indicate that the injected carriers diffused around the injection point and successfully induced a chain reaction of the adjacent cells.

Figure 7.16 shows the result of an excitatory reaction-diffusion device without external injections of minority carriers. With an appropriate initial pattern of minority-carrier densities, the reaction-diffusion device produced rotating spiral patterns of minority carriers. Notice that the wave disappears at the collision points (Fig. 7.16c and d) because of the depletion of minority carriers. This is the same phenomenon as observed in natural reaction-diffusion systems.

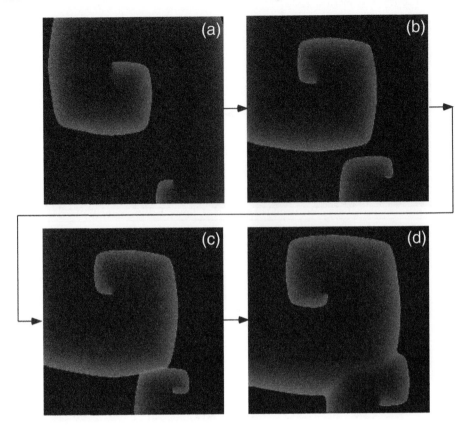

Figure 7.16: Simulation results of the two-dimensional reaction-diffusion device producing spiral waves.

7.3   Computing in reaction-diffusion semiconductor devices

Now we understand how reaction-diffusion semiconductor devices behave but we still need to specify their operational characteristics in computational terms. A reaction-diffusion computation, particularly in active chemical media, is implemented by travelling waves which interact with, or collide with, each other and form a dynamical — oscillating pattern, or stationary — precipitate concentration profile — structure. This structure represents the result of the computation, while the data are naturally represented by the initial configurations of the wave generators.

Before starting any computation one should input information into the reaction-diffusion medium. A parallel input is an essential feature of a cutting edge parallel computing architecture. Serial inputs, so common for the vast majority of massively parallel processors, dramatically decrease the performance of the computing devices, particularly those operating in transducer mode, where information is constantly fed into the processor (e.g. in tasks of image processing). Experimental reaction-diffusion chemical computers, at least in some specific cases, may well have analogues of parallel inputs. It has been demonstrated widely that by applying light of varying intensity we can control excitation dynamics in the Belosuov–Zhabotinsky (BZ) medium [47, 218, 117, 100], wave velocity [243] or pattern formation [299]. Of particular interest is experimental evidence of light-induced back-propagating waves, wave-front splitting and phase shifting [308]; we can also manipulate the medium's excitability by varying the intensity of the illumination [64]. This proved to be particularly important for experiments concerned with image processing using BZ-medium-based computing devices [154, 230, 232, 233]. We are not aware of any rigorous experimental results demonstrating the possibility of optical inputs of reaction-diffusion semiconductor devices. However, there is simulation-related evidence of the possibility of optical parallel inputs. Thus, Mohajerzadeh *et al.* [184] discuss the particulars of such a photoresponse, via photon-induced generation of electron–hole pairs in p-n-p-n devices, dependent on the primary colour components in the stimulating light input as elementary processors that can also act as colour sensors.

Thus, there exists the possibility of parallel optical outputs on semiconductor devices. Technologies for integrating optoelectronic devices and electronic circuitry are fully developed, but limited hybrid integration is available commercially at present. An important problem of such integration is that pursuing it involves the simultaneous development of sophisticated technologies for optoelectronic devices (III-V semiconductors) and silicon integrated circuits. Indeed, recent development in optoelectronic integrated circuits (OEICs) enables us to implement light-emitting devices (LEDs) on silicon substrates by controlling defects at the III-V–silicon interface. For example, Furukawa et al. [107] demonstrated that lattice-matched and defect-free GaPN epilayers can be grown on silicon with a thin GaP buffer layer. The task is enormous and as a practical matter only small-scale OEICs have been demonstrated. While the integration levels of III-V OEICs have remained low, the degree of integration in commercial GaAs integrated circuits has reached VLSI levels in recent years. These advances offer a route to achieving much higher levels of optoelectronic integration through epitaxial growth of III-V heterostructures on GaAs-based VLSI electronics.

Most experimental prototypes of chemical excitable computing devices suffer from difficulties in representing effectively the results of wave interaction. This is because, being excited, a medium's micro-volume becomes refractory and then reverts to the resting state. Excitation wave fronts usually annihilate as a result of collision. Therefore, experimental excitable reaction-diffusion computers required to date some external devices, like a digital camera, to record their spatio-temporal dynamics. Thus, for example, to compute a collision-free path around obstacles in a thin-layer BZ medium one must record snapshots of the BZ-medium's activity and then analyse this series of snapshots to extract the results of the computation [28, 11]; thus, a degree of conventional processing is also presently required. This is the cost we pay for re-usable reaction-diffusion processors. Another option of preserving the results of the computations may be to employ a non-excitable reaction-diffusion medium, where a precipitate is formed as a result of diffusive wave interactions on a substrate (or competition of several wave fronts for the same substrate) [18, 80, 15, 12]. Precipitation is an analogue of infinite memory. The feature is very useful;

however, it also makes the experimental prototypes simply disposable — in contrast to excitable media the precipitate-forming media can be used just once.

In contrast, reaction-diffusion semiconductor computers, because they are essentially man-made devices, may allow us to combine the re-usability and rich space–time dynamics of the excitable reaction-diffusion media with the low post-processing costs of the precipitate-forming reaction-diffusion media. This can be accomplished by embedding a lattice of oscillatory p-n-p-n elements into a lattice of excitatory p-n-p-n elements. The excitatory elements (EEs) will form a substrate to support travelling excitation waves while oscillatory elements (OEs) will play the role of a re-writable memory. For example, to represent sites of wave-front collision we must adjust an activation threshold of OEs in such a manner that the wave front of a single wave will not trigger these OEs; however, when two or more wave fronts collide the OEs at the sites of collision are triggered and continue to oscillate even when the wave fronts have annihilated.

What computational tasks can be realised in reaction-diffusion semiconductor devices? Most primitive operations of image processing (see overview in [7]), e.g. detection of a contour and enhancement, can be mapped straight onto the silicon architecture. However, silicon implementation of reaction-diffusion (precipitate formation based) algorithms for computational geometry — Voronoi diagrams [18, 80, 81] and skeletonisation [19] — would require the above-mentioned embedding of oscillatory elements in a lattice of excitatory p-n-p-n devices; the oscillating elements will represent the bisectors of a Voronoi diagram and the line segments of a skeleton.

Universal computation can also be realised in reaction-diffusion semiconductor devices using two approaches. Firstly, by employing a collision-based mode of computation such as that discussed for excitable media in [5], where functionally complete sets of Boolean gates are implemented by colliding waves. In collision-based mode, however, we must force the medium to be in a subexcitable regime which may pose some problems from a fabrication point of view. Secondly, we can 'physically' embed logical circuits, namely their diagram-based representations, into the excitable medium [283, 269]. In this we employ particulars of the photoresponse of p-n-p-n elements and project the logical circuit onto the medium as a pattern of heterogeneous illumination.

7.4   Conclusion

In this chapter, a massively parallel computing device based on principles of information processing in reaction-diffusion chemical media [6, 7] was introduced. This novel silicon device imitates autocatalytic and dissipative phenomena of the chemical reaction-diffusion systems; however, when compared to the real chemical medium, this semiconductor analogue of the reaction-diffusion processors functions much faster. We studied the operational characteristics of the reaction-diffusion silicon devices and demonstrated the feasibility of the approach in several computational tasks. The results indicate that the proposed reaction-diffusion device will be a useful tool for developing novel hardware architectures based on the reaction-diffusion principles of information processing.

Chapter 8

Single-electron reaction-diffusion devices

This chapter introduces a single-electron device that is analogous to a reaction-diffusion system [209]. This electrical reaction-diffusion device consists of a two-dimensional array of single-electron non-linear oscillators that are combined with one another through diffusive coupling. The device produces animated spatio-temporal patterns of node voltages, e.g. rotating spiral patterns similar to those found in colonies of cellular slime moulds and a dividing-and-multiplying pattern that reminds us of cell division. A method of fabricating actual devices by using self-organised crystal growth technology is also described. The following is an excerpt from [209]. For details, see the reference.

8.1 Constructing electrical analogue of reaction-diffusion systems

The behaviour of reaction-diffusion systems, or the spatio-temporal patterns of chemical concentration, can be expressed by the reaction-diffusion equation, a partial differential equation with chemical concentrations as variables:

$$\frac{\partial \mathbf{u}}{\partial t} = \mathbf{f}(\mathbf{u}) + D\Delta\mathbf{u} \qquad [\mathbf{u} = (u_1, u_2, u_3, \dots)], \tag{8.1}$$

where t is time, $\mathbf{u}$ is the vector of chemical concentrations, u_i is the concentration of the ith substance and D is the diagonal matrix of diffusion coefficients. Non-linear function $\mathbf{f}(\mathbf{u})$ is the reaction term that represents the reaction kinetics of the system. Spatial derivative $D\Delta\mathbf{u}$ is the diffusion term that represents the change of $\mathbf{u}$ due to the diffusion of the substance. A greater number of variables results in more complex dynamics and a more complicated dissipative structure. A simple reaction-diffusion system with few variables, however, will still exhibit quite complex dynamics similar to those observed in biological systems.

A reaction-diffusion system can be considered an aggregate of coupled chemical oscillators, or a chemical cellular automaton, as described in Fig. 8.1. Each oscillator represents the local reaction of chemical substances and generates non-linear dy-

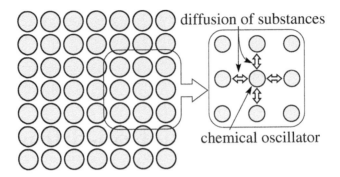

Figure 8.1: Simplified model of reaction-diffusion systems, consisting of many chemical oscillators. Each oscillator has variables corresponding to chemical concentrations u_1, u_2, u_3, ... in Eq. (8.1) and interacts with its neighbours through diffusion of substances.

namics $\mathrm{d}\mathbf{u}/\mathrm{d}t = \mathbf{f}(\mathbf{u})$ that corresponds to reaction kinetics in Eq. (8.1). The oscillator interacts with its neighbours through non-local diffusion of substances; this corresponds to the diffusion term in Eq. (8.1) and produces dynamics $\mathrm{d}\mathbf{u}/\mathrm{d}t = D\Delta\mathbf{u}$. Because of diffusion, all oscillators correlate with one another to generate synchronisation and entrainment. Consequently, the system as a whole produces orderly dissipative structures on a macroscopic level. The size of each oscillator, or the size of the local space in which chemical concentrations are roughly uniform, depends on the diffusion coefficients and reaction velocities in the system. It is several micrometres in diameter in many liquid reaction-diffusion systems; therefore, even a tiny reaction-diffusion system in a test tube contains many millions of oscillators.

An electrical analogue of reaction-diffusion systems can be created by using electrical oscillation circuits instead of chemical oscillators and coupling these circuits with one another in a way that imitates diffusion. Variables are the electrical potentials of nodes in the oscillation circuits in this electrical reaction-diffusion system. The system will produce electrical dissipative structures, i.e. orderly spatio-temporal patterns of node potentials, under appropriate conditions.

The key to building an electrical reaction-diffusion system is to integrate a large number of oscillation circuits on a chip with coupled subcircuits. A large arrangement of oscillators (e.g. 1000×1000 or more) is needed to generate complex, varied dissipative structures as observed in chemical reaction-diffusion systems. Oscillators constructed by single-electron circuits are thus useful to achieve such large-scale integration. A single-electron circuit can generate non-linear oscillation through a simple circuit structure, so it can effectively be used in producing small oscillators for electrical reaction-diffusion systems.

8.1.1   Non-linear oscillator using a single-electron circuit

Figure 8.2 shows an electrical oscillator constructed by a single-electron circuit. It consists of a tunnelling junction C_j and a high resistance R connected in series

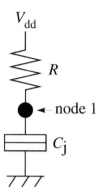

Figure 8.2: Single-electron oscillator (a SET cell) consisting of tunnelling junction C_j, high resistance R connected at node 1 and positive bias voltage V_{dd}.

at node 1 and biased by a positive voltage V_{dd}. This circuit is an elementary component of single-electron circuits known as the SET cell, see [114] for detailed explanation. A SET cell only has a single variable, voltage V_1 of node 1, but it can be oscillatory or excitatory in operation — which is indispensable in creating reaction-diffusion systems — because the node voltage can produce a discontinuous change because of electron tunnelling. In continuous-variable systems such as chemical reaction systems, two or more variables are needed for oscillatory and excitatory operations.

The SET cell operates as a non-linear oscillator at the low temperatures at which the Coulomb-blockade effect occurs. It is oscillatory (astable) if $V_{dd} > e/(2C_j)$ (e is the elementary charge) and produces non-linear oscillation in voltage at node 1 (Fig. 8.3a). The node voltage gradually increases as junction capacitance C_j is charged through resistance R, then drops discontinuously because of electron tunnelling through the junction and again gradually increases to repeat the same cycles. In contrast, the oscillator is excitatory (monostable) if $V_{dd} < e/(2C_j)$ and produces single-pulse operation excited by an external trigger (Fig. 8.3b). A modified Monte Carlo method is used for the simulation. Kuwamura *et al.* [158] have given details of this method; also see the appendix in [277]. For the task of constructing electrical reaction-diffusion systems, oscillatory and excitatory ones, or both, can be used.

The oscillator exhibits discontinuous, probabilistic kinetics resulting from electron tunnelling. The kinetics is given in the form of

$$\frac{dV_1}{dt} = \frac{V_{dd} - V_1}{R\,C_j} - \frac{e}{C_j}\delta\left(V_1 - \frac{e}{2C_j} - \Delta V\right),$$

where $\delta()$ represents a discontinuous change in node voltage caused by electron tunnelling. Probabilistic operation arises from the stochastic nature of tunnelling, i.e. a time lag or delay exists between the time when the junction voltage exceeds the tunnelling threshold $e/(2C_j)$ and when tunnelling actually occurs. This effect is represented by the delay term ΔV in the equation. Because the value of ΔV has probabilistic fluctuations in every tunnelling event and cannot be expressed in an

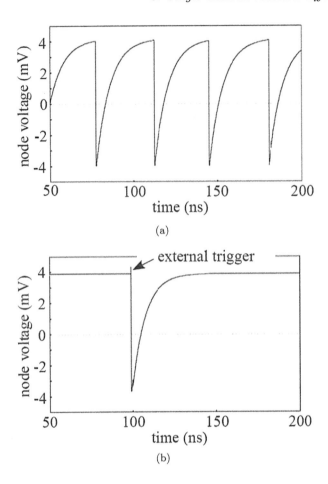

Figure 8.3: Operation of the oscillator. Waveforms of node voltage are shown for (a) self-induced oscillation and (b) monostable oscillation, simulated with the following set of parameters: tunnelling junction capacitance $C_j = 20$ aF, tunnelling junction conductance 1 μS, high resistance $R = 400$ MΩ and zero temperature. Bias voltage $V_{dd} = 4.2$ mV for self-induced oscillation and $V_{dd} = 3.8$ mV for monostable oscillation.

analytical form, the Monte Carlo simulation is necessary for studying the behaviour of the oscillator.

8.1.2 Single-electron oscillator with a multiple tunnelling junction

In fabricating actual oscillators, a high resistance of hundreds of mega-ohms or more is not easy to implement on an LSI chip. A better way is to use a multiple tunnelling junction, i.e. a series of many tunnelling junctions, instead of a high resis-

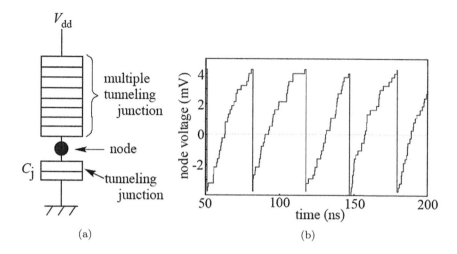

(a)   (b)

Figure 8.4: Single-electron oscillator with a multiple tunnelling junction: (a) circuit configuration,(b) simulated self-induced oscillation. Parameters are as follows: capacitance of a single tunnelling junction $C_j = 10$ aF, conductance of the single tunnelling junction 1 μS, 30 tunnelling junctions in the multiple tunnelling junction, capacitance and conductance of a tunnelling junction in the multiple tunnelling junction are 300 aF and 50 nS, bias voltage $V_{dd} = 8.6$ mV and zero temperature.

tance (Fig. 8.4a). This structure also enables oscillatory and excitatory operations to be obtained because sequential electron tunnelling through a multiple tunnelling junction has a similar effect to current flowing at high resistance (Fig. 8.4b). In the following sections, however, the high-resistance SET cell (Fig. 8.2) is used to construct electrical reaction-diffusion systems because less computing time is required in simulating the reaction-diffusion operation. One can expect that the knowledge obtained from high-resistance reaction-diffusion systems will be able to be easily applied to reaction-diffusion systems consisting of multiple-junction oscillators.

8.1.3   Diffusive coupling of oscillators

To construct reaction-diffusion systems, oscillators have to be connected with one another so that they will interact through 'diffusive' coupling to generate synchronisation and entrainment. To do this, the oscillators are connected by means of intermediary oscillation cells and coupling capacitors. Figure 8.5a illustrates the method of connection with a one-dimensional chain of oscillators. The oscillators — SET cells denoted by A1, A2, ..., with their nodes represented by closed circles — are connected with their neighbouring oscillators through intermediary oscillation cells — SET cells denoted by B1, B2, ..., with their nodes represented by open circles — and coupling capacitors C. We use an excitatory SET cell biased with a negative voltage $-V_{ss}$ as the intermediary oscillation cell.

When electron tunnelling occurs in an oscillator in this structure, the node volt-

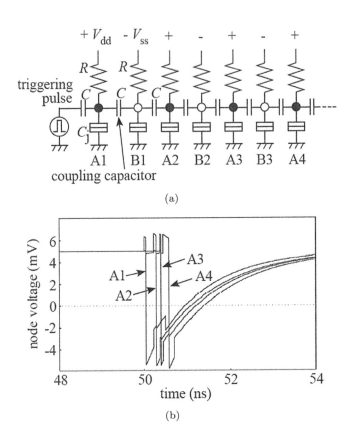

(a)

(b)

Figure 8.5: Diffusive connection of oscillators. (a) One-dimensional chain of oscillators (A1, A2, …) with intermediary cells (B1, B2, …) and coupling capacitors C. For the study of the transmission of tunnelling, a triggering pulse generator is connected to the left-hand end. (b) Transmission of tunnelling through the chain of excitatory oscillators. The waveform of the node voltage is plotted for each oscillator. A triggering pulse was applied to the oscillator on the far left, and tunnelling started at this oscillator to transmit along the chain with delay. Jumps in curves A1–A4 result from electron tunnelling in oscillators A1–A4. Simulated with a set of parameters: $C_j = 10$ aF, $C = 2$ aF, $R = 77$ MΩ, tunnelling junction conductance $= 5$ μS, $V_{dd} =$ mV, $-V_{ss} = -5$ mV and zero temperature.

age of the oscillator changes from positive to negative, and this induces, through the coupling capacitor C, electron tunnelling in an adjacent intermediary cell. The induced tunnelling changes the node voltage of the intermediary cell from negative to positive, and this induces electron tunnelling in an adjacent oscillator. In this way, electron tunnelling is transmitted from one oscillator to another along the oscillator chain. There is a time lag between two tunnelling events in two neighbouring oscillators as if these oscillators interacted via diffusion. This phenomenon is not diffusion itself and cannot be expressed in the form $D\Delta \mathbf{u}$ in Eq. (8.1), but can be used as a substitute for diffusion.

The transmission of tunnelling with delay is illustrated in Fig. 8.5b with simulated results for a chain of excitatory oscillators with intermediary cells. Electron tunnelling was induced in the oscillator on the far left by a triggering pulse, and it was transmitted to the right along the chain with delay. In other words, an excitation wave of tunnelling travelled from left to right along the chain. Its delay in travelling from one oscillator to a neighbour has probabilistic fluctuations because of the stochastic nature of tunnelling, but this is not a problem for applications to produce reaction-diffusion systems.

8.1.4 Electrical reaction-diffusion system with single-electron oscillators

An electrical reaction-diffusion system can be constructed by connecting oscillators into a network by means of intermediary cells and coupling capacitors (Fig. 8.6). Each oscillator is connected to its neighbouring four oscillators by means of four intermediary cells and coupling capacitors. This is a two-dimensional reaction-diffusion system. A 3D reaction-diffusion system can also be constructed in a similar way by arranging oscillators into a cubic structure and connecting each oscillator with its six neighbouring oscillators by means of six intermediary cells and coupling capacitors.

8.2 Spatio-temporal dynamics produced by the single-electron system

In the single-electron reaction-diffusion system, the node voltage of each oscillator changes temporally as the oscillators operate through mutual interactions. Consequently, a two-dimensional spatio-temporal pattern of the node voltages is produced in the reaction-diffusion system. Since this voltage pattern corresponds to the dissipative structure in chemical reaction-diffusion systems, it can be called an 'electrical dissipative structure'.

A variety of electrical dissipative structures are produced from different sets of system parameters. To understand the behaviour of an electrical reaction-diffusion system entirely, a phase diagram for the system must be drawn, i.e. a diagram that depicts — in the multidimensional space of system parameters — what kind of dissipative structure will appear for each set of parameter values. However, a phase diagram for our reaction-diffusion system cannot be drawn without a long numerical computer simulation because its reaction-diffusion kinetics cannot be expressed

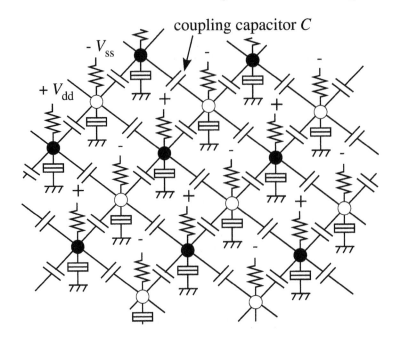

Figure 8.6: Two-dimensional reaction-diffusion system consisting of a network of single-electron oscillators. Each oscillator (closed-circle node) is connected with four neighbouring oscillators by means of four intermediary cells (open-circle nodes) and coupling capacitors.

in analytical form. Instead, a few examples of electrical dissipative structures simulated with a few sample sets of parameter values are demonstrated.

Although a single-electron reaction-diffusion system differs greatly from chemical reaction-diffusion systems in terms of reaction-diffusion kinetics, it can produce dissipative structures similar to those of chemical reaction-diffusion systems. We show three examples — an expanding circular pattern, a rotating spiral pattern and a dividing-and-multiplying pattern. The following sections detail the results simulated for a reaction-diffusion system consisting of 201×201 excitatory oscillators and 200×200 intermediary cells.

A. Expanding circular pattern

A single-electron reaction-diffusion system consisting of excitatory oscillators is in a stable uniform state as it stands. Once a triggering signal is applied to an oscillator in the system, an excitation wave of tunnelling starts at the oscillator and propagates in all directions to form an expanding circular pattern. This can be seen in Fig. 8.7, where the node voltage of each oscillator is represented by a grey scale: the light shading means high voltage, and the dark means low voltage. The front, labelled 'F' in Fig. 8.7, of the wave is the region where tunnelling just occurred and, therefore, the node voltage of the oscillators is at the lowest negative

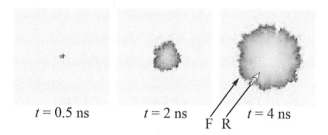

$t = 0.5$ ns $t = 2$ ns $t = 4$ ns
                                              F R

Figure 8.7: Snapshots of expanding circular pattern in the single-electron reaction-diffusion system, simulated with the following parameters: tunnelling junction capacitance $C_j = 1$ aF, tunnelling junction conductance 1 μS, high resistance $R = 137.5$ MΩ, coupling capacitance $C = 1$ aF, bias voltage $V_{dd} = 16.5$ mV, $-V_{ss} = -16.5$ mV and zero temperature.

value. The front line is uneven or irregular because the velocity of the travelling wave fluctuated in each direction throughout the process because of the stochastic delay of tunnelling.

After the excitation wave has passed through, the node voltage of each oscillator gradually increases and returns to its initial value, the positive bias voltage. This is indicated in Fig. 8.7 by the light shading on the rear, labelled 'R', of the wave. If a triggering signal is applied repeatedly to one oscillator, a concentric circular wave — called a target pattern in chemical reaction-diffusion systems — will be generated.

B. Rotating spiral pattern

This pattern appears when an expanding circular wave is broken by an external disturbance, thereby making an endpoint appear in the wave front. With this endpoint acting as a centre, the wave begins to curl into itself and forms a rotating spiral pattern (Fig. 8.8). The principle of curling is similar to that in chemical reaction-diffusion systems.

In this example, a triggering signal was applied to the middle oscillator on the left of the reaction-diffusion system. When an excitation wave started and expanded a little, the lower half of the wave was broken by re-setting the node voltage of the oscillators to zero (Fig. 8.8, $t = 3$ nsec). After that, the reaction-diffusion system was left to operate freely, and a rotating spiral pattern of node voltages was automatically generated as can be seen in Fig. 8.8, $t \geq 6$ nsec.

C. Dividing-and-multiplying pattern

This pattern appears when the coupling between oscillators is weak (i.e. small coupling capacitance or low bias voltage). When this happens, electron tunnelling in an oscillator cannot be transmitted to all four adjacent intermediary cells, e.g. tunnelling can be transmitted to the right and left cells but not to the upper and lower cells. As a result, an expanding node-voltage pattern splits into many pieces,

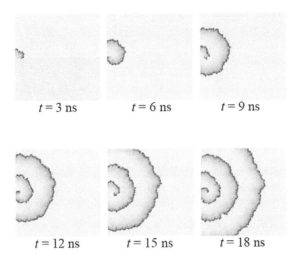

Figure 8.8: A series of snapshots of rotating spiral pattern, simulated with the same parameters as used for Fig. 8.7.

or wave fragments, and each wave fragment expands and splits again. This produces dividing-and-multiplying patterns (Fig. 8.9). The principle of division is different from that in chemical reaction-diffusion systems, but the behaviour of the created patterns is somewhat similar. In a way, we may consider that there are electrical 'microbes' consisting of negative voltages 'living' on the reaction-diffusion system, these eating positive charges on the nodes, and thus propagating, enabling them to spread through the system.

8.3 Towards actual reaction-diffusion devices

The unit element in this reaction-diffusion system is a single-electron oscillator coupled with four neighbours. The multiple-tunnelling-junction oscillator (Fig. 8.4) is preferable for this element because it can be made without high resistance, which is difficult to implement on an LSI chip. Arranging such oscillators into a two-dimensional array produces a reaction-diffusion system, so the next task is to fabricate many identical oscillators on a substrate. Figure 8.10a shows the 3D and cross-sectional schematics for the structure of the device. Each oscillator consists of a conductive nano-dot (minute dot) with four coupling arms, and there is a tunnelling junction between the nano-dot and the conductive substrate beneath it. Many series-connected junctions run between the nano-dot and a positive-bias or a negative-bias electrode. Capacitive coupling between neighbouring oscillators can be achieved by laying their coupling arms close to each other.

The key in this construction is to prepare a large arrangement of nano-dots with coupling arms and tunnelling junctions. A process technology that could be used to fabricate the reaction-diffusion-system structure was previously proposed and demonstrated [208]. This technology uses self-organised crystal growth achieved

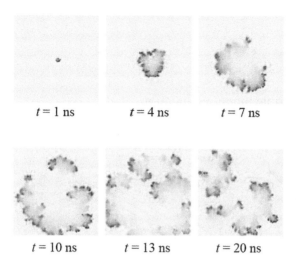

$t = 1$ ns $t = 4$ ns $t = 7$ ns

$t = 10$ ns $t = 13$ ns $t = 20$ ns

Figure 8.9: Snapshots of dividing-and-multiplying pattern, simulated with the same parameters as used in producing patterns shown in Fig. 8.7 apart from $R = 150.5$ MΩ, $V_{dd} = 15.8$ mV and $-V_{ss} = -15.8$ mV.

by selective-area metalorganic vapour-phase epitaxy (SA-MOVPE), and it can be used to fabricate GaAs nano-dots with arms and tunnelling junctions on a GaAs substrate by making use of the dependence of the crystal-growth rate on crystal orientation; see details in [156, 155]. With this technology, a nano-dot with four coupling arms can be formed automatically in a self-organising manner. This technology can also be used to automatically create the structure for multiple tunnelling junctions on nano-dots simply by repeating the growth of an n-type GaAs layer and an insulating AlGaAs layer. Using such a process, the formation of GaAs nano-dots with their arms and tunnelling junctions beneath them was successful in the form of a two-dimensional array on a substrate (Fig. 8.10b and c), though the technology is not yet perfect and a complete device has not yet been fabricated. An improved process technology to form GaAs nano-dots with arms and multiple tunnelling junctions is now under development, where the arms and multiple tunnelling junctions are arranged regularly with a smaller pitch of 100 nm or less, corresponding to 1010 oscillators/cm^2. With the improved process technology, we will be able to integrate coupled single-electron oscillators on a chip and proceed from there to develop reaction-diffusion-based LSIs.

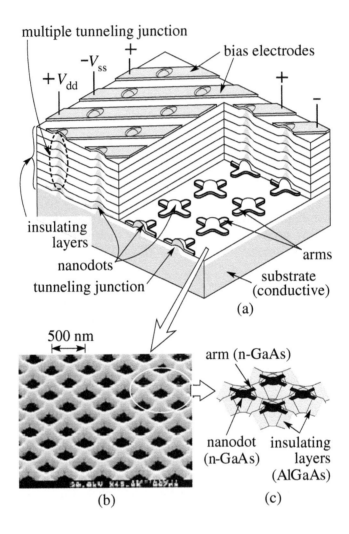

Figure 8.10: Device structure for the single-electron reaction-diffusion system: (a) 3D and cross-sectional schematics, (b) SEM photograph of a two-dimensional array of GaAs nano-dots with coupling arms and tunnelling junctions, (c) schematic diagram of the nano-dots with coupling arms.

Chapter 9

Non-constructibility: from devil's advocate

By the end of this part of the book dealing predominantly with chemical systems and models thereof, we decided to make an unusual move and to cool down the reader's excitement by demonstrating that reaction-diffusion computers are not 'as perfect' as we want them to be (that is, if you hadn't guessed already!). In this chapter we show some limitations of the chemical processors in relation to two problems: construction of a Voronoi diagram — given $\mathbf{P}$, compute VD($\mathbf{P}$), and inverting a Voronoi diagram — given VD($\mathbf{P}$), compute $\mathbf{P}$. The first problem was discussed in detail in Chap. 2, the second one – inversion of a Voronoi diagram, is a much more exotic problem (when compared to the construction of a Voronoi diagram) and just a handful of algorithms are available, see [246] for a brief account of existing (conventional) techniques of Voronoi-diagram inversion.

In experiments, discussed in this chapter, we used the 'palladium processor', a reaction-diffusion chemical processor based on the reaction of palladium chloride loaded gel with potassium iodide; this was originally discussed in [282, 18] and significantly improved and modified in [79, 81, 13]; see details in Chap. 2.

The processor starts its computation as soon as drops/shapes with potassium iodide are applied; the computation is finished when no more precipitate is formed. Configurations of drops/filter-paper templates of potassium iodide correspond to the input states of the processor; the resulting two-dimensional concentration profile of the precipitate (iodo-palladium species) is the output state of the processor. The processor can be seen as a pre-programmed or hard-wired device because the way in which the chemical species react could not be changed.

9.1 Computing with singularities

As previously, we represent data points (to be separated) using a clear solution of potassium iodide. The potassium iodide diffuses from the sites of the drops or from the edges of planar shapes into the palladium chloride loaded gel. The potassium iodide reacts with the palladium chloride to form iodo-palladium species. Palladium chloride gel is lightly yellow coloured while iodo-palladium species are dark brown.

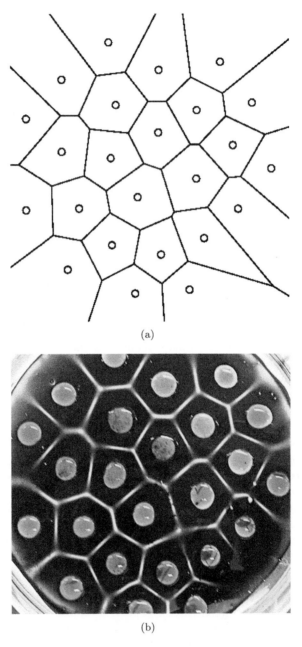

(a)

(b)

Figure 9.1: (a) Voronoi diagram of the planar set **P** (elements of **P** are shown by small circles) and (b) Voronoi diagram of the same planar set **P** implemented in an experimental reaction-diffusion processor (elements of **P** are light-coloured discs).

Thus, during the process of computation we observe the growth of dark-brown patterns emanating from the initial data sites. At sites where two or more diffusive fronts meet each other almost no precipitate is formed; these sites therefore remain uncoloured, and they represent the bisectors of a Voronoi diagram, generated by the initial geometrical configuration of data objects (Fig. 9.1).

The concentration of the precipitate at a given site of the processor represents the likelihood of that site belonging to the edges of a Voronoi diagram — due to the specific mechanism of the chemical computation, the sites corresponding to initial data objects also remain uncoloured. The degree of lightness, or lack of precipitate, of the processor's colouration can be represented by a grey-scale image $g(VD(\mathbf{P}))$ of the processor taken at the end of the computation. To extract a Voronoi diagram from the image of the chemical processor, we determine a threshold θ such that if $g(x) > \theta$ then $x \in VD(\mathbf{P})$.

To produce an image of $VD(\mathbf{P})$ we 'switch' all pixels above the threshold to white and all others to black. A series of $VD(\mathbf{P})$ for various thresholds θ is shown in Fig. 9.2.

For these examples and for all other computations utilising this type of processor, it is almost impossible to find a threshold value θ which is maximal enough to cut off regions not belonging to $VD(\mathbf{P})$ and minimal enough to keep the edges of $VD(\mathbf{P})$ connected.

It is common for very short segments rather than an unbroken bisector to be formed at the bisecting sites of certain sets of points. Why does this occur? Is the experimental technique at fault or is the underlying mechanism of the reaction-diffusion processor responsible?

To prove the latter we will analyse the dynamics of wave-front interactions and its influence on the precipitation kinetics. Possible physico-chemical mechanisms of bisector formation were tackled in [282, 18, 13, 15, 79] and in Chap. 2. It is thought that diffusing wave fronts repel and annihilate, which results in the formation of a depleted zone with no (negligible amounts of) precipitate (light sites representing bisectors) [81].

We could assume that the repulsive interaction of the fronts depends on the speed of the advancing fronts, which in turn depends on the fronts' curvature: $v = w - D/c$, where v is the velocity of the curved front, w is the velocity of the planar wave front, D is a diffusion coefficient and c is the curvature, see e.g. [179]. There are three identifiable phases of bisector formation.

In the first phase, two fronts approach each other at high velocity; due to the velocities of the fronts they are able to overcome repulsive forces and thus annihilate at very close range (minimal distance between the points). A cusp is formed, and this cusp-like bound state of the two fronts accelerates (due to negative curvature), causing maximal depletion of the substrate and the formation of very light-coloured segments at the sites of the fronts' initial interaction (Fig. 9.3a, phase 1).

In the second phase, the curvature value of the cusp is lowered and the depletion of the substrate slows down, causing erosion of the bisector (Fig. 9.3a, phase 2).

In the third phase, a wave-front structure formed of three or more cusps smooths down and transforms itself to a contracting circle-shaped front; depletion of the substrate is high at this stage and a disc-like locus of very light-coloured sites is produced (Fig. 9.3a, phase 3). If the palladium processor starts its development

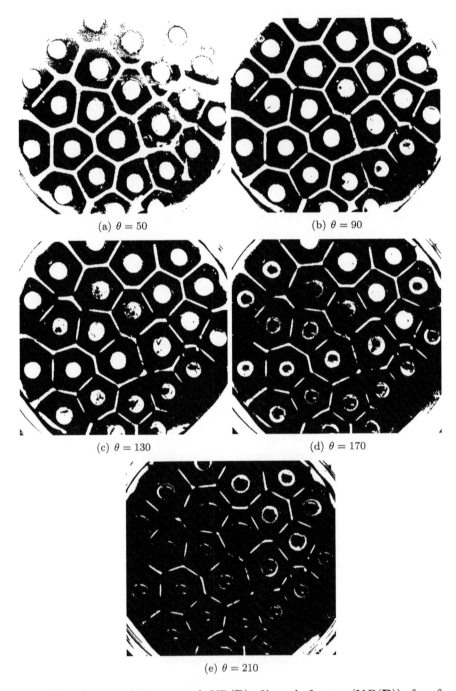

Figure 9.2: Series of images of VD(**P**) filtered from $g(VD(\mathbf{P}))$ for $\theta = 50, 90, \ldots, 210$.

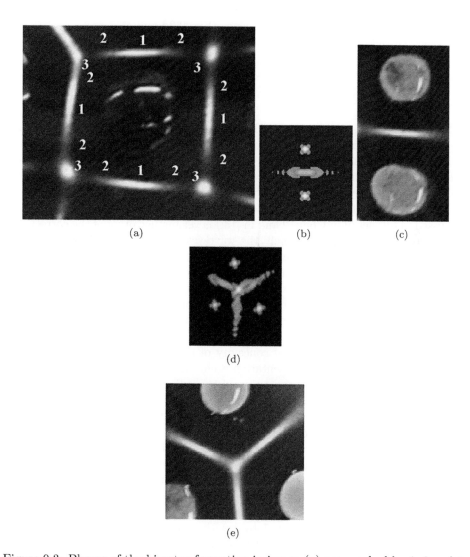

Figure 9.3: Phases of the bisector formation in image (a) are marked by 1, 2 and 3. Bisecting segments of two (b) and (c), and three (d) and (e) planar points calculated in an experimental reaction-diffusion processor (c) and (e); and, its cellular-automaton model, with neighbourhood radius 7 and cell-state transition functions described in [15], (b) and (d). Images (c) and (e) are provided for illustrative purposes only (because they are parts of images of larger diagrams).

with just two data points then the third phase of bisector formation is never reached, the cusp is transformed to a planar wave and no substrate is depleted after this point, so the bisector segment disappears (Fig. 9.3b and c). For three or more points the third phase happens at least once (Fig. 9.3d and e).

The palladium chemical processor computes by singularities. Travelling wave fronts are used to transmit information whilst singularities of the wave fronts implement the computation.

Singularities are so important in the mechanism of bisector formation because the 'density' of the front's states is much higher in the singularities than in parts of the front with positive curvature. Actually, all our previous cellular-automaton models of constructing Voronoi diagrams and skeletonisation were based on this assumption (and they worked perfectly!) [2, 4, 13, 15].

Skeletonisation of the planar shapes in a reaction-diffusion chemical processor (discussed in full detail in Chap. 2 and in [13]) is yet further proof of the correctness of our proposed mechanism, namely the wave-velocity-dependent bisector formation. During skeletonisation there are no interactions of wave fronts emanating from different sources; there is only one front, generated at the edges of the closed planar shape. Concave parts of the shape produce bisectors; however, the further the bisecting sites are from the given shape the more precipitate they contain (this is because the velocity of the fronts corresponding to the concave parts reduces with time) [13].

Let us look at the evolution of concavity of the planar wave front during the front's evolution; a cartoon representation is shown in Fig. 9.4.

We could speculate that competition of the adjacent parts of the wave front is proportional to the area of intersection of normals to the curve. With higher rates of competition (Fig. 9.4a and b) very narrow (but highly visible) zones of gel without precipitate will be formed. However, when the curvature becomes more positive (Fig. 9.4c) the bisector becomes less visible because of the scattering of loci without precipitate. Finally, when all parts of the front have positive curvature no bisector at all is formed (Fig. 9.4d). Shown in Fig. 9.4d is the extinction of a singularity on an inflating wave front, roughly (wave velocity is not considered) illustrated by cusp evolution on a limacon, $r = b + 2a\cos(t)$, in polar coordinates, where $2a = 1$ and b evolves from 1 to 2 with increment 1.1. The width of the bisector constructed by this singularity may be seen as proportional to b and the concentration of the precipitate in the bisector sites as inversely proportional to b.

The palladium reaction-diffusion chemical processor only partially constructs a Voronoi diagram of a planar point set.

This follows from experiments on the computation of a Voronoi diagram in the palladium processor and analysis of filtering transformations for various values of θ. Also, in a more simple example of just two planar points we see that neither arbitrarily long, or even a reasonably long, bisecting segment is constructed, see e.g. Fig. 9.3b and c.

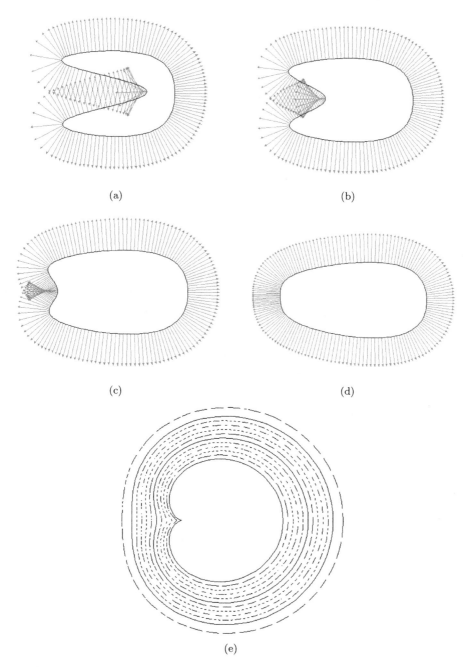

Figure 9.4: Development of the pattern of normals to the curve during transformation of the front with concavity (a), (b) and (c) to the convex shape (d). Drawn using a two-dimensional curve simulator [310]. Shown in (e) is an inflating limacon.

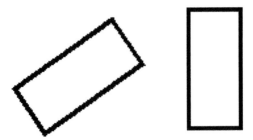

Figure 9.5: Example of 'worst' orientation of convex shapes to be separated by a bisector.

These results show that a Voronoi diagram is constructed, but there are some imperfections due to the mechanisms of formation controlled by the underlying chemical processes. Further on we detail a differing case which highlights more concerns about the accuracy of Voronoi diagrams constructed in chemical processors.

Voronoi diagrams of non-point geometrical objects are not constructed correctly in the palladium-based reaction-diffusion processor.

We demonstrate that edges of the Voronoi diagram of planar shapes, constructed in the chemical processor, are not necessarily equidistant from the objects they separate. For shapes with concavities the demonstration is simple: wave fronts generated at the edges of shapes with concavities will become completely convex after some period of time. This is because any concave front tends to become planar at higher velocity. Therefore, at the beginning of the computation the wave front exactly corresponds to the original contour; however, concave parts of the front move faster than convex ones, so after some time — depending on diffusion coefficients and reactant interaction — the wave front evolves towards a more circular morphology. At this stage the wave front does not represent the original shape any more and the further it travels from the object the more smoothed its appearance becomes. So, information about the geometry of the shape is lost progressively during the development of the wave front. What is it about convex objects specifically that causes this effect?

Given two convex shapes the exact bisector computed depends on the relative orientation of the shapes. Consider, for example, two rectangles positioned as shown in Fig. 9.5.

The almost planar wave front generated by the left-hand edge of the right-hand rectangle (Fig. 9.5) moves with a higher velocity than the wave front (with positive curvature) emanating from the opposed corner (bottom-right corner) of the left-hand rectangle (Fig. 9.5); therefore, some parts of the bisector will be positioned more closely (than the theoretical or actual computation) to the left-hand rectangle. This is experimentally demonstrated in the palladium processor, which computes an approximated Voronoi diagram of rectangular planar shapes (Fig. 9.6).

Figure 9.6: Voronoi diagram of the set of rectangles approximated in reaction-diffusion palladium processor. Positions of 'corner-to-side' orientation of rectangle pairs are shown by arrows.

9.2 Voronoi diagram is not invertible

Previously, we discussed minor inaccuracies of chemical processors in relation to the computation of a Voronoi diagram. In this case the processors do not always compute perfect results which correspond to the theoretical ideal; however, the results of their computations give good and usable approximations. In the present section we will consider a specific problem which cannot be solved by the chemical processor, namely the inversion of a Voronoi diagram.

To invert VD(**P**) experimentally we must prepare a contour of the completed diagram from filter paper, saturate this template with potassium iodide and place it on the gel containing palladium chloride. Diffusive wave fronts start to develop and they travel inwards from the edges of the Voronoi cells. Ideally the travelling waves would annihilate at the original points of VD(**P**), so that for every Voronoi cell one point was computed. However, from experiments we know that a point will only be computed exactly by a circular cell. Possibly given a very large cell the concavities would be nullified, yielding a point-like computation; however, this is not practical from a computing point of view and reactants may be depleted prior to the completion of the computation. This leads us to the following proposition.

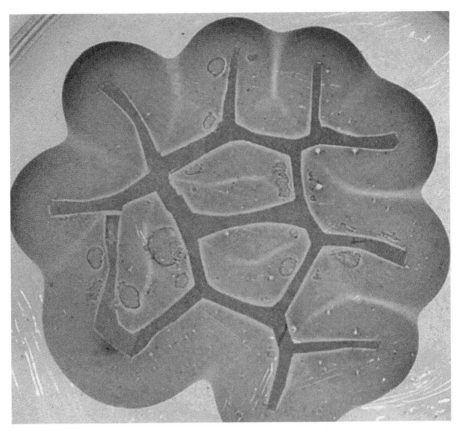

Figure 9.7: An attempt to invert a Voronoi diagram (prepared using filter paper soaked with potassium iodide) in an experimental chemical reaction-diffusion palladium processor. The chemical processor computed skeletons for each Voronoi cell. The skeletons are visible as light (less coloured) loci inside the dark-coloured domain. The rough template of the Voronoi diagram is visible as a dark-brown network.

A Voronoi diagram cannot be inverted in a reaction-diffusion chemical processor.

We have shown that given a planar set **P** we can construct VD(**P**); however, if we are given VD(**P**) (and not **P**) we will never re-construct **P** in a reaction-diffusion chemical processor.

When a single Voronoi cell is applied as a data set to the chemical processor, a skeleton of the cell is generated (more information about skeletonisation can be found in [18, 13]). This happens because any Voronoi cell — in a diagram built up of at least three planar points — has at least one singular point, which is a seed of bisector growth during self-interaction of the wave front emanating from the contour of the cell. An example showing how 'inversion' of a Voronoi diagram is implemented in a chemical processor is shown in Fig. 9.7.

As mentioned a skeleton is formed, i.e. an 'internal' Voronoi diagram of the original Voronoi cells. Naively it may be considered possible to treat the intersection points of the internal skeleton as a good approximation of **P**; however, attempting to re-construct experimentally the Voronoi diagram from these points yields another different Voronoi diagram. It would be a massive computational task (maybe impossible) for a reaction-diffusion chemical processor to find a Voronoi diagram which when its internal skeleton's intersection points were used to approximate **P** would re-construct the same Voronoi diagram. This would be an exceptional case and the processor would have required such a degree of pre-processing to make the processing step inconsequential. This is especially so considering that even to utilise the internal skeleton requires a significant level of analysis (pre-processing for experimental re-construction) in the first instance.

Could we re-construct any points from 'very smooth' Voronoi cells? The answer to this question is 'No' given the following reasoning.

A chemical processor re-constructs the centre of a circle (Fig. 9.8a and b) and approximates the centre of an ellipse (Fig. 9.8c). This happens because the skeleton passes through the singularities of all consecutive contours of the shrinking wave front. There are no singularities in a shrinking circle (Fig. 9.8a and b); however, if ellipsoidal fronts are considered an elongated segment not a point is produced (Fig. 9.8c).

This shows that the computation becomes irreversible at the moment when at least two wave fronts, emanating from different sources, interact with each other. At this stage the processor partly 'forgets' its past. Effectively, we treated the chemical processor in a very simple manner — for the computation of any Voronoi diagram the results obtained were intuitive, a good approximation of the theoretical ones and usable in the context of the set tasks. However, a more detailed analysis of the processor and related interactions (computation) showed a more complex underlying mechanism of bisector construction where there is still scope for further study. Eventually, this complexity could be utilised to significantly increase the processing power of this class of chemical processor. After all, the results of the current processors have encoded information which represents much more than the simple geometric relationship (bisector patterns show the number of converging waves, distance waves have travelled, related to velocity, and thus distance from source, possibly relative concentrations of reactant/substrate — none of which are utilised fully in the current analysis). It is this complexity which causes 'faults' — incomplete bisectors, curvature, widening of bisectors (where the geometric ideal is minimally thin bisectors) — in the target pattern when considering the construction of a Voronoi diagram. It is then impossibly complex to re-construct these 'faults' effectively to approximate the inversion of a Voronoi diagram. The chemical process alone for this class of processor is irreversible and thus the same processor must be used for inversion and subsequent re-construction. However, even if a reversible reaction could be utilised (and bisectors could be accurately constructed) it seems likely that the irreversibility of the wave dynamics in this class and other classes of chemical processor is a more insurmountable problem in the inversion of a Voronoi diagram.

Let us further substantiate the proposition concerning the irreversibility of a Voronoi diagram in reaction-diffusion processors. A point p of a Voronoi cell $vor(p)$

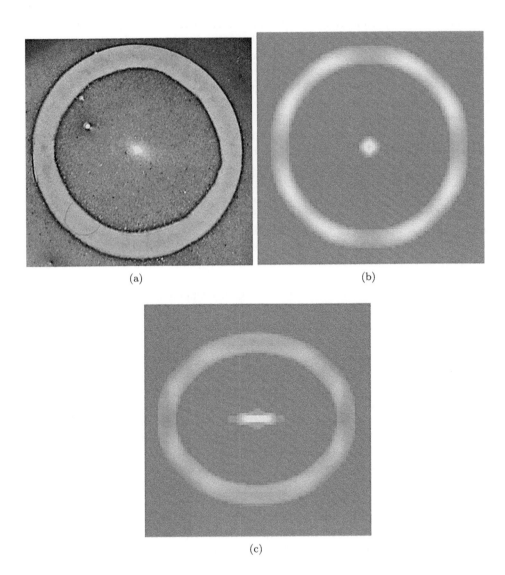

(a)   (b)

(c)

Figure 9.8: Centres of circles (a) and (b) and ellipse computed in reaction-diffusion chemical processor: (a) shows an experimental palladium processor, (b) and (c) are results of simulation in two-dimensional cellular automaton with neighbourhood radius 7 and cell-state transition functions defined in [13, 15].

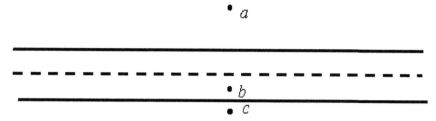

Figure 9.9: Voronoi diagram VD(**P**) is represented by two bisectors.

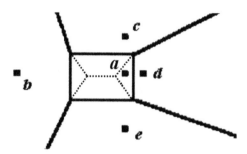

Figure 9.10: Closed Voronoi cell.

cannot be re-constructed using distance-travelled techniques because the point p, around which $vor(p)$ was constructed, does not lie at the same distance from the opposite edges of $vor(p)$. Consider several examples.

Look at the planar set $\mathbf{P} = \{a, b, c\}$ in Fig. 9.9. Its Voronoi diagram VD(**P**) is represented by two bisectors, shown by solid lines in Fig. 9.9. The diffusive waves generated by these bisectors meet at a dotted line (Fig. 9.9); because the waves travel in a uniform medium and with the same speed, they meet at an equal distance from their sites of origin. The point b does not belong to the dotted line. One may say that this example is not a typical one because not every open Voronoi cell can be inverted even using conventional techniques [246]. So, let us look at a closed cell in Fig. 9.10.

The closed cell $vor(a)$ is part of VD(**P**), $\mathbf{P} = \{a, b, c, d, e\}$; cells $vor(b)$, $vor(c)$, $vor(d)$ and $vor(e)$ are open (Fig. 9.10). When the edges of $vor(a)$ are inputted to the reaction-diffusion processor a skeleton, shown by the dotted lines in Fig. 9.10, is constructed. Point a does not belong to the skeleton of $vor(a)$. This means that we could not even approximately re-construct the positions of the original planar points when inverting the Voronoi diagram. As soon as the demonstration is purely geometrical we can be sure that, however many reagents (even a unique reagent for each edge of the Voronoi diagram!) we use, we will never invert the Voronoi diagram in a reaction-diffusion processor.

9.3   Conclusion

In the chapter we highlighted 'faults' in the palladium processor — a passive reaction-diffusion processor — in the context of Voronoi-diagram construction and inversion. The palladium processor proved to be fairly 'omnivorous': in that it constructs Voronoi diagrams of planar point sets [282, 79, 15], generalised Voronoi diagrams of arbitrary geometric shapes [14] and it can be used in the skeletonisation of planar shapes [13], computation of a shortest path and control of robot navigation [13] and computation of logical functions [12]. With this increase in the number of problems solvable in the reaction-diffusion processor, it became important to highlight the limitations of the processors.

Using the example of inverting a Voronoi diagram, we showed that there are problems that — due to their intrinsic structure — cannot be solved in reaction-diffusion processors. All findings discussed in the book were based on the palladium chemical processor. Are they true for other implementations of reaction-diffusion computers? Yes, provided that the architecture of the chemical computer remains the same — a gel-based thin-layer film of substrate and solution-based application of geometrical data objects [15].

The palladium processor considered in the chapter is a 'passive chemical processor'. How do 'active chemical processors', based on excitable chemical media such as the Belousov–Zhabotinsky reaction, deal with the problem? In [11] we demonstrated that a Voronoi diagram is approximated in excitable chemical media; this approximation however requires substantial computational resources to extract edges of the diagram from snapshots of excitation wave dynamics. The lack of any stationary output is therefore the major disadvantage of active chemical processors. However, the 'active chemical processors' overcome one limitation of the 'passive processors' as they can be initiated at a point, rather than a circular structure (reagent drop), which is mediated by the gel viscosity and surface tension of the reactant solution causing imperfections in the circular waves and thus the Voronoi diagrams formed. The fast dynamics of the 'active processors' is an obvious advantage in any computational task; the re-usability due to the annihilation of wave fronts is also an advantage in selected tasks, for example in the dynamic control of a robot where a collision-free path may need to be calculated and then updated on-line. However, the 'active' and 'passive' processors share many common characteristics, and it is difficult to produce a case for the active processors overcoming the more fundamental limitations of the chemical processors *per se* as outlined in this chapter, particularly with regard to the inversion of a Voronoi diagram.

The results of the chapter exhibit some minor drawbacks of reaction-diffusion chemical computers but in no way dismiss the efficiency and high prospects for reaction-diffusion computing in general. The impurities discovered will help us to improve existing prototypes and design more advanced chemical computers. The undecidability of the inversion of a Voronoi-diagram problem in reaction-diffusion processors offers a great challenge in the theory of computation, including unbounded action machines, local computation and unconventional computing.

Glossary

- **Analogue circuit** is an electronic circuit that operates with currents and voltages that vary continuously with time and have no abrupt transitions between levels. Since most physical quantities, e.g. velocity and temperature, vary continuously, as does audio, an analogue circuit provides the best means of representing them.

- **Belousov–Zhabotinsky (BZ) reaction** is a term applied to a group of chemical reactions in which an organic substrate (typically malonic acid) is oxidised by bromate ions in the presence of acid and a one electron transfer redox catalyst (e.g. ferroin, or the light-sensitive ruthenium complex). During the BZ reaction there are three major interlinked processes — firstly the reduction of the inhibitor (bromide ions) via reaction with bromate ions, secondly autocatalysis in bromous acid and the oxidation of the redox catalyst and finally reduction of the redox catalyst and production of the inhibitor (bromide ions) via a reaction with the organic substrate and its brominated derivative. The reaction produces oscillations in well-stirred reactors and travelling waves in thin layers, which may be visualised if the redox behaviour of the catalyst is accompanied by a change of colour (e.g. the colour is changed from orange to blue when ferroin is oxidised to ferriin).

- **Capacitor** is a device for the storage of electrical charge. Simple capacitors consist of two plates made of an electrically conducting material (e.g. a metal) and separated by a non-conducting material or dielectric (e.g. glass, paraffin, mica, oil, paper, tantalum or air).

- **Cellular automaton** is an array of locally connected finite automata, which update their discrete states in discrete time depending on the states of their neighbours; all automata of the array update their states in parallel.

- **Chip** is a piece of semiconductor (or other material) in which an integrated circuit is embedded.

- **Chua's circuit** is the simplest electronic device which exhibits complex and chaotic behaviour. The circuit consists of basic elements (capacitor, resistor, conductor, inductor) and one non-linear resistor (Chua's diode).

- **CMOS, Complementary metal oxide semiconductor** is a **semiconductor** that uses both negative and positive polarities (each at any given

289

time) and thus CMOS chips require less power than those using just one type of **transistor**.

- **CNN, Cellular neural network** is an array of elements (cells); all cells take continuous-valued states, operate in continuous or discrete time and update their states in parallel depending on the states of their closest neighbours. The cell state update rules are described by several parametric functionals.

- **Collision-based computer** is a uniform homogeneous medium which employs mobile compact patterns (particles, wave fragments) which travel in space and perform computation (e.g. implement logical gates) when they collide with each other. Truth values of logical variables are given by either the absence or presence of a localisation or by various types of localisations. In contrast to conventional architectures, collision-based computers have no pre-determined components, there are no stationary wires; a trajectory of the travelling pattern is a momentary wire and almost any part of the medium's space can be used as a wire.

- **Comparator** is an electrical circuit with positive and negative voltage inputs and one voltage (or current) output. The output is set at 1 (or 0) when the positive voltage is larger (or smaller) than the negative one.

- **Copper ferrocyanide (hydroxide) processor** is a **reaction-diffusion** chemical medium, where the substrate includes potassium ferrocyanide (copper(II) chloride), and data are represented by drops of copper(II) chloride, which when it reacts with potassium ferrocyanide forms a coloured precipitate of copper ferrocyanide. Interaction of the chemical waves in the medium leads to the formation of uncoloured domains, which — depending on the particulars of the problem — represent the result of the computation (e.g. **Voronoi diagram**).

- **Current mirror** is a circuit that copies a single input current to single (or multiple) output nodes. Two types of current mirrors exist; nMOS for current sinks and pMOS for current sources. Combining both types of current mirrors, one can invert the direction of currents; e.g. sink to source or source to sink.

- **Digital circuit** is an electronic circuit that takes a finite number of states. Binary (two-state) digital circuits are the most common. The two possible states of a binary circuit are represented by the binary digits, or bits, 0 and 1. The simplest forms of digital circuits are built from logic gates, the building blocks of the digital computer.

- **Diode** is a device that allows current flow only in one direction. A chemical diode allows for propagation of chemical waves only in one direction.

- **Excitable medium** is spatially distributed assembly of coupled **excitable systems**; spatial distribution and coupling allow for propagation of excitation waves.

- **Excitable system** is a system with a single steady quiescent state that is stable to small perturbations, but responds with an excursion from its quiescent state (excitation event) if the perturbation is above a critical threshold level. After excitation the system enters a refractory period during which time it is insensitive to further excitation before returning to its steady state.

- **FET, Field-effect transistor**, see **nMOS FET**.

- **FitzHugh–Nagumo** model is a system of two coupled differential equations, representing the dynamics of excitable and recovery variables, aimed to simulate excitable systems; the model is a reduction of the four-variable Hodgkin–Huxley model of excitability.

- **Flip-flop circuit** is a synchronous bistable device where the output changes its state only when the clock input is triggered, i.e. changes in the output of the circuit occur in synchronisation with the clock.

- **Floating-gate transistor** consists of a control gate, a floating gate and a thin oxide layer; when the floating gate is given an electrical charge, the charge is trapped in the insulating thin oxide layer. The transistors are used as non-volatile storage devices because they store electrical charge for a long time without powering.

- **Glider**, as related to **cellular automaton**, is a compact (neither infinitely expanding nor collapsing) pattern of non-quiescent states that travels along the cellular-automaton lattice.

- **Image processing** is a transformation of an input image to an output image with desirable properties, using manipulation of images to enhance or extract information.

- **Inversion of a Voronoi diagram** is a re-construction of the original data set of planar points from which the **Voronoi diagram** was computed.

- **Light-sensitive Belousov–Zhabotinsky reaction** is a **Belousov–Zhabotinsky reaction** where the ruthenium catalyst is excited by 460-nm light (blue) and reacts with bromomalonic acid to produce bromine, an inhibitor of autocatalysis. By varying the light levels the excitability of the system can be controlled.

- **Logical gate** is an elementary building block of a digital, or logical, circuit, which represents (mostly) binary logical operations, e.g. AND, OR, XOR, with two input terminals and one output terminal. In Boolean logic terminals are in one of two binary conditions (e.g. low voltage and high voltage) corresponding to TRUE and FALSE values of logical variables.

- **Logically universal processor** is a system which can realise a functionally complete set of logical operations in its development, e.g. conjunction and negation.

- **LSI, Large-scale integration, circuit** is an electronic circuit built on a semiconductor substrate, usually one of single-crystal silicon. It contains from 100 to 1000 transistors. Some LSI circuits are analogue devices; an operational amplifier is an example. Other LSI circuits, such as the microprocessors used in computers, are digital devices.

- **Minority-carrier transport** is produced in forwardly biased semiconductor p-n junctions. Minority carriers are generated in both areas of p- and n-type semiconductors. Minority carriers are electrons in p-type semiconductors and holes n-type semiconductors. Once minority carriers are generated, they diffuse among the semiconductor and finally disappear by the re-combination of electrons and holes.

- **MOSIS** is an integrated circuit fabrication service where you can purchase prototype and small-volume production quantities of integrated circuits. www.mosis.org

- **Multi-tasking chemical processor** is a **reaction-diffusion** processor that implements computations over two separate sets of data and produces two separate results, while both the sets of data and results share the same physical medium (e.g. substrate and diffusion space); in an ideal case the tasks do not interfere with each other.

- **nMOS FET** is an abbreviation of n-type metal–oxide–semiconductor field-effect transistor, where the semiconductor is negatively charged so that the transistors are controlled by movement of electrons; these transistors have three modes of operation: cut-off, triode and saturation (active).

- **n-type semiconductor** is a **semiconductor** produced by adding a certain type of atoms to it to increase the amount of free negative charges.

- **Operational transconductance amplifier** is a transconductance-type device, where input voltage controls an output current by means of the device transconductance. Transconductance is the control of a current through two output points by a voltage at two input points, as if the conductance is transferred from the input points to the output points.

- **Oregonator** is a system of three (or two in a modified version) coupled differential equations aimed to simulate oscillatory phenomena in the **Belousov–Zhabotinsky reaction**.

- **Oscillator** is a system exhibiting periodic behaviour even without external stimulation. In chemical oscillators concentrations of several intermediary species are changed periodically.

- **Palladium processor** is a **reaction-diffusion** chemical medium, where the substrate includes palladium chloride, and data are represented by drops of potassium iodide, which forms a coloured precipitate of iodo-palladium species when it reacts with palladium chloride. Interaction of chemical waves in the medium leads to formation of uncoloured domains, which — depending on particulars of the problem — represent the result of the computation (e.g. **Voronoi diagram**).

- **Pixbot** is a virtual pixel-sized robot travelling on digitised images navigating by colour values of image pixels.

- **pMOS FET** is a device which works by analogy with an nMOS FET but the transistors are moved on and off by movement of electron vacancies.

- **p-n-p-n device** is a coupling of transistors which can be switched into the high-current, low-voltage state, capable of 'accidental' triggering and thus can act analogously to 'excitable' elements. These are typically fabricated as a four-layer device with alternate n-type and p-type layers.

- **Precipitation reaction** is a reaction where soluble components mixed together form an insoluble compound that drops out of solution as a solid (called a precipitate).

- **Prussian-blue processor** is a **reaction-diffusion** chemical medium, where the substrate includes potassium ferrocyanide, and data are represented by drops of ferric ion solution (e.g. iron(III) nitrate), which forms a coloured precipitate of ferric ferrocyanide (Prussian blue) when it reacts with potassium ferrocyanide. Interaction of chemical waves in the medium leads to formation of uncoloured domains, which — depending on particulars of the problem — represent the result of the computation (e.g. **Voronoi diagram**).

- **Reaction-diffusion processor** is a thin layer of a reagent mixture which reacts to changes of one reagent's concentration — data configuration — in a predictable way to form a stationary pattern corresponding to the concentration of the reagent — result configuration. A computation in the chemical processor is implemented via the spreading and interaction of diffusive or phase waves.

- **Reagent** is a substance used in a chemical reaction.

- **Semiconductor** is an element that is intermediate in electrical conductivity between conductors and insulators, through which conduction takes place by means of holes and electrons.

- **Shift register** is an electronic circuit including several storage locations (registers); during each clock cycle the information (e.g. bits) in each location moves (is shifted) into the adjacent location.

- **Shortest-path problem** is the problem of finding a path between two sites (e.g. vertices of the graph, locations in the space) such that the length (sum of the weights of the graph edges or travel distances) of the path is minimised.

- **Single-electron circuit** is an electrical circuit that is functionally constructed by controlling movements of single electrons. Single-electron circuits consists of tunnelling junctions and electrons are controlled by using a physical phenomenon called the Coulomb blockade.

- **Singularity** is a point in space–time at which the space–time curvature becomes infinite.

- **Skeleton** of a planar contour is a set of centres of bi-tangent circles lying inside the contour.

- **Subexcitable medium** is a medium whose steady state lies between the excitable and the unexcitable domains. In excitable media waves initiated by perturbations of a sufficient size propagate throughout the media. In an unexcitable medium no perturbation is large enough to trigger a wave. In a subexcitable medium **wave fragments** with open ends are formed.

- **Substrate** is any stratum lying underneath another; in chemical processors it is commonly the substance impregnated into a gel layer and acted upon by diffusing species.

- **Substrate leaching (mechanism)** is a competitive process for the substrate initiated between two advancing chemical wave fronts which is the primary reason for the lack of precipitation at the collision points of said wave fronts.

- **Taxis** is a locomotor response towards or away from an external stimulus by a motile system.

- **Transistor** is a solid-state **semiconductor** device acting as a variable valve which controls the amount of current to flow from the voltage supply depending on its input current and voltage values; this can be used for amplification, switching, signal modulation, etc.

- **Turing (chemical) pattern** is a spatially periodic concentration profile of chemical species formed as a result of instabilities caused by differences in the diffusion rates of activating (slow) and inhibiting (fast) chemical species.

- **Unconventional computing** is a 'buzz' word representing novel, emerging or re-emerging computing paradigms, architectures and implementations that do not fit in the mainstream framework of classical computer science. Also known as non-classical, novel and lateral computing.

- **Unstable chemical processor** is a **reaction-diffusion** chemical medium with an above-threshold substrate-reagent concentration where travelling circular wavefronts become unstable and spontaneously split, thus preventing formation of desirable results of computation.

- **VLSI, Very large scale integration, circuit** is a circuit having in excess of 1000 digital gates, or equivalent elements.

- **Voronoi diagram** (also known as Dirichlet tessellation) of a planar set **P** of planar points is a partition of the plane into regions, each for any element of **P**, such that a region corresponding to a unique point p contains all those points of the plane that are closer to p than to any other node of **P**.

- **Wave fragment** is an excitation wave formed in a **subexcitable medium**; this is a segment with free ends, which either expand or contract, depending on their size and the medium's excitability. In a subexcitable medium with lower excitability waves with free ends contract and eventually disappear.

- **Wilson–Cowan (neural) oscillator** indexoscillator!Wilson–Cowan is derived from interacting populations of excitatory and inhibitory cortical neurons.

Colour insert

Figure 9.11: The reaction of cobalt and manganese salts with ammonia ($CoCl_2$ 0.05 M, $MnCl_2$ 0.25 M, agar gel 1%). From left to right ammonia solution concentration 4.5 M, 9 M, 13.5 M, 18 M.

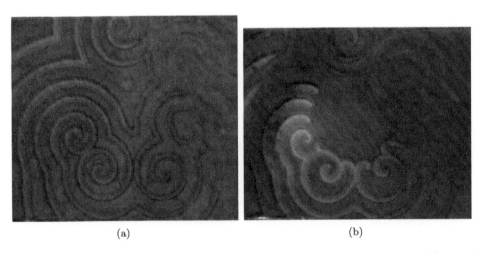

(a)   (b)

Figure 9.12: Effects of light on waves in the light-sensitive BZ reaction: (a) spiral wave formation where light from a blue LED was used to break the original waves in (b). Wave-free area caused by illumination with a blue LED at one spot (~ 10 sec). The formation of a number of free ends can be observed — these will naturally generate additional spirals.

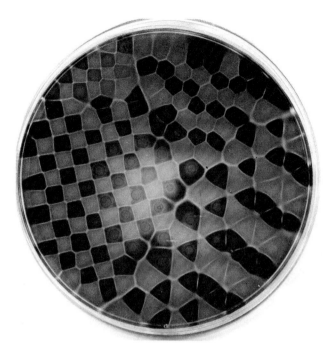

Figure 9.13: Mixed cell Voronoi diagram (a) computed when ferric ions (dark) and cupric ions (light) were reacted on a potassium ferrocyanide agar gel substrate (2.5 mg/ml). Drop size 5 mm.

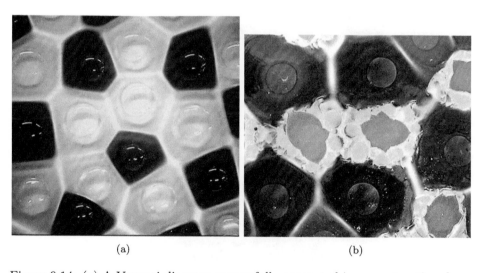

Figure 9.14: (a) A Voronoi diagram successfully computed in a constructive chemical processor. (b) The result of the same experiment where the substrate concentration has been increased and the formation of copper ferrocyanide has become unstable.

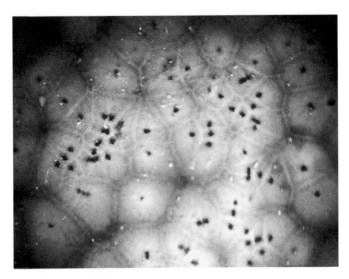

Figure 9.15: An example of the structures formed in the copper ferrocyanide gel reactor at potassium ferrocyanide concentrations between 20 mg/ml and 40 mg/ml. It shows that in the unstable system Voronoi diagrams are formed spontaneously when the reactant is applied to the gel substrate. The dark points at the Voronoi cells' centres correspond to the original points of instability.

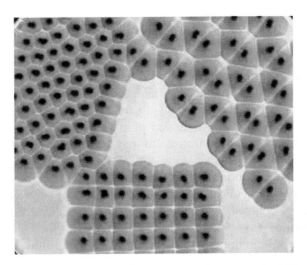

Figure 9.16: Consecutive stages of development of a controlled Voronoi diagram, representing triangular, rectangular and hexagonal arrangements. Elapsed time was 878 sec after the initiation of the reaction.

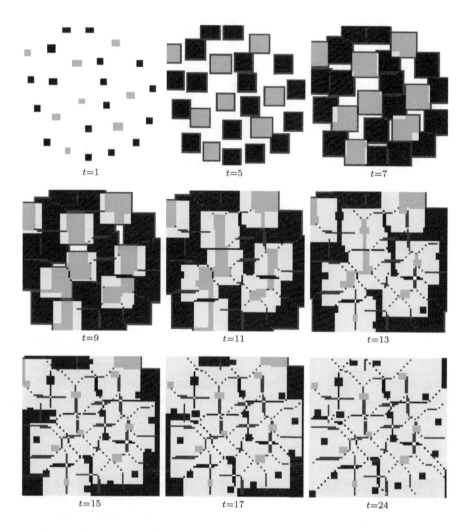

Figure 9.17: A series of configurations of a two-dimensional cellular automaton that simulates the reaction-diffusion construction of two Voronoi diagrams. The initial drops of two separate reagents are shown in brown (the reagent R_1) and green (the reagent R_2). The Voronoi diagrams are shown in cyan — VD($\mathbf{P}_1$) and magenta — VD($\mathbf{P}_2$).

Figure 9.18: A multitasking parallel chemical processor near to completing the computation of three Voronoi diagrams. The primary bisectors evident in the single processors are very clear, but in this example two additional sets of bisectors are formed so that the plane of the Petri dish is approximately split into 1-cm squares.

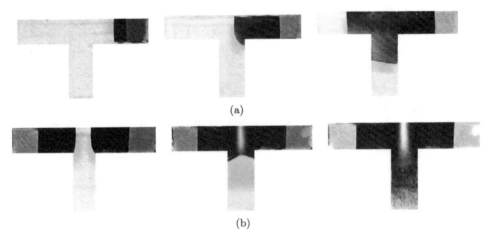

Figure 9.19: Photographs of experimental implementation of the XOR gate. (a) The progression of the reaction-diffusion gate for the inputs $x = F$ and $y = T$. (b) The progression of the reaction-diffusion gate for the inputs $x = T$ and $y = T$.

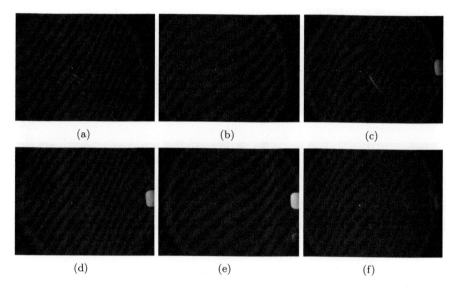

Figure 9.20: Spatio-temporal dynamics of BZ bullets.

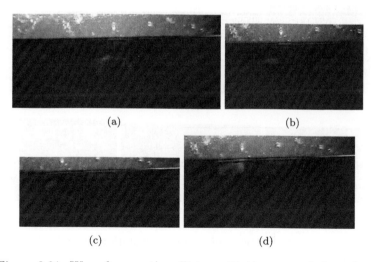

Figure 9.21: Wave-fragment's collision with the reactor's boundary.

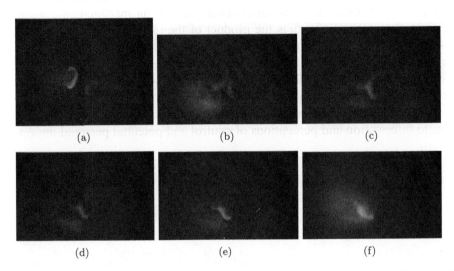

Figure 9.22: Snapshots of experimental implementation of collision gate $\langle x, y \rangle \rightarrow \langle x$ AND y, x AND NOT y, NOT x AND $y \rangle$ (see Fig. 3.12) in BZ medium, $x =$ TRUTH, y = TRUTH.

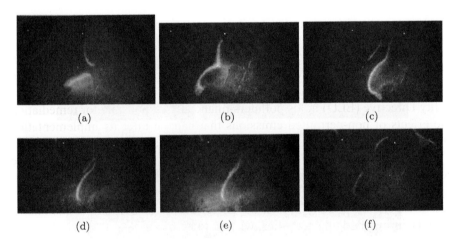

Figure 9.23: Snapshots of experimental implementation of multiple collision between wave fragments. The dynamics realised gate $\langle x, y \rangle \rightarrow \langle x$ AND y, NOT x AND y, x AND NOT y, NOT x OR NOT $y \rangle$, shown in Fig. 3.15.

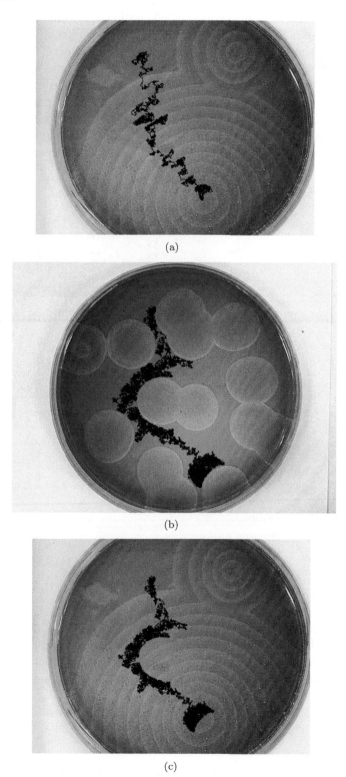

(a)

(b)

(c)

Figure 9.24: Trajectories of pixbot moving towards the target (a) without obstacles and with obstacles (b) and (c).

Figure 9.25: Colloidal silver on the finger 'nails' excites target waves in the BZ medium [307].

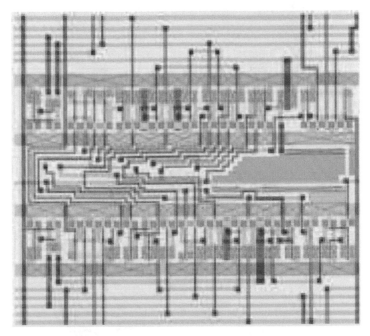

Figure 9.26: Cell layout with 1.5-μm double-poly double-metal n-well CMOS process (MOSIS, vendor: AMIS). The circuit size is $261\lambda \times 299\lambda$, $\lambda = 0.8$ μm.

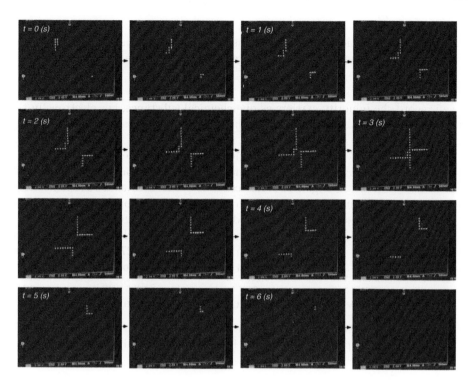

Figure 9.27: Snapshots of recorded movie obtained from fabricated reaction-diffusion chip: white dots represent excited cells, where EXC is logical '1'.

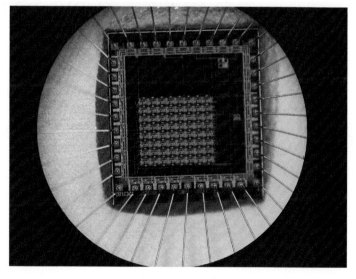

Figure 9.28: Micrograph of fabricated reaction-diffusion chip.

Bibliography

[1] Abdel-Hamid G. and Yang Y.-H. Multiscale skeletonization: an electrostatic field-based approach. VR-94-6, 14 (1994). Available at citeseer.nj.nec.com/201707.html.

[2] Adamatzky A. Reaction-diffusion algorithm for constructing discrete generalized Voronoi diagram. Neural Network World 6 (1994) 635–643.

[3] Adamatzky A. Computation of shortest path in cellular automata. Math. Comput. Modelling 23 (1996) 105–113.

[4] Adamatzky A. Voronoi-like partition of lattice in cellular automata. Math. Comput. Modelling 23 (1996) 51–66.

[5] Adamatzky A. Universal dynamical computation in multi-dimensional excitable lattices. Int. J. Theor. Phys. 37 (1998) 3069–3108.

[6] Adamatzky A. Reaction-diffusion and excitable processors: a sense of the unconventional. Parallel Distrib. Comput.: Theory Pract. (Spec. Issue: Unconventional Parallel Architectures) 3 (2000) 113–132.

[7] Adamatzky A. Computing in Nonlinear Media and Automata Collectives. IoP Publishing, Bristol, 2001.

[8] Adamatzky A. (ed.). Collision-Based Computing. Springer, London, 2002.

[9] Adamatzky A. Collision-based computing in Belousov–Zhabotinsky medium. Chaos Solitons Fractals 21 (2004) 1259–1264.

[10] Adamatzky A. Programming reaction-diffusion processors. In: Unconventional Programming Paradigms Workshop, Le Mont Saint Michel, France, 15–17 September 2004.

[11] Adamatzky A. and De Lacy Costello B. P. J. Collision-free path planning in the Belousov–Zhabotinsky medium assisted by a cellular automaton. Naturwissenschaften 89 (2002) 474–478.

[12] Adamatzky A. and De Lacy Costello B. P. J. Experimental logical gates in a reaction-diffusion medium: the XOR gate and beyond. Phys. Rev. E 66 (2002) 046112.

[13]   Adamatzky A. and De Lacy Costello B. Experimental reaction-diffusion pre-processor for shape recognition. Phys. Lett. A 297 (2002) 344–352.

[14]   Adamatzky A. and De Lacy Costello B. Reaction-diffusion path planning in a hybrid chemical and cellular-automaton processor. Chaos Solitons Fractals 16 (2003) 727–736.

[15]   Adamatzky A. and De Lacy Costello B. P. J. On some limitations of reaction-diffusion computers in relation to a Voronoi diagram and its inversion. Phys. Lett. A 309 (2003) 397–406.

[16]   Adamatzky A. and Melhuish C. Towards the design of excitable lattice controllers for (nano) robots. Smart Eng. Syst. Design 3 (2001) 265–277.

[17]   Adamatzky A. and Melhuish C. Phototaxis of mobile excitable lattices. Chaos Solitons Fractals 13 (2002) 171–184.

[18]   Adamatzky A. and Tolmachiev D. Chemical processor for computation of skeleton of planar shape. Adv. Mater. Opt. Electron. 7 (1997) 135–139.

[19]   Adamatzky A., De Lacy Costello B. and Ratcliffe N. M. Experimental reaction-diffusion pre-processor for shape recognition. Phys. Lett. A 297 (2002) 344–352.

[20]   Adamatzky A., De Lacy Costello B., Melhuish C. and Ratcliffe N. Experimental reaction-diffusion chemical processors for robot path planning. J. Intell. Robot. Syst. 37 (2003) 233–249.

[21]   Adamatzky A., De Lacy Costello B., Melhuish C. and Ratcliffe N. Experimental implementation of mobile robot taxis with onboard Belousov–Zhabotinsky chemical medium. Mater. Sci. Eng. C 24 (2004) 541–548.

[22]   Adamatzky A., De Lacy Costello B., Melhuish C., Rambidi N., Ratcliffe N. and Wessnitzer J. Excitable chemical controllers for robots. In: EP-SRC/BBSRC Int. Workshop Biologically Inspired Robotics, UK, 2002, p. 78.

[23]   Adamatzky A., Arena P., Basile A., Carmona-Galán R., De Lacy Costello B., Fortuna L., Frasca M. and Rodríguez-Vázquez A. Reaction-diffusion navigation robot control: from chemical to VLSI analogic processors. IEEE Trans. Circuits Syst. I 51 (2004) 926–938.

[24]   Adelman L. M. Molecular computation of solutions to combinatorial problems. Science 266 (1994) 1021–1024.

[25]   Agladze K., Obata S. and Yoshikawa K. Phase-shift as a basis of image processing in oscillating chemical medium. Physica D 84 (1995) 238–245.

[26]   Agladze K., Aliev R. R., Yamaguhi T. and Yoshikawa K. Chemical diode. J. Phys. Chem. 100 (1996) 13895–13897.

[27] Agladze K., Toth A., Ichino T. and Yoshikawa K. Propagation of chemical waves at the boundary of excitable and inhibitory fields. J. Phys. Chem. A 104 (2000) 6677–6680.

[28] Agladze K., Magome N., Aliev R., Yamaguchi T. and Yoshikawa K. Finding the optimal path with the aid of chemical wave. Physica D 106 (1997) 247–254.

[29] Alamgir M. and Epstein I. R. Experimental study of complex dynamics behavior in coupled chemical oscillators. J. Phys. Chem. 88 (1984) 2848–2851.

[30] Ammelt E., Astro Y. and Purwins H. G. Hexagon structures in a two-dimensional DC-driven gas discharge system. Phys. Rev. E 58 (1998) 7109–7117.

[31] Ammelt E., Schweng D. and Purwins H. G. Spatio-temporal pattern formation in a lateral high-frequency glow discharge system. Phys. Lett. A 179 (1993) 348–354.

[32] Andreou A. G., Boahen K. A., Pouliquen P. O., Pavasović A., Jenkins R. E. and Strohbehn K. Current-mode subthreshold MOS circuits for analog VLSI neural systems. IEEE Trans. Neural Networks 2 (1991) 205–213.

[33] Asai T., Adamatzky A. and Amemiya Y. Towards reaction-diffusion computing devices based on minority-carrier transport in semiconductors. Chaos Solitons Fractals 20 (2004) 863–876.

[34] Asai T., De Lacy Costello B. and Adamatzky A. Silicon implementation of a chemical reaction-diffusion processor for computation of Voronoi diagram. Int. J. Bifurcat. Chaos 15 (2005), in press.

[35] Asai T., Kanazawa Y. and Amemiya Y. A subthreshold MOS neuron circuit based on the Volterra system. IEEE Trans. Neural Networks 14 (2003) 1308–1312.

[36] Asai T., Nishimiya Y. and Amemiya Y. A CMOS reaction-diffusion circuit based on cellular-automaton processing emulating the Belousov–Zhabotinsky reaction. IEICE Trans. Fundam. Electron. Commun. Comput. E85-A 9 (2002) 2093–2096.

[37] Asai T., Ohtani M., and Yonezu H. Analog integrated circuits for the Lotka–Volterra competitive neural networks. IEEE Trans. Neural Networks 10 (1999) 1222–1231.

[38] Asai T., Ikebe M., Hirose T. and Amemiya Y. A quadrilateral-object composer for binary images with reaction-diffusion cellular automata. Int. J. Parallel Emergent Distrib. Syst. 20 (2005) 57–68.

[39] Asai T., Kanazawa Y., Hirose T. and Amemiya Y. Analog reaction-diffusion chip imitating the Belousov–Zhabotinsky reaction with hardware Oregonator model. Int. J. Unconv. Comput. 1 (2005) 123–147.

[40]   Asai T., Sunayama T., Amemiya Y. and Ikebe M. A νMOS vision chip based on the cellular-automaton processing. Jpn. J. Appl. Phys. 40 (2001) 2585–2592.

[41]   Aurenhammer F. Voronoi diagrams — a survey of a fundamental geometric data structure. Comput. Surv. 23 (1991) 34.

[42]   Bakkum D. J., Shkolnik A. C., Ben-Ary G., Gamblen P., De Marse T. B. and Potter S. M. Removing some 'A' from AI: embodied cultured networks, embodied artificial intelligence. Lect. Notes AI 3139 (2004) 130–145.

[43]   Bar-Eli K. On the stability of coupled oscillators. Physica D 14 (1985) 242–252.

[44]   Bar-Eli K. and Reuveni S. Stable stationary-states of coupled chemical oscillators. Experimental evidence. J. Phys. Chem. 89 (1985) 1329–1330.

[45]   Barlow G. W. Hexagonal territories. Anim. Behav. 22 (1974) 876–878.

[46]   Barraquand J., Langlois B. and Latombe J. C. Numerical potential field techniques for robot path planning. IEEE Trans. Syst. Man Cybern. 22 (1992) 224–241.

[47]   Beato V. and Engel H. Pulse propagation in a model for the photosensitive Belousov–Zhabotinsky reaction with external noise. In: Noise in Complex Systems and Stochastic Dynamics, ed. by Schimansky-Geier L., Abbott D., Neiman A. and Van den Broeck C. Proc. SPIE 5114 (2003) 353–362.

[48]   Beaver D. Computing with DNA. J. Comput. Biol. 2 (1995) 1–7.

[49]   Benenson Y., Adar R., Paz-Elizur T., Livneh Z. and Shapiro E. DNA molecules provide a computing machine with both data and fuel. PNAS 100 (2003) 2191–2196.

[50]   Ben-Jacob E. and Garik P. The formation of patterns in non-equilibrium growth. Nature 343 (1990) 523.

[51]   Berlekamp E. R., Conway J. H. and Guy R. L. Winning Ways for your Mathematical Plays, Vol. 2. Academic Press, 1982.

[52]   Blackburn C. G. and Dunckley L. The application of Voronoi tessellations in the development of 3D stochastic models to represent tumour growth. Z. Angew. Math. Mech. 76 (1996) 335–338.

[53]   Blittersdorf R., Müller J. and Schneider F. M. Chemical visualization of Boolean functions: a simple chemical computer. J. Chem. Educ. 72 (1995) 760–763.

[54]   Blum H. A transformation for extracting new descriptors of shape. In: Wathen-Dunn W. (ed.), Models for the Perception of Speech and Visual Form. MIT Press, Cambridge, MA, 1967, pp. 362–380.

[55] Blum H. Biological shape and visual science. J. Theor. Biol. 38 (1973) 205–287.

[56] Boccaletti S., Grebogi C., Lai Y.-C., Mancini H. and Maza D. The control of chaos: theory and applications. Phys. Rep. 329 (2000) 103–197.

[57] Bode M. and Purwins H. G. Pattern formation in reaction diffusion systems — dissipative solitons in physical systems. Physica D 86 (1995) 53–63.

[58] Bode M., Liehr A. W., Schenk C. P. and Purwins H. G. Interaction of dissipative solitons: particle-like behaviour of localized structures in a three-component reaction-diffusion system. Physica D 161 (2000) 45–66.

[59] Boga E., Kadar S., Peintler G. and Nagypal I. Effect of magnetic fields on a propagating reaction front. Nature 347 (1990) 749–751.

[60] Bonaiuto V., Maffucci A., Miano G., Salerno M., Sargeni F., Serra P. and Visone C. Hardware implementation of a CNN for analog simulation of reaction-diffusion equations. Proc. IEEE Int. Conf. Circuits Syst. 3 (2001) 485–488.

[61] Bouzat S. and Wio H. S. Pattern dynamics in inhomogeneous active media. Physica A 293 (2001) 405–420.

[62] Brady M. Criteria for representations of shape. In: Beck J., Hope B. and Rosenfeld A. (eds.), Human and Machine Vision. Academic Press, 1983, pp. 39–84.

[63] Braich R. S., Chelyapov N., Johnson C., Rothemund P. W. K. and Adelman L. Solution of a 20-variable 3-SAT problem on a DNA computer. Science 296 (2002) 499–502.

[64] Brandtstädter H., Braune M., Schebesch I. and Engel H. Experimental study of the dynamics of spiral pairs in light-sensitive Belousov–Zhabotinskii media using an open-gel reactor. Chem. Phys. Lett. 323 (2000) 145–154.

[65] Bray N. H., Anderson J. B., Devine J. D. and Kwasnik J. M. Topological properties of random crack networks. Math. Geol. 8 (1976) 617–626.

[66] Byers J. A. Dirichlet tessellation of bark beetle spatial attack points. J. Anim. Ecol. 61 (1992) 759–768.

[67] Calabi L. and Hartnett W. E. Shape recognition, prairie fires, convex deficiencies and skeletons. Am. Math. Mon. 75 (1968) 335–342.

[68] Castets V., Dulos E., Boissonade J. and De Kepper P. Experimental evidence of a sustained standing Turing-type nonequilibrium chemical pattern. Phys. Rev. Lett. 64 (1990) 2953–2956.

[69] Chen G. and T. Ueta (eds.). Chaos in Circuits and Systems. World Scientific, New Jersey, 2002.

[70] Ciletti, M. D. Advanced Digital Design with the Verilog HDL. Prentice-Hall, New Jersey, 2002.

[71] Conrad M. Molecular information processing in the central nervous system. Part I. Selection circuits in the brain. In: Lect. Notes Biomathematics, Physics and Mathematics of the Nervous System. Springer, Berlin, 1974.

[72] Conrad M. and Zauner K.P. Molecular computing with artificial neurons. Commun. Korea Inf. Sci. Soc. 18 (2000) 78–89.

[73] Crounse R. K., Chua O. L., Thiran P. and Setti G. Characterization and dynamics in cellular neural networks. Int. J. Bifurcat. Chaos 6 (1996) 1703–1724.

[74] Crowley M. F. and Epstein I. R. Experimental and theoretical studies of a coupled chemical oscillator: phase death, multistability and in-phase and out of phase entrainment. J. Phys. Chem. 93 (1989) 2496–2502.

[75] Crowley M. F. and Field R. J. Electrically coupled Belousov–Zhabotinskii oscillators 1. Experiments and simulations. J. Phys. Chem. 90 (1986) 1907–1915.

[76] Dahlem M. A. and Muller S. C. Self-induced splitting of spiral-shaped spreading depression waves in chicken retina. Exp. Brain Res. 115 (1997) 319–324.

[77] Daikoku T., Asai T. and Amemiya Y. An analog CMOS circuit implementing Turing's reaction-diffusion model. In: Proc. Int. Symp. Nonlinear Theory and its Applications, 2002, pp. 809–812.

[78] De Lacy Costello B. Double chemical processor. Unpublished (2002).

[79] De Lacy Costello B. P. J. Constructive chemical processors — experimental evidence that shows that this class of programmable pattern forming reactions exist at the edge of a highly non-linear region. Int. J. Bifurcat. Chaos 13 (2003) 1561–1564.

[80] De Lacy Costello B. and Adamatzky A. On multitasking in parallel chemical processors: experimental findings. Int. J. Bifurcat. Chaos 13 (2003) 521–533.

[81] De Lacy Costello B., Adamatzky A., Ratcliffe N., Zanin A. L., Liehr A. W. and Purwins H. G. The formation of Voronoi diagrams in chemical and physical systems: experimental findings and theoretical models. Int. J. Bifurcat. Chaos 14 (2004) 2187–2210.

[82] De Marse T. B., Wagenaar D. A. and Blau A. W. The neurally controlled Animat: biological brains acting with simulated bodies. Auton. Robot. 11 (2001) 305–310.

[83] Deussen O., Hiller S., van Overveld C. and Strothotte T. Floating points: a method for computing stipple drawings. Computer Graphics Forum. 19 (2000) 40–51.

[84]   Deutsch D. Quantum theory, the Church–Turing principle and the universal quantum computer. Proc. R. Soc. A 400 (1985) 97–117.

[85]   Ditto W. L. and Showalter K. Introduction: control and synchronization of chaos. Chaos 7 (1997) 509–511.

[86]   Dolnik M. and Epstein I. R. Coupled chaotic oscillators. Phys. Rev. E 54 (1996) 3361–3368.

[87]   Dori D. and A. Bruckstein (eds.). Shape, Structure and Pattern Recognition. World Scientific, Singapore, 1994.

[88]   Drysdale S. Voronoi Diagrams: Applications from Archeology to Zoology. Regional Geometry Institute, Smith College, 19 July 1993.

[89]   Dylla R. J. and Korgel B. A. Temporal organization of nanocrystal self-assembly directed by a chemical oscillator. ChemPhysChem 2 (2001) 62–64.

[90]   Earnshaw R. A. (ed.). Theoretical Foundations of Computer Graphics and CAD. NATO ASI Ser. F: Computer and Systems Sciences, Vol. 40. Springer, 1998.

[91]   Elwakil A. S. and Soliman A. M. Two twin-T based op amp oscillators modified for chaos. J. Franklin Inst. 335B (1998) 771–787.

[92]   Elwakil A. S. and Soliman A. M. Two modified for chaos negative impedance converter op amp oscillators with symmetrical and antisymmetrical nonlinearities. Int. J. Bifurcat. Chaos 8 (1998) 1335–1346.

[93]   Feeney R., Schmidt S. L. and Ortoleva P. Experiments of electric field–BZ chemical wave interactions: annihilation and the crescent wave. Physica D 2 (1981) 536–544.

[94]   Fenton F. H., Cherry E. M., Hastings H. M. and Evans S. J. Multiple mechanisms of spiral wave breakup in a model of cardiac electrical activity. Chaos 12 (2002) 852–892.

[95]   Fiasconard A., Valenti D. and Spagnolo B. Nonmonotonic behavior of spatiotemporal pattern formation in a noisy Lotka–Volterra system. Acta Phys. Pol. B 35 (2004) 1491–1500.

[96]   Field R. J. and Burger M. Oscillation and Travelling Waves in Chemical Systems. Wiley Interscience, New York, 1985.

[97]   Field R. J. and Noyes R. M. Oscillations in chemical systems. IV. Limit cycle behavior in a model of a real chemical reaction. J. Chem. Phys. 60 (1974) 1877–1884.

[98]   Field R. J. and Winfree A. T. Travelling waves of chemical activity in the Zaikin–Zhabotinsky–Winfree reagent. J. Chem. Educ. 56 (1979) 754.

[99]   FitzHugh R. Impulses and physiological states in theoretical models of nerve membrane. Biophys. J. 1 (1961) 445–466.

[100] Flesselles J.-M., Belmonte A. and Gáspár V. Dispersion relation for waves in the Belousov–Zhabotinsky reaction. J. Chem. Soc. Faraday Trans. 94 (1998) 851–855.

[101] Forsyth D., Mundy J., Cipolla R. and Goos G. (eds.). Shape, Contour and Grouping in Computer Vision. Springer, 2000.

[102] Fourie C. J. Intelligent path planning for a mobile robot using a potential field algorithm. In: Proc. 29th Int. Symp. Robotics. Advanced Robotics: Beyond 2000. DMG Business Media, Redhill, UK, 1998, pp. 221–224.

[103] Fredkin F. and Toffoli T. Conservative logic. Int. J. Theor. Phys. 21 (1982) 219–253.

[104] Fromherz P. and Gaede V. Exclusive-OR function of single arborized neuron. Biol. Cybern. 69 (1993) 337–344.

[105] Fukuda H., Morimura H. and Kai S. Global synchronisation in two-dimensional lattices of discrete Belousov–Zhabotinsky oscillators. Physica D (2005) in press.

[106] Fukuda H., Nagano H. and Kai S. Stochastic synchronisation in two-dimensional coupled lattice oscillators in the Belousov–Zhabotinsky reaction. J. Phys. Soc. Jpn. 72 (2003) 487–490.

[107] Furukawa Y., Yonezu H., Ojima K., Samonji K., Fujimoto Y., Momose K. and Aiki K. Control of N content of GaPN grown by molecular beam epitaxy and growth of GaPN lattice-matched to Si(100) substrate. Jpn. J. Appl. Phys. 41 (2001) 528–532.

[108] Garfinkel A., Spano M. L., Ditto W. L. and Weiss J. N. Controlling cardiac chaos. Science 257 (1992) 1230–1235.

[109] Gerhardt M., Schuster H. and Tyson J. J. A cellular automaton model of excitable media. Physica D 46 (1990) 392–415.

[110] Goel S. N., Maitra C. S. and Montroll W. E. On the Volterra and other nonlinear models of interacting populations. Rev. Mod. Phys. 43 (1971) 231–276.

[111] Gorecka J. and Gorecki J. T-shaped coincidence detector as a band filter of chemical signal frequency. Phys. Rev. E 67 (2003) 067203.

[112] Gorecki J., Yoshikawa K. and Igarashi Y. On chemical reactors that can count. J. Phys. Chem. A 107 (2003) 1664–1669.

[113] Graham R. and Yao F. A whirlwind tour of computational geometry. Am. Math. Mon. 97 (1990) 687–701.

[114] Gravert H. and Devoret M. H. Single Charge Tunneling — Coulomb Blockade Phenomena in Nanostructures. Plenum Press, New York, 1992.

[115] Gray C. M., Konig P., Engel A. K. and Singer W. Oscillatory responses in cat visual-cortex exhibit inter-columnar synchronization which reflects global stimulus properties. Nature 338 (1989) 334.

[116] Grigorishin T., Abdel-Hamid G. and Yang Y.-H. Skeletonization: an electrostatic field-based approach. VR-96-1 1 (1996).

[117] Grill S., Zykov V. S. and Müller S. C. Spiral wave dynamics under pulsatory modulation of excitability. J. Phys. Chem. 100 (1996) 19082–19088.

[118] Grossberg S. Nonlinear neural networks: principles, mechanisms and architectures. Neural Networks 1 (1988) 17–61.

[119] Hanna A., Saul A. and Showalter K. Detailed studies of propagating fronts in the iodate oxidation of arsenous acid. J. Am. Chem. Soc. 104 (1982) 3838–3844.

[120] Hantz P. Pattern formation in the $NaOH + CuCl_2$ reaction. J. Phys. Chem. B 104 (2000) 4266–4272.

[121] Hantz P. Regular microscopic patterns produced by simple reaction-diffusion systems. Phys. Chem. Chem. Phys. 4 (2002) 1262–1267.

[122] Haralick P. and Shapiro L. Computers and Robots. Addison Wesley, 1992.

[123] Hargittai I. (ed.). Symmetry: Unifying Human Understanding. Pergamon, 1986.

[124] Hassinger T. D., Guthrie P. B., Atkinson P. B., Bennett M. V. L. and Kater S. B. An extracellular signalling component in propagation of astrocytic calcium waves. PNAS 93 (1996) 13268–13273.

[125] Heitkotter J. and Beasley D. (eds.). The hitchhikers guide to evolutionary computation: a list of FAQs (2000). Available at `ftp://rtfm.mit.edu/pub/usenet/news.answers/ai-faq/genetic/`.

[126] Henisch H. K. Crystals in Gels and Liesegang Rings. Cambridge University Press, Cambridge, 1988.

[127] Hjelmfelt A. and Ross J. Mass-coupled chemical systems with computational properties. J. Phys. Chem. 97 (1993) 7988–7992.

[128] Hjelmfelt A. and Ross J. Implementation of logic functions and computations by chemical kinetics. Physica D 84 (1995) 180–193.

[129] Hjelmfelt A., Schneider F. W. and Ross J. Pattern recognition in coupled chemical kinetic systems. Science 260 (1993) 335–337.

[130] Hjelmfelt A., Weinberger E. D. and Ross J. Chemical implementation of neural networks and Turing machines. Proc. Natl. Acad. Sci. USA 88 (1991) 10983–10987.

[131] Hjelmfelt A., Weinberger E. D. and Ross J. Chemical implementation of finite-state machines. Proc. Natl. Acad. Sci. USA 89 (1992) 383–387.

[132] Holz R. and Schneider F. W. Control of dynamic states with time-delay between 2 mutually flow-rate coupled reactors. J. Phys. Chem. 97 (1993) 12239.

[133] Honda H. and Eguchi G. J. How much does the cell boundary contract in a monolayered cell sheet? J. Theor. Biol. 84 (1980) 575–588.

[134] Hou E. S. H. and Zheng D. Mobile robot path planning based on hierarchical hexagonal decomposition and artificial potential fields. J. Robot. Syst. 11 (1994) 605–614.

[135] Hussien B. and McLaren R. W. Real-time robot path planning using the potential function method. Autom. Construct. 2 (1993) 241–250.

[136] Hwang Y. K. and Ahuja N. A potential field approach to path planning. IEEE Trans. Robot. Autom. 8 (1992) 23–32.

[137] Hwang Y. K. and Ahuja N. Gross motion planning: a survey. ACM Comput. Surv. 2 (1992) 219–291.

[138] Ichino T., Igarashi Y., Motoike I. N. and Yoshikawa K. Different operations on a single circuit: field computation on an excitable chemical system. J. Chem. Phys. 118 (2003) 8185–8190.

[139] Ishikawa Y., Yu W., Yokoi H. and Kakazu Y. Development of robot hands with an adjustable power transmitting mechanism. In: Dagli C. H. et al. (eds.). Proc. Intell. Eng. Syst. Through Artificial Neural Networks, 2000, Vol. 10, pp. 631–636.

[140] Jornè J. The diffusive Lotka–Volterra oscillating system. J. Theor. Biol. 65 (1977) 1330–1339.

[141] Jung P., Cornell-Bell A., Madden K. S. and Moss F. J. Noise-induced spiral waves in astrocyte syncytia show evidence of self-organized criticality. Neurophysiology 79 (1998) 1098–1101.

[142] Kador S., Amemiya T. and Showalter K. Reaction mechanism for light sensitivity of the Ru (bby)32+ catalysed BZ reaction. J. Phys. Chem. 101 (1997) 8200–8206.

[143] Kador S., Wang J. and Showalter K. Noise-supported travelling waves in sub-excitable media. Nature 391 (1998) 770–772.

[144] Karahaliloglu K. and Balkir S. An MOS cell circuit for compact implementation of reaction-diffusion models. In: Proc. 2004 Int. Joint Conf. Neural Networks, p. 1222.

[145] Kassim A. A. and Kumar B. V. Path planners based on the wave expansion neural network. Robot. Auton. Syst. 26 (1999) 1–22.

[146] Kastánek P., Kosek J., Snita D., Schreiber I. and Marek M. Reduction waves in the BZ reaction: circles, spirals and effects of electric field. Physica D 84 (1995) 79–94.

[147] Kawato M. and Suzuki R. Two coupled neural oscillators as a model of the circadian pacemaker. J. Theor. Biol. 86 (1980) 547–575.

[148] Keener J. P. and Sneyd J. Mathematical Physiology. Springer, New York, 1998.

[149] Kennedy B., Melhuish C. and Adamatzky A. Biologically inspired robots. In: Bar-Cohen Y. (ed.). Electroactive Polymer (EAP) Actuators as Artificial Muscles — Reality, Potential and Challenges. SPIE Press, 2001.

[150] Klein R. Concrete and Abstract Voronoi Diagrams. Springer, Berlin, 1990.

[151] Kobayashi K. and Sugihara K. Crystal Voronoi diagram and its applications. Future Generat. Comput. Syst. 18 (2002) 681–692.

[152] Krug H. J., Pohlmann L. and Kuhnert L. Analysis of the modified complete Oregonator (MCO) accounting for oxygen- and photosensitivity of Belousov–Zhabotinsky systems. J. Phys. Chem. 94 (1990) 4862–4866.

[153] Kuhnert L. Photochemische Manipulation von chemischen Wellen. Naturwissenschaften 76 (1986) 96–97.

[154] Kuhnert L., Agladze K. L. and Krinsky V. I. Image processing using light-sensitive chemical waves. Nature 337 (1989) 244–247.

[155] Kumakura K., Motohisa J. and Fukui T. Formation and characterization of coupled quantum dots (CQDs) by selective area metalorganic vapor phase epitaxy. J. Cryst. Growth 170 (1997) 700–704.

[156] Kumakura K., Nakakoshi K., Motohisa J., Fukui T. and Hasegawa H. Novel formation method of quantum dot structures by self-limited selective area metalorganic vapor phase epitaxy. Jpn. J. Appl. Phys. 34 (1995) 4387–4389.

[157] Kurin-Csorgei K., Epstein I. R. and Orban M. Systematic design of chemical oscillators using complexation and precipitation equilibria. Nature 433 (2005) 139–142.

[158] Kuwamura N., Taniguchi K. and Hamakawa C. Simulation of single-electron logic circuits. IEICE Trans. Electron. J77-C-II (1994) 221–228.

[159] Laplante J. P., Pemberton M., Hjelmfelt A. and Ross J. Experiments on pattern recognition by chemical kinetics. J. Phys. Chem. 99 (1995) 10063–10065.

[160] Larter R. Understanding complexity in biophysical chemistry. J. Phys. Chem. B 107 (2003) 415–429.

[161] Lebender D. and Schneider F. W. Logical gates using a nonlinear chemical reaction. J. Phys. Chem. 98 (1994) 7533–7537.

[162] Lee K. J., McCormick W. D., Swinney H. L. and Pearson J. E. Experimental observation of self-replicating spots in a reaction-diffusion system. Nature 369 (1994) 215.

[163] Lemmon M. D. 2-degree-of-freedom robot path planning using cooperative neural fields. Neural Comput. 3 (1991) 350–362.

[164] Lin A. L., Hagberg A., Meron E. and Swinney H. L. Resonance tongues and patterns in periodically forced reaction-diffusion systems. Phys. Rev. E 69 (2004) 066217.

[165] Liu C. S., Kramer J., Indiveri G., Delbrück T. and Douglas R. Analog VLSI: Circuits and Principles. MIT Press, Cambridge, MA, 2002.

[166] Loncaric S. A survey of shape analysis techniques. Pattern Recogn. 31 (1998) 983–1001.

[167] Louste C. and Liegeois A. Near optimal robust path planning for mobile robots: the viscous fluid method with friction. J. Intell. Robot. Syst. 27 (2000) 99–112.

[168] Maniatty W. and Szymanski B. Fine-grain discrete Voronoi diagram algorithms in L_1 and L_∞ norms. Math. Comput. Modelling 26 (1997) 71–78.

[169] Mao C., LaBean T. H., Relf J. H. and Seeman N. C. Logical computation using algorithmic self-assembly of DNA triple-crossover molecules. Nature 407 (2000) 493–496.

[170] Maragos P. A. Tutorial on advances in morphological image processing and analysis. Opt. Eng. 26 (1987) 623–632.

[171] Margolus N. Physics-like models of computation. Physica D 10 (1984) 81–95.

[172] Marr D. Vision. Freeman, 1982.

[173] Marshall G. F. and Tarassenko L. Robot path planning using VLSI resistive grids. IEE Proc. Vision Image Signal Proc. 141 (1994) 267–272.

[174] Masia M., Marchettini N., Zambranoa V. and Rustici M. Effect of temperature in a closed unstirred Belousov–Zhabotinsky system. Chem. Phys. Lett. 341 (2001) 285–291.

[175] Matsubara Y., Asai T., Hirose T. and Amemiya Y. Reaction-diffusion chip implementing excitable lattices with multiple-valued cellular automata. IEICE Electron. Express 1 (2004) 248–252.

[176] Matsumoto T., Chua L. O. and Komuro G. M. The double scroll. IEEE Trans. Circuits Syst. I CS-32 (1985) 798–818.

[177] McCulloch W. J. and Pitts W. A logical calculus of the ideas immanent in nervous activity. Bull. Math. Biophys. 5 (1943) 115–133.

[178] Meinhardt H. Biological pattern formation — new observations provide support for theoretical predictions. Bioessays 16 (1994) 627–632.

[179] Mikhailov A. S.. Foundations of Synergetics I. Distributed Active Systems. Springer, Berlin, 1990.

[180] Miller J. F. et al. (eds.). In: Proc. 3rd Int. Conf. Evolvable Systems: From Biology to Hardware, Edinburgh, 2000.

[181] Mills J. W. The new computer science and its unifying principle: complementarity and unconventional computing. In: Int. Workshop The Grand Challenges in Non-Classical Computation, York, UK, 18–19 April 2005. Available at http://www.cs.york.ac.uk/nature/workshop/papers/Mills.pdf.

[182] Mimura M. and Kan-on Y. Predation-mediated coexistence and segregation structures. In: Nishida T., Mimura M. and Fujii H. (eds.). Patterns and Waves: Qualitative Analysis of Nonlinear Differential Equations. Kinokuniya, Tokyo, 1986, pp. 129–155.

[183] Mirkin C. A. and Ratner M. A. Molecular electronics. Annu. Rev. Phys. Chem. 43 (1992) 719–754.

[184] Mohajerzadeh S., Nathan A. and Selvakumar C. R. Numerical simulation of a p-n-p-n color sensor for simultaneous color detection. Sens. Actuators A 44 (1994) 119–124.

[185] Moorthy S. and Ghosh S. A Voronoi cell finite element model for particle cracking in composite materials. Comput. Methods Appl. Mech. Eng. 151 (1998) 377–400.

[186] Moreu P. and Braquelaire J.-P. Two-dimensional thick-skeleton morphing (1996). Available at citeseer.nj.nec.com/87262.html.

[187] Morita K., Margenstern M. and Imai K. Universality of reversible hexagonal cellular automata. Theor. Inf. Appl. 33 (1999) 535–550.

[188] Motoike I. N. and Adamatzky A. Three-valued logic gates in reaction-diffusion excitable media. Chaos Solitons Fractals 24 (2004) 107–114.

[189] Motoike I. and Yoshikawa K. Information operations with an excitable field. Phys. Rev. E 59 (1999) 5354–5360.

[190] Motoike I. N. and Yoshikawa K. Information operations with multiple pulses on an excitable field. Chaos Solitons Fractals 17 (2003) 455–461.

[191] Motoike I. N., Yoshikawa K., Iguchi Y. and Nakata S. Real-time memory on an excitable field. Phys. Rev. E 63 (2001) 036220.

[192] Muenster A. F., Watzl M. and Schneider F. W. Two-dimensional Turing-like patterns in the PA-MBO-system and effects of an electric field. Phys. Scr. T 67 (1996) 58–62.

[193] Muller S. C., Mair T. and Steinbock O. Traveling waves in yeast extract and in cultures of *Dictyostelium discoideum*. Biophys. Chem. 72 (1998) 37.

[194] Muñuzuri A. P., Davydov V. A., Pérez-Muñuzuri V., Gómez-Gesteira M. and Pérez-Villar V. General properties of the electric-field-induced vortex drift in excitable media. Chaos Solitons Fractals 7 (1996) 585–595.

[195] Murray J. D. Mathematical Biology. Springer, Berlin, 1989.

[196] Murray, J. D. Mathematical Biology I: An Introduction. Springer, New York, 2000.

[197] Murphy R. An Introduction to AI Robotics. MIT Press, Cambridge, MA, 2000.

[198] Nagypal I., Bazsa G. and Epstein I. R. Gravity-induced anisotropies in chemical waves. J. Am. Chem. Soc. 108 (1986) 3635–3640.

[199] Nicolis G. and Prigogine I. Self-organization in Nonequilibrium Systems — From Dissipative Structures to Order through Fluctuations. John Wiley, New York, 1977.

[200] Niedernostheide F. J., Kreimer M., Kukuk B., Schulze H. J. and Purwins H. G. Travelling current density filaments in multilayered silicon devices. Phys. Lett. A 191 (1994) 285–290.

[201] Nishiyama N. Eccentric motions of spiral cores in aggregates of *Dictyostelium* cells. Phys. Rev. E 57 (1998) 4622–4626.

[202] Noyes R. M., Field R. J. and Koros E. Oscillations in chemical systems I. Detailed mechanism in a system showing temporal oscillations. J. Am. Chem. Soc. 94 (1972) 1394–1395.

[203] Ogniewicz R. L. and Ilg M. Voronoi skeletons. Theory and applications. In: Proc. IEEE Conf. Computer Vision and Pattern Recognition (CVPR). IEEE Press, 1992, pp. 63–69.

[204] Ogniewicz R. L. and Kübler O. Hierarchic Voronoi skeletons. Pattern Recogn. 28 (1995) 343–359.

[205] Okabe A., Boots B., Sugihara K. and Chiu S. N. Spatial Tesselations: Concepts and Applications of Voronoi Diagrams. John Wiley, New York, 2000.

[206] Ortoleva P. Chemical wave–electrical field interaction phenomena. Physica D 26 (1987) 67–84.

[207] Ott E., Grebogi C. and Yorke J. A. Controlling chaos. Phys. Rev. Lett. 64 (1990) 1196–1199.

[208] Oya T., Asai T., Fukui T. and Amemiya Y. A majority-logic nanodevice using a balanced pair of single-electron boxes. J. Nanosci. Nanotechnol. 2 (2002) 333–342.

[209] Oya T., Asai T., Fukui T. and Amemiya Y. Reaction-diffusion systems consisting of single-electron circuits. Int. J. Unconv. Comput. 1 (2004) 177–194.

[210] Painter K. J. In: Mathematical Models for Biological Pattern Formation. The IMA Volumes in Mathematics and Applications, ed. by Maini P. K. and Othmer H. G. Springer, New York, 2001, Vol. 121, pp. 59-81.

[211] Park M. G., Jeon K. H. and Lee M. C. Obstacle avoidance for mobile robots using artificial potential field approach with simulated annealing. In: Proc. 2001 IEEE Int. Symp. Industrial Electronics. IEEE, Piscataway, NJ, 2001, Vol. 3, pp. 1530–1535.

[212] Pasztor L. Astronomical data analysis software and systems. In: 3rd ASP Conf. Ser. (1994) 61.

[213] Paun Gh. Computing with membranes. J. Comput. Syst. Sci. 61 (2000) 108–143.

[214] Paun Gh. and Rozenberg G. A guide to membrane computing. Theor. Comput. Sci. 287 (2002) 73–100.

[215] Pavlidis T. Algorithms for shape analysis of contours and waveforms. IEEE Trans. Pattern Anal. Mach. Intell. 2 (1980) 301–312.

[216] Pearce A. R., Caelli T., Sestito S., Goss S., Selvestrel M. and Murray G. Skeletonizing topographical regions for navigational path planning. Tech. Rep. CVPRL and ARL, Australia, 1993.

[217] Petrov V. and Showalter K. Nonlinear prediction, filtering and control of chemical systems from time series. Chaos 7 (1997) 614–620.

[218] Petrov V., Ouyang Q. and Swinney H. L. Resonant pattern formation in a chemical system. Nature 388 (1997) 655–657.

[219] Petrov V., Peng B. and Showalter K. A map based algorithm for controlling low-dimensional chaos. J. Chem. Phys. 96 (1992) 7506–7513.

[220] Petrov V., Gaspar V., Masere J. and Showalter K. Controlling chaos in the Belousov–Zhabotinsky reaction. Nature 361 (1993) 240–243.

[221] Philippedes A., Husbands P. and O'Shea M. Four dimensional neuronal signalling by nitric oxide: a computational analysis. J. Neurosci. 20 (2000) 1199–1207.

[222] Pontius J., Richelle J. and Wodak S. J. Deviations from standard atomic volumes as a quality measure for protein crystal structures. J. Mol. Biol. 264 (1996) 121–136.

[223] Pornprompanya M., Muller S. C. and Sevcikova H. Pulse waves under an electric field in the Belousov–Zhabotinsky reaction with pyrogallol as substrate. PCCP 4 (2002) 3370–3375.

[224] Preparata F. P. and Shamos M. I. Computational Geometry: An Introduction. Springer, Berlin, 1991.

[225] Press W. H, Teukolsky S. A., Vetterling W. T. and Flannery B. P. Numerical Recipes in C. Cambridge University Press, New York, 1992.

[226] Preston K. and Duff M. Modern Cellular Automata. Plenum Press, 1984.

[227] Qian H. and Murray J. D. A simple method of parameter space determination for diffusion-driven instability with three species. Appl. Math. Lett. 14 (2001) 405–411.

[228] Radwan A. G., Soliman A. M. and El-Sedeek A.-L. MOS realization of the double-scroll-like chaotic equation. IEEE Trans. Circuits Syst. I 50 (2003) 285–288.

[229] Rambidi N. G. Biomolecular computer: roots and promises. Biosystems 44 (1997) 1–15.

[230] Rambidi N. G. Neural network devices based on reaction-diffusion media: an approach to artificial retina. Supramol. Sci. 5 (1998) 765–767.

[231] Rambidi N. Chemical-based computing and problems of high computational complexity: the reaction-diffusion paradigm. In: Seinko T., Adamatzky A., Rambidi N. and Conrad M. (eds.). Molecular Computing. MIT Press, Cambridge, MA, 2003.

[232] Rambidi N. G. and Yakovenchuk D. Chemical reaction-diffusion implementation of finding the shortest paths in a labyrinth. Phys. Rev. E 63 (2001) 026607.

[233] Rambidi N. G., Shamayaev K. R. and Peshkov G. Yu. Image processing using light-sensitive chemical waves. Phys. Lett. A 298 (2002) 375–382.

[234] Ramos J. J. Oscillatory dynamics of inviscid planar liquid sheets. Appl. Math. Comput. 143 (2003) 109–144.

[235] Rekeczky T., Roska R., Carmona F., Jiménez-Garrido A. and Rodríguez-Vázquez A. Exploration of spatial-temporal dynamic phenomena in a 32× 32-cell stored program two-layer CNN universal machine chip prototype. J. Circuits Syst. Comput. 12 (2003) 691–710.

[236] Reyes D. R., Ghanem M. M., Whitesides G. M. and Manz A. Glow discharge in microfluidic chips for visible analog computing. Lab on a Chip 2 (2002) 113–116.

[237] Rosenfeld A. and Pfaltz J. L. Distance functions on digital pictures. Pattern Recogn. 1 (1968) 33–61.

[238] Rossler O. E. In: Lecture Notes in Biomathematics 4, eds. Conrad M., Guttinger W. and Dal Cin M. Springer, Berlin, 1974, pp. 399-418, 546–582.

[239] Sakai T. and Yoshida R. Self oscillating nanogel particles. Langmuir 20 (2004) 1036–1038.

[240] Sakurai T., Mihaliuk E., Chirila F. and Showalter K. Design and control of wave propagation patterns in excitable media. Science 296 (2002) 2009–2012.

[241] Sakurai T., Miike H., Yokoyama E. and Muller S. C. Initiation front and annihilation center of convection waves developing in spiral structures of Belousov–Zhabotinsky reaction. J. Phys. Soc. Jpn. 66 (1997) 518–521.

[242] Schaudt B. F. and Drysdale R. L. Multiplicatively weighted crystal growth Voronoi diagram. In: Proc. 7th Annu. Symp. Computational Geometry, 1991, pp. 214–223.

[243] Schebesch I. and Engel H. Wave propagation in heterogeneous excitable media. Phys. Rev. E 57 (1998) 3905–3910.

[244] Schenk C. P., Or-Guil M., Bode M. and Purwins H.-G. Interacting pulses in three-component reaction-diffusion systems on two-dimensional domains. Phys. Rev. Lett. 78 (1997) 3781–3784.

[245] Schiff S. J., Jerger K., Duong D. H., Chang T., Spano M. L. and Ditto W. L. Controlling chaos in the brain. Nature 370 (1994) 615–620.

[246] Schoenberg F. P., Ferguson T. and Li C. Inverting Dirichlet tessellations. The Computer J. 46 (2003) 76–83.

[247] Schmidt G. K. and Azarm K. Mobile robot path planning and execution based on a diffusion equation strategy. Adv. Robot. 7 (1993) 479–490.

[248] Scott S. K. Oscillations, Waves and Chaos in Chemical Kinetics. Oxford University Press, Oxford, 1994.

[249] Seipel M., Schneider F. W. and Münster A. F. Control and coupling of spiral waves in excitable media. Faraday Discuss. 120 (2001) 395–405.

[250] Sendiña-Nadal I., Mihaliuk E., Wang J., Pérez-Muñuzuri V. and Showalter K. Wave propagation in subexcitable media with periodically modulated excitability. Phys. Rev. Lett. 86 (2001) 1646–1649.

[251] Serradilla F. and Maravall D. A navigation system for mobile robots using visual feedback and artificial potential fields. In: Cybernetics and Systems '96, Proc. 13th Eur. Meet. Cybernetics and Systems Research. Austrian Soc. Cybernetic Studies, Vienna, 1996, Vol. 2, pp. 1159–1164.

[252] Serrano-Gotarredona T. and Linares-Barranco B. Log-domain implementation of complex dynamics reaction-diffusion neural networks. IEEE Trans. Neural Networks 14 (2003) 1337–1355.

[253] Sevćikova H. and Marek M. Chemical waves in electric field. Physica D 9 (1983) 140–156.

[254] Sevćikova H. and Marek M. Chemical front waves in an electric field. Physica D 13 (1984) 379–386.

[255] Sevcikova H., Marek M. and Muller S. C. The reversal and splitting of waves in an excitable medium caused by an electric field. Science 257 (1992) 951.

[256] Shahaf G. and Marom S. Learning in networks of cortical neurons. J. Neurosci. 21 (2001) 8782–8788.

[257] Shahinpoor M. and Kim K. J. Ionic polymer–metal composites: IV. Industrial and medical applications. Smart Mater. Struct. 14 (2005) 197–214.

[258] Shi B. E. and Luo B. T. Spatial pattern formation via reaction-diffusion dynamics in $32 \times 32 \times 4$ CNN chip. IEEE Trans. Circuits Syst. I 51 (2004) 939–947.

[259] Shibata T. and Ohmi T. A functional MOS transistor featuring gate-level weighted sum and threshold operations. IEEE Trans. Electron Devices 39 (1992) 1444–1455.

[260] Shor P. W. Algorithms for quantum computation: discrete logarithms and factoring. In: Proc. 35th Annu. Symp. Foundations of Computer Science. IEEE Computer Society Press, 1994, pp. 124–134.

[261] Sielewiesiuka J. and Górecki J. On the response of simple reactors to regular trains of pulses. Phys. Chem. Chem. Phys. 4 (2002) 1326–1333.

[262] Song H., Cai Z. S., Zhao X. Z., Li Y. J., Xi B. M. and Li Y. N. A new method of controlling chemical chaos — nonlinear artificial neural network (ANN)–occasional perturbation feedback control in the whole chaotic region. Sci. China Ser. B: Chem. 42 (1999) 624–630.

[263] Sprott J. C. A new class of chaotic circuit. Phys. Lett. A 266 (2000) 19–23.

[264] Sprott J. C. Algebraically simple chaotic flows. Int. J. Chaos Theory Appl. 5 (2000) 3–22.

[265] Sprott J. C. Simple chaotic systems and circuits. Am. J. Phys. 68 (2000) 758–763.

[266] Stan M. R., Burleson W. P., Connolly C. I. and Grupen R. A. Analog VLSI for robot path planning. J. VLSI Signal Process. 8 (1994) 61–73.

[267] Steinbock O., Kettunen P. and Showalter K. Chemical wave logic gates. J. Phys. Chem. 100 (1996) 18970.

[268] Steinbock O., Schutze J. and Muller S. C. Electric-field-induced drift and deformation of spiral waves in an excitable medium. Phys. Rev. Lett. 68 (1992) 248–251.

[269] Steinbock O., Tóth A. and Showalter K. Navigating complex labyrinths: optimal paths from chemical waves. Science 267 (1995) 868–871.

[270] Stevens P. S. Patterns in Nature. Little Brown and Co., 1979.

[271] Strümpel C., Astrov Yu. A., Ammelt E. and Purwins H.-G. Dynamics of zigzag destabilized solitary stripes in a dc-driven pattern-forming semiconductor gas-discharge system. Phys. Rev. E 61 (2000) 4899–4905.

[272] Stuchl I. and Marek M. Dissipative structures in coupled cells: experiments. J. Phys. Chem. 77 (1982) 2956–2963.

[273] Sunayama T., Ikebe M., Asai T. and Amemiya Y. Cellular νMOS circuits performing edge detection with difference-of-Gaussian filters. Jpn. J. Appl. Phys. 39 (2000) 399–407.

[274] Sze S. M. Physics of Semiconductor Devices. John Wiley, New York, 1981.

[275] Tabata O., Hirasawa H., Aoki S., Yoshida R. and Kokufuta E. Ciliary motion using self-oscillating gel. Sens. Actuators A 95 (2002) 234–238.

[276] Takahashi O. and Schilling R. J. Motion planning in a plane using generalized Voronoi diagram. IEEE Trans. Robot. Autom. 5 (1989) 143–150.

[277] Tamada T., Akazawa M., Asai T. and Amemiya Y. Boltzmann machine neural network devices using single-electron tunneling. Nanotechnology 12 (2001) 60–67.

[278] Taylor A. F., Armstrong G. R., Goodchild N. and Scott S. K. Propagation of waves across inexcitable gaps. PCCP 5 (2003) 3928–3932.

[279] Thiran P., Crounse R. K., Chua O. L. and Hasler M. Pattern formation properties of autonomous cellular neural networks. IEEE Trans. Circuits Syst. I 42 (1995) 757–774.

[280] Toffoli T. Programmable matter methods. Future Generat. Comput. Syst. 16 (1998) 187–201.

[281] Toffoli T. and Margolus N. Cellular Automata Machines. MIT Press, Cambridge, MA.

[282] Tolmachiev D. and Adamatzky A. Chemical processor for computation of Voronoi diagram. Adv. Mater. Opt. Electron. 6 (1996) 191–196.

[283] Tóth A. and Showalter K. Logic gates in excitable media. J. Chem. Phys. 103 (1995) 2058–2066.

[284] Tóth A., Gáspár V. and Showalter K. Propagation of chemical waves through capillary tubes. J. Phys. Chem. 98 (1994) 522–531.

[285] Tyson J. J. and Fife P. C. Target patterns in a realistic model of the Belousov–Zhabotinskii reaction. J. Chem. Phys. 73 (1980) 2224–2237.

[286] Tzafestas C. S. and Tzafestas S. G. Recent algorithms for fuzzy and neuro-fuzzy path planning and navigation of autonomous mobile robots. Syst. Sci. 25 (1999) 25–39.

[287] Vadakkepat P., Tan K. C. and Liang W. M. Evolutionary artificial potential fields and their application in real time robot path planning. In: Proc. 2000 Congr. Evolutionary Computation (CEC00). IEEE, Piscataway, NJ, 2000, Vol. 1, pp. 256–263.

[288] Vanag V. K. and Epstein I. R. Pattern formation in a tunable medium. Phys. Rev. Lett. 87 (2001) 228301.

[289] Vanag V. K. and Epstein I. R. Inwardly rotating spiral waves in a reaction diffusion system. Science 294 (2001) 835–837.

[290] Vanag V. K. and Epstein I. R. Packet waves in a reaction-diffusion system. Phys. Rev. Lett. 88 (2002) 088303.

[291] Vanag V. K. and Epstein I. R. Dash-waves in a reaction diffusion system. Phys. Rev. Lett. 90 (2003) 098301.

[292] Vanag V. K. and Epstein I. R. Segmented spirals in reaction diffusion systems. PNAS 100 (2003) 14635–14638.

[293] Vanag V. K. and Epstein I. R. Stationary and oscillatory localized patterns, and subcritical bifurcations. Phys. Rev. Lett. 92 (2004) 128301.

[294] Vergis A., Steiglitz K. and Dickinson B. The complexity of analog computation. Math. Comput. Simul. 28 (1986) 91–113.

[295] Vinson M., Mironov S., Mulvey S. and Pertsov A. Control of spatial orientation and lifetime of scroll rings in excitable media. Nature 386 (1997) 477–480.

[296] Vittoz E. A. Micropower techniques. In: Tsividis Y. et al. (eds.). Design of MOS VLSI Circuits for Telecommunications. Prentice-Hall, New Jersey, 1985, pp. 104–144.

[297] Von Neumann J. The Computer and the Brain. Yale University Press, 1958.

[298] Walter M., Fournier A. and Reimers M. Clonal mosaic model for the synthesis of mammalian coat pattern. In: Proc. Graphics Interface 1998. Available at http://www.graphicsinterface.org/proceedings/1998/118/.

[299] Wang J. Light-induced pattern formation in the excitable Belousov–Zhabotinsky medium. Chem. Phys. Lett. 339 (2001) 357–361.

[300] Wang J., Kadar S., Jung P. and Showalter K. Noise driven avalanche behaviour in subexcitable media. Phys. Rev. Lett. 82 (1999) 855–858.

[301] Wang J., Yang S., Cai R., Lin Z. and Liu Z. A new method for determination of uric acid by the lactic acid-acetone-BrO^{3-}-Mn^{2+}-H_2SO_4 oscillating reaction using the analyte pulse perturbation technique. Talanta 65 (2005) 799–805.

[302] Wang Y. F. and Chirikjian G. S. A new potential field method for robot path planning. In: Proc. IEEE Int. Robotics and Automation Conf., San Francisco, CA. IEEE, Piscataway, NJ, 2000, pp. 977–982.

[303] Wiesenfeld K. and Moss F. Stochastic resonance and the benefits of noise from ice ages to crayfish and SQUIDs. Nature 373 (1995) 33–36.

[304] Wilson H. R. and Cowan J. D. Excitatory and inhibitory interactions in localized populations of model neurons. Biophys. J. 12 (1972) 1–24.

[305] Wuensche A. Self-reproduction by glider collisions: the beehive rule. Available at http://www.cogs.susx.ac.uk/users/andywu/multi_value/self_rep.html.

[306] Wuensche A. Glider dynamics in 3-value hexagonal cellular automata: the beehive rule. Int. J. Unconv. Comput. (2005) in press.

[307] Yokoi H., Adamatzky A., De Lacy Costello B. and Melhuish C. Excitable chemical medium controlled by a robotic hand: closed loop experiments. Int. J. Bifurcat. Chaos 14 (2004) 3347–3354.

[308] Yoneyama M. Optical modification of wave dynamics in a surface layer of the Mn-catalyzed Belousov–Zhabotinsky reaction. Chem. Phys. Lett. 254 (1996) 191–196.

[309] Yoshikawa K., Aihara R. and Magome N. Mode locking in coupled oscillators as is exemplified in chemical and hydrodynamic systems. ACH-Models Chem. 135 (1998) 417–423.

[310] Yoshizawa S. 2D curve simulator (2000). Available at http://www.mpi-sb.mpg.de/~shin/Research/CCurve/CCurve.html.

[311] Young D. A local activator–inhibitor model of vertebrate skin patterns. Math. Biosci. 72 (1984) 51.

[312] Zaikin A. N. and Zhabotinsky A. M. Concentration wave propagation in two-dimensional liquid-phase self-oscillating system. Nature 225 (1970) 535.

[313] Zanin A. L., Liehr A. W., Moskalenko A. S. and Purwins H. G. Voronoi diagrams in barrier gas discharge. Appl. Phys. Lett. 81 (2002) 3338–3340.

[314] Zauner K.-P. and Conrad M. Enzymatic computing. Biotechnol. Prog. 17 (2001) 553–559.

[315] Zhou L. Q., Jia X. and Ouyang Q. Experimental and numerical studies of noise-induced coherent patterns in a subexcitable system. Phys. Rev. Lett. 88 (2002) 138301.

Index

331

Printed and bound by CPI Group (UK) Ltd, Croydon, CR0 4YY

03/10/2024

01040419-0017